D1423497

The Behaviour and Design of Steel Structures to BS5950

Third edition – British

N.S. TRAHAIR
Emeritus Professor of Civil Engineering
The University of Sydney

M.A. BRADFORD
Professor of Civil Engineering
The University of New South Wales

and

D.A. NETHERCOT
Professor of Civil Engineering
Imperial College of Science and Technology

London and New York

First published 2001 by Spon Press
11 New Fetter Lane, London EC4P 4EE

Simultaneously published in the USA and Canada
by Spon Press
29 West 35th Street, New York, NY 10001

Spon Press is an imprint of the Taylor & Francis Group

© 2001 N.S. Trahair, M.A. Bradford and D.A. Nethercot

Typeset in 10/12pt Times by Wearset, Boldon, Tyne and Wear

Printed and bound in Great Britain by Biddles Ltd, Guildford and
King's Lynn

British Library Cataloguing in Publication Data
A catalogue record for this book is available
from the British Library

Library of Congress Cataloging in Publication Data
Trahair, N.S.
The behaviour and design of steel structures to BS5950 / N.S. Trahair,
M.A. Bradford, and D.A. Nethercot. – 3rd ed. – British
p.   cm.
Rev. ed. of: The behaviour and design of steel structures. 1988.
Includes bibliographical references and index.
1. Building, Iron and steel. 2. Structural design. I. Bradford, M.A.,
1955– II. Nethercot, D.A. III. Trahair, N.S. Behaviour and design of
steel structures. IV. Title.
TA684 .T7 2001
624.1'821–dc21   2001017040

ISBN 0-419-26190-7 (hbk)
ISBN 0-419-23820-4 (pbk)

Contents

Preface

This third British edition has been directed specifically to the design of steel structures in accordance with the British standard BS5950: Part 1: 2000. The removal of material on Australian and American methods of design has allowed the inclusion of additional material relevant to British practice, and of more detail in the worked examples. Thus designers, teachers, and students using BS5950 will find greater clarity and more helpful material.

The previous Chapter 7 has been divided into two new chapters, Chapter 7 on Beam-Columns, and Chapter 8 on Frames. The latter has been significantly expanded, both with new material and worked examples, and also with material on frame buckling from the previous Chapter 3. Torsion is now dealt with in Chapter 10. This includes new material on designing for torsion and for combined torsion and bending which is based on recent research. Chapter 9 on Connections has been expanded by including material on common connections.

The preparations of this third British edition has provided an opportunity to revise the text generally to incorporate the results of recent findings and research. This is in accordance with the principal objective of the book, to provide students and practising engineers with an understanding of the relationships between structural behaviour and the design criteria implied by the rules of design codes such as BS5950.

N.S. Trahair, M.A. Bradford and D.A. Nethercot
December 2000

Preface to the second edition revised

Since publication of the second edition of this book in 1988, the British Standard has been revised as BS5950: Part 1: 1990, and the new Australian Standard AS4100–1990 (referred to in this book as the AS) based on the draft revision DR87164 has been published. As a consequence of these revisions, and using the opportunity of a reprint, a number of changes have been made to this book to reflect the changes in the British and Australian Standards.

N.S. Trahair and M.A. Bradford
June 1991

Preface to the second edition

The second edition of this book has been prepared following the very significant revisions made to the British, American and Australian methods of designing steel structures. The British BS449:1969 has been replaced by the BS5950: Part 1: 1985 'Structural Use of Steelwork in Building', the American AISC has published its 1986 'Load and Resistance Factor Design Specification for Structural Steel Buildings' to be used as an alternative to its updated 1969 Specification, and a draft Australian limit state revision of the updated AS1250–1975 'SAA Steel Structures Code' has been issued. The second edition of this book deals specifically with these revisions, which are referred to as the BS5950, the AISC, and the AS in the text.

The principal change in these revisions is the replacement of the Working Stress Method of Design discussed in the first edition by the more logical and efficient *Limit State (or Load and Resistance Factor) Method of Design*. The general basis of this new method is explained in section 1.7.3.4, and the details of the method are fully discussed in each chapter.

The revisions are also changed substantially and expanded, reflecting the better understanding of the behaviour of steel structures which has developed from more than a decade of intensive research and investigation. The bases of these changes and expansions are explained in this second edition as a continuation of the objective of the first edition, which sought to provide students and structural engineers with an understanding of the relationships between structural behaviour and the design criteria implied by the rules of design codes, and of the bases and limitations of these rules.

The opportunity provided by the preparation of the second edition has been used to expand the material on connections into a full chapter (Chapter 9), and to add to the material on frames (sections 7.5–7.7). Modifications have also been made based on further teaching experiences in Australia, New Zealand and Canada, at the suggestions of interested users, while unworked examples have been suggested in Chapters 2–9.

The preparation of this second edition was made possible by the significant contributions made by my co-author, Dr M.A. Bradford of the University of New South Wales. We would both like to thank all those who assisted in the preparation of this second edition, particularly Dr R.Q. Bridge, my secretary Jean Whittle and the tracing staff led by Ron Brew at the University of Sydney.

Nicholas Trahair

Preface to the first edition

The designer of a steel structure must make a proper choice of a method to analyse the structure and of the design criteria to be used to proportion it. To do this he needs a sound knowledge of structural steel behaviour, including the material behaviour of the steel itself, and the structural behaviour of the individual members and of the complete structure. He also needs to understand the relationships between structural behaviour and the design criteria implied by the rules of design codes, and the bases and limitations of these rules. Thus, the basic training of a student of structural engineering has as its object the promotion of this understanding.

Structural knowledge is continually increasing, and techniques for analysis, design, fabrication, and erection of structures are being extended or revised, while new types of structure are being introduced. These changes are reflected by the continual changes being made to the design criteria given in steel design codes, and by their growing detail and sophistication. The structural engineer who does not keep pace with these increases in knowledge may come to use design codes blindly by accepting the rules at their face value and without question, by interpreting them rigidly, and by applying them incorrectly in situations beyond their scope. On the other hand, the structural engineer who has kept abreast of these increases in knowledge and who understands the bases of the code rules will be able to design routine structures more efficiently, and to determine appropriate design criteria for unusual structures and for structures to which the design codes do not directly apply.

There are many excellent textbooks available on the analysis of frame structures, while there are a number of texts which demonstrate the application of design code rules to the proportioning of structural steel members. This book, which is concerned with the behaviour and design of steel structures, is not one of either of these types, although it is assumed that it will be used in courses preceded by appropriate courses in structural analysis and accompanied by suitable assignments on the design of steel structures. Instead, the purpose of this book is to promote the understanding of the behaviour of steel structures by summarizing the present state of knowledge, and to facilitate design by relating this behaviour to the criteria adopted for design, with particular reference to the Australian AS 1250–1975 (SAA Steel Structures Code), the British Standard BS449: Part 2: 1969 (The Use of Structural Steel in Building), and the American Institute of Steel Construction 1969 Specification (for the Design, Fabrication and Erection of Structural Steel for Buildings). The book is written in metric (SI) units, which are now in use in England and Australia, and which are being introduced into technical papers published in the USA. However, most of the material is presented in a non-dimensional format, and those readers who wish to continue using Imperial units will not be unduly inconvenienced.

This book is for use by undergraduate and graduate engineering students and by practising structural engineers. Because of this wide range of usage, the level of the material presented varies somewhat, and the user of the book will need to select material suitable to his purpose. Thus, for a first undergraduate course, some of the topics treated in the book should be passed over until they can be presented in a later course, while the teacher may also simplify other topics by omitting some of the finer details. On the other hand, the simpler subject matters already treated can be omitted from advanced undergraduate or postgraduate courses, while some teachers may wish to expand some of the material presented in the book, or even to add material on additional specialist topics. Finally, the book is intended to be of interest and use to research workers and practising designers. It is anticipated that designers will find it of value in updating their knowledge of structural behaviour, and in furthering their understanding of the code rules and their bases.

The book is not intended, however, to be an advanced treatise for the exclusive use of research workers, so many of the analytical details have been omitted in favour of more descriptive explanations of the behaviour of steel structures. However, where a solution of a characteristic problem can be simply derived, this has been done in an abbreviated form (usually in an appendix which the reader can pass over, if desired) in order to indicate to those interested the rigour of the results presented. Further details may be obtained from the references quoted.

Chapter 1 – Introduction deals with the scope of the book, with the role of structural design in the complete design process, and with the relationships between the behaviour and analysis of steel structures and their structural design. It also presents relevant information on the material properties of structural steels under static, repeated, and dynamic loads. The behaviour of connections, members, and structures is summarized, while the dead and live loads and the forces of nature which act on structures are discussed.

Chapters 2, 3, 5, and 8 are concerned primarily with the more common behaviour of structural steel members under simple loading conditions. Chapter 2 – Tension Members and Chapter 3 – Compression Members deal with axially loaded members and frames, while Chapter 5 – In-Plane Bending of Beams deals with transversely loaded members and frames, and Chapter 8 – Torsion Members deals with the twisting of steel members. Some of the material given in these chapters is at an advanced level, and might well be taught in a postgraduate course. This includes the analysis of shear due to transverse forces given in Chapter 5, and the analysis of torsion and distortion given in Chapter 8. The material on plastic analysis given in Chapter 5 is available in many textbooks, but is included here because of its relevance to the ultimate strength of beams and flexural frames.

Chapters 4, 6, and 7 contain material much of which is often omitted from a first course in structural steel design, and some of which might well be taught in a postgraduate course. Chapter 4 – Local Buckling of Thin Plate Elements discusses the local strength of thin plates under in-plane loading, and the design of the flanges and webs of steel members. Chapter 6 – Lateral Buckling of Beams deals with the flexural–torsional buckling of laterally unsupported beams and rigid-jointed flexural frames. Chapter 7 – Beam-Columns and Frames discusses the in-plane behaviour, the flexural–torsional buckling, and the biaxial bending of members subjected to both axial and transverse loads, and of rigid-jointed frames composed of these members.

The author has been greatly influenced in his preparation of this book by his own teaching and research experiences in Australia at the University of Sydney, in the USA at Washington University, and in the UK at the University of Sheffield, and also by his work with the Standards Association of Australia. The discussions that he has had with his own teachers, with his colleagues and other structural engineers, and with his students have also been important. Thus, while a significant proportion of the material in this book has been developed by the author, much of the material is not original, but has been gathered from many sources. Unfortunately, it is very difficult or even impossible to acknowledge individual sources, and so the references given in this book are restricted to those which the general reader may wish to consult for further information.

The author would like to thank all those who assisted in the preparation of this book. These include the University of Sydney and the University of Sheffield for the facilities made available, including the typing of the manuscript and the tracing of the figures, and the author's colleagues for their helpful advice and criticism, especially Dr D.A. Nethercot of the University of Sheffield.

Units and conversion factors

Units

While most expressions and equations used in this book are arranged so that they are non-dimensional, there are a number of exceptions. In all of these, SI units are used which are derived from the basic units of kilogram (kg) for mass, metre (m) for length, and second (s) for time.

The SI unit of force is the newton (N), which is the force which causes a mass of 1 kg to have an acceleration of 1 m/s^2. The acceleration due to gravity is 9.807 m/s^2 approximately, and so the weight of a mass of 1 kg is 9.807 N.

The SI unit of stress is the pascal (Pa), which is the average stress exerted by a force of 1 N on an area of 1 m^2. The pascal is too small to be convenient in structural engineering, and it is common practice to use either the mega-pascal (1 MPa = 10^6 Pa) or the identical newton per square millimetre (1 N/mm^2 = 10^6 Pa). The newton per square millimetre (N/mm^2) is used generally in this book.

Table of conversion factors

To Imperial (British) units			To SI units		
1 kg	=	0.068 53 slug	1 slug	=	14.59 kg
1 m	=	3.281 ft	1 ft	=	0.3048 m
	=	39.37 in.	1 in.	=	0.0254 m
1 mm	=	0.003 281 ft	1 ft	=	304.8 mm
	=	0.039 37 in.	1 in.	=	25.4 mm
1 N	=	0.2248 lb	1 lb	=	4.448 N
1 kN	=	0.2248 kip	1 kip	=	4.448 kN
	=	0.100 36 ton	1 ton	=	9.964 kN
1 N/mm^{2*†}	=	0.1450 kip/in.^{2} (ksi)	1 kip/in.^{2}	=	6.895 N/mm^2
	=	0.064 75 ton/in.^{2}	1 ton/in.^{2}	=	15.44 N/mm^2
1 kNm	=	0.737 6 kip ft	1 kip ft	=	1.356 kNm
	=	0.329 3 ton ft	1 ton ft	=	3.037 kNm

* 1 N/mm^2 = 1 MPa

$\dagger$ There are some dimensionally inconsistent equations used in this book which arise because a numerical value (in N/mm^2) is substituted for the Young's modulus of elasticity E while the yield stress f_y remains algebraic. The value of the yield stress f_y used in these equations should therefore be expressed in N/mm^2. Care should be used in converting these equations from SI to Imperial units.

Glossary of terms

Beam A member which supports transverse loads or moments only.

Beam-column A member which supports transverse loads or moments which cause bending and axial loads which cause compression.

Biaxial bending The general state of a member which is subjected to bending actions in both principal planes together with axial compression and torsion actions.

Brittle fracture A mode of failure under a tension action in which fracture occurs without yielding.

Buckling A mode of failure in which there is a sudden deformation in a direction or plane normal to that of the loads or moments acting.

Cleat A short length component (often of angle cross-section) used in a connection.

Column A member which supports axial compression loads.

Compact section A section capable of reaching the full plastic moment.

Connection The means by which members are connected together and through which forces and moments are transmitted.

Dead load The weight of all permanent construction.

Deformation capacity A measure of the ability of a structure to deform as a plastic collapse mechanism develops without otherwise failing.

Design resistance The capacity of the structure or element to resist the design load.

Design load A combination of factored nominal loads which the structure is required to resist.

Distortion A mode of deformation in which the cross-section of a member changes shape.

Effective length The length of an equivalent simply supported member which has the same elastic buckling load as the actual member.

Effective width That portion of the width of a flat plate which has a non-uniform stress distribution (caused by local buckling or shear lag) which may be considered as fully effective when the non-uniformity of the stress distribution is ignored.

Factor of safety The factor by which the strength is divided to obtain the working load capacity and the maximum permissible stress.

Fastener A bolt, pin, rivet, or weld used in a connection.

Fatigue A mode of failure in which a member fractures after many applications of load.

First-order analysis An analysis in which equilibrium is formulated for the undeformed position of the structure, so that the moments caused by products of the loads and deflections are ignored.

Flexural–torsional buckling A mode of buckling in which a member deflects and twists.

Friction–grip joint A joint in which forces are transferred by friction forces generated between plates by clamping them together with tensioned high tensile bolts.

Girt A horizontal member between columns which supports wall sheeting.

Gusset A short plate element used in a connection.

Imposed load The load assumed to act as a result of the use of the structure, but excluding wind load.

Inelastic behaviour Deformations accompanied by yielding.

In-plane behaviour The behaviour of a member which deforms only in the plane of the applied loads.

Joint A connection.

Lateral buckling Flexural–torsional buckling of a beam.

Limit states design A method of design in which the performance of the structure is assessed by comparison with a number of limiting conditions of usefulness. The most common conditions are the strength limit state and the serviceability limit state.

Load factor A factor used to multiply a nominal load to obtain part of the design load.

Local buckling A mode of buckling which occurs locally (rather than generally) in a thin plate element of a member.

Mechanism A structural system with a sufficient number of frictionless and plastic hinges to allow it to deform indefinitely under constant load.

Member A one-dimensional structural element which supports transverse or longitudinal loads or moments.

Nominal load The load magnitude determined from a loading code or specification.

Non-uniform torsion The general state of torsion in which the twist of the member varies non-uniformly.

Plastic analysis A method of analysis in which the ultimate strength of a structure is computed by considering the conditions for which there are sufficient plastic hinges to transform the structure into a mechanism.

Plastic hinge A fully yielded cross-section of a member which allows the member portions on either side to rotate under constant moment (the plastic moment).

Plastic section A section capable of reaching and maintaining the full plastic moment until a plastic collapse mechanism is formed.

Post-buckling strength A reserve of strength after buckling which is possessed by some thin plate elements.

Purlin A horizontal member between main beams which supports roof sheeting.

Reduced modulus The modulus of elasticity used to predict the buckling of inelastic members under constant applied load, so called because it is reduced below the elastic modulus.

Residual stresses The stresses in an unloaded member caused by uneven cooling after rolling, flame cutting, or welding.

Second-order analysis An analysis in which equilibrium is formulated for the deformed position of the structure, so that the moments caused by products of the loads and deflections are included.

Semi-compact section A section which can reach the yield stress, but which does not have sufficient resistance to inelastic local buckling to allow it to reach or to maintain the full plastic moment while a plastic mechanism is forming.

Service loads The design loads appropriate for the serviceability limit state.

Shear centre The point in the cross-section of a beam through which the resultant transverse force must act if the beam is not to twist.

Shear lag A phenomenon which occurs in thin wide flanges of beams in which shear straining causes the distribution of bending normal stresses to become sensibly non-uniform.

Slender section A section which does not have sufficient resistance to local buckling to allow it to reach the yield stress.

Splice A connection between two similar collinear members.

Squash load The value of the compressive axial load which will cause yielding throughout a short member.

Strain-hardening A stress–strain state which occurs at stresses which are greater than the yield stress.

Strength limit state The state of collapse or loss of structural integrity.

Tangent modulus The slope of the inelastic stress–strain curve which is used to predict buckling of inelastic members under increasing load.

Tensile strength The maximum nominal stress which can be reached in tension.

Tension-field A mode of shear transfer in the thin web of a stiffened plate girder which occurs after elastic local buckling takes place. In this mode, the tension diagonal of each stiffened panel behaves in the same way as does the diagonal tension member of a parallel chord truss.

Tension member A member which supports axial tension loads.

Ultimate load design A method of design in which the ultimate load capacity of the structure is compared with factored loads.

Uniform torque That part of the total torque which is associated with the rate of change of the angle of twist of the member.

Uniform torsion The special state of torsion in which the angle of twist of the member varies linearly.

Warping A mode of deformation in which plane cross-sections do not remain plane.

Warping torque The other part of the total torque (than the uniform torque). This only occurs during non-uniform torsion, and is associated with changes in the warping of the cross-sections.

Working load design A method of design in which the stresses caused by the service loads are compared with maximum permissible stresses.

Yield strength The average stress during yielding when significant straining takes place. Usually, the minimum yield strength in tension specified for the particular steel.

Notation

The following notation is used in this book. Usually, only one meaning is assigned to each symbol, but in those cases where more meanings than one are possible, the correct one will be evident from the context in which it is used.

A	Area of cross-section, or
	Area of weld group
A_e	Area enclosed by hollow section, or
	Effective area of cross-section
A_{eff}	Effective cross-sectional area
A_{fc}	Flange area at critical section
A_{fm}	Flange area at minimum section, or
	Lesser flange effective area
A_g	Gross area of cross-section
A_h	Area of hole reduced for stagger
A_i	Area of ith connector
A_n	Area of nth rectangular element
A_n	Net area of a cross-section
A_r	Reduced area of an angle tension member connected through one leg
A_{sn}	Bolt area in nth shear plane
A_s	Area of stiffener, or
	Area defined by distance s around section, or
	Shear area of a bolt
A_t	Tensile stress area of a bolt
A_v	Shear area of a member
A_1, A_2, A_3	Constants
B	Bimoment, or
	Flange width
B^*	Design bimoment
B_e	Bimoment capacity
B_p	Fully plastic bimoment
B_Y	First yield bimoment
D	Plate rigidity $Et^3/12(1-v^2)$, or
	Diameter of a circular hollow section, or
	Overall depth of section
$\{D\}$	Vector of nodal deformations
E	Young's modulus of elasticity
E_r	Reduced modulus
E_{st}	Strain-hardening modulus

E_t	Tangent modulus
F	Buckling factor for beam-columns with unequal end moments
F_b^*	Design bearing force
F_c^*	Design compression force
$F_L^*, F_{Tx}^*, F_{Ty}^*$	Weld design forces
F_q^*	Design axial force in transverse stiffener
F_s^*	Design shear force on a bolt
F_t^*	Design tension force
F_v^*	Design shear force in a member
G	Shear modulus of elasticity
$[G]$	Stability matrix
G_{st}	Strain-hardening shear modulus
H	Height, or
	Horizontal reaction
H_q^*	Design longitudinal anchor force
I	Second moment of area
I_b, I_c	Second moments of area of beam and column
I_{cy}	Second moment of area of compression flange
I_e	Effective second moment of area
I_{eff}	Effective second moment of area
I_f	Second moment of area of a flange $= I_y/2$
I_m	Second moment of area of member
I_n	$= b_n^3 t_n/12$
I_r	Second moment of area of restraining member or rafter
I_s	Second moment of area of stiffener
I_w	Warping section constant
I_x, I_y	Second moments of area about the x, y axes
I_{x1y1}	Product second moment of area about the x_1, y_1 axes
I_{ym}	Value of I_y for critical segment
I_{yr}	Value of I_y for restraining segment
J	Torsion section constant
K	Beam or torsion constant $= \sqrt{(\pi^2 EI_w/GJL^2)}$, or
	Fatigue life constant, or
	Brittle fracture thickness factor, or
	Weld transverse capacity coefficient
$[K]$	Elastic stiffness matrix
K_b	Beam stiffness coefficient
K_e	Effective net area coefficient
K_m	$= \sqrt{(\pi^2 EI_y d_f^2/4GJL^2)}$
K_s	Slip resistance factor for hole shape and size
L	Length of member, or
	Length of weld
L_b, L_c	Lengths of beam and column
L_c	Length of column which fails under N alone

L_{Ex}, L_{Ey}	Effective lengths for buckling about the x, y axes
L_{Ez}	Effective length for torsional buckling
L_j	Joint length
L_{LT}	Lateral buckling segment length
L_m	Length of critical segment, or
	Member length, or
	Brace spacing
L_r	Length of reduced cross-section, or
	Length of restraining segment or rafter
L_v	Length of shear failure path
LF	Load factor
M	Moment
M^*	Design bending moment
M_A, M_B	End moments
M_B, M_T	Bottom and top flange end moments
M_b	Lateral buckling resistance moment
M_{btx}	Out-of-plane moment resistance of a tension member in bending
M_{bx}, M_{by}	Major and minor axis beam moment resistances
M_{cw}	Moment capacity of a web
M_{cx}, M_{cy}	Section moment capacities
M_E	$= (\pi/L)\sqrt{(EI_y GJ)}$, or
	Elastic lateral buckling moment for uniform bending
M_f	Flange moment, or
	Flanges only moment capacity
M_f^*	Design first order end moment of frame member
M_{fb}^*	Braced component of M_f^*
M_{fp}	Major axis moment resisted by plastic flanges
M_{fY}	First yield moment of a flange
M_{fs}^*	Sway component of M_f^*
M_I	Inelastic beam buckling moment
M_{Iu}	Value of M_I for uniform bending
M_L	Limiting end moment on a crooked and twisted beam at first yield
M_m, M_{max}	Maximum moment
M_m^*	Design first order maximum moment in frame member
M_{ob}	Maximum moment at elastic buckling
M_{ob0}	Value of M_{ob} for centroidal loading
M_{oc}	Elastic buckling moment of a beam-column
M_{ox}	Out-of-plane member moment resistance
M_{pf}, M_{pw}	Flange and web full plastic moments
M_{px}, M_{py}	Full plastic moments
M_{prx}, M_{pry}	Reduced full plastic moments
M_{rx}, M_{ry}	Section moment capacities reduced for axial load

M_S	Simple beam moment
M_{sx}, M_{sy}	Section moment capacities
M_{tx}	Lesser of M_{rx} and M_{ox}
M_u	Ultimate moment, or
	Uniform torque
M_u^*	Design uniform torque
M_{ue}	Uniform torque capacity
M_{up}	Fully plastic uniform torque
M_{urx}, M_{ury}	Values of M_{ux}, M_{uy} reduced for axial load
M_{ux}, M_{uy}	Major and minor axis bending strengths of a beam
M_{uxu}, M_{uyu}	Values of M_{ux}, M_{uy} for uniform bending
M_{uY}	First yield uniform torque
M_w	Warping torque, or
	Moment in a web
M_x^*, M_y^*	Design moments about x, y axes
M_x, M_y, M_z	Moments about x, y, z axes
M_Y	First yield moment $= f_y Z$
M_{yz}	Value of M_{ob} for a simply supported doubly symmetric beam in uniform bending
M_{yzr}	Value of M_{yz} reduced for incomplete torsional end restraint
M_z	Total torque
N	Applied axial load
N_c	Average column force
$\{N_i\}$	Vector of initial axial forces
N_{im}	Initial member axial force
N_{Iy}	Inelastic minor axis compression buckling load
N_L	Limiting compression force on a crooked column at first yield
N_m	Maximum load
N_o	Elastic buckling load
N_{oc}	Elastic buckling compression load
N_{ol}	Elastic local buckling compression load
N_{oL}	Euler buckling load $= \pi^2 EI/L^2$
N_{omb}, N_{oms}	Elastic buckling compression loads of braced and sway members
N_{ox}, N_{oy}, N_{oz}	Loads at elastic buckling about the x, y, z axes
N_r	Reduced modulus buckling load, or
	Average rafter force
N_s	Axial force in stiffener
N_t	Tension force capacity, or
	Tangent modulus buckling load
N_{tf}	Tension force capacity of fastener
N_u	Strength of a concentrically loaded member, or
	Ultimate load
N_{ux}, N_{uy}	Strengths of a compression member about the x, y axes
N_Y	Squash load

N_{Yr}	Reduced squash load
N_z	Force in z direction
N_{zi}	Axial force in ith fastener
P_{b}	Bearing capacity of a web and its stiffener
P_{bb}	Bearing capacity of a bolt
P_{bs}	Plate bearing capacity
P_{bw}	Bearing capacity at the flange of an unstiffened web
P_{c}	Compression resistance
P_{o}	Bolt preload
P_{q}	Buckling resistance of an intermediate web stiffener
P_{s}	Shear capacity of a bolt
P_{sL}	Slip resistance
P_{t}	Tension capacity
P_{t}^{*}	Design tension force
P_{v}	Shear capacity of a member
P_x	Bearing buckling resistance of an unstiffened web
Q	Concentrated load
Q^{*}	Design concentrated load
Q_{D}	Concentrated dead load
Q_{f}	Flange force
Q_{I}	Concentrated imposed load
Q_{m}	Upper bound mechanism estimate of Q_{u}
Q_{ms}	Value of Q_{s} for the critical segment
Q_{o}	Value of Q at elastic buckling
Q_{ob}	Value of Q at elastic beam buckling
Q_{rs}	Value of Q_{s} for an adjacent restraining segment
Q_{s}	Buckling load for an unrestrained segment, or
	Lower bound static estimate of Q_{u}
Q_{u}	Value of Q at plastic collapse
R	Ratio of minimum to maximum stress, or
	Radius of circular cross-section, or
	Reaction force, or
	Ratio of column and rafter stiffnesses
R^{*}	Design reaction force
$R, R_1, R_2,$	
R_3, R_4	Restraint parameters
R_{f}	Ratio of flange areas at points of minimum and maximum moment
SF	Factor of safety
S_{eff}	Effective plastic modulus
S_{f}	Flanges only plastic modulus
S_{r}	Reduced plastic modulus
S_{v}	Plastic modulus of the shear area
S_x, S_y	Major and minor axis plastic section moduli

T	Applied torque, or
	Flange thickness
T^*	Design torque
T_M	Torque exerted by bending moment
T_P	Torque exerted by axial load
T_0	End torque
T_u	Uniform torsion torque capacity
T_w	Warping torsion torque capacity
U	Strain energy, or
	Lesser of flange and flange plate widths
U_b	Minimum tensile strength of a bolt
U_e	Minimum tensile strength of a weld electrode
U_s	Minimum tensile strength of steel
V	Shear force
V^*	Design shear force, or
	Design horizontal storey shear
V_b	Shear buckling resistance of a web
V_{cr}	Critical shear buckling resistance of a web
V_f	Flange shear force, or
	Flange dependent shear buckling resistance
V_L	Longitudinal shear force in a fillet weld
V_L^*	Design longitudinal shear force in a fillet weld
V_R	Resultant shear force
V_{Tx}, V_{Ty}	Transverse shear forces in a fillet weld
V_{Tx}^*, V_{Ty}^*	Design transverse shear forces in a fillet weld
V_{vi}	Shear force in ith fastener
V_w	Simple shear buckling resistance
V_x, V_y	Shear forces in x, y directions
W	Work
Y_b	Minimum yield strength of a bolt
Y_s	Minimum yield strength of steel
Z	Elastic section modulus
Z_e	Effective section modulus
Z_{eff}	Effective elastic section modulus
Z_x, Z_y	Major and minor axis elastic section moduli
Z_{xB}, Z_{xT}	Values of Z_x for bottom and top flanges
a	$= \sqrt{(EI_w/GJ)}$, or
	Distance along member, or
	Distance from web to shear centre, or
	Robertson imperfection constant, or
	Spacing of transverse stiffeners, or
	Effective throat size of a weld
a_0	Distance from shear centre
a_1	Gross area of the connected leg of an angle tension member

a_2	$= A_g - a_1$
b	Width, or
	Width of rectangular section
b_b	Bearing length at neutral axis
b_{bf}	Bearing length at flange–web junction
b_e	Effective width
b_{eff}	Effective width
b_f	Flange width
b_n	Net width of a tension member (see Fig. 2.8), or
	Width of nth rectangular element
b_1	Stiff bearing length
c_{mx}, c_{my}	Bending coefficients for beam-columns with unequal end moments
d	Depth of section, or
	Depth of rectangular section, or
	Diameter of hole, or
	Depth of web, or
	Bolt diameter
d_c, d_m	Values of d at critical and minimum sections
d_e	Depth of elastic core
d_{eff}	Effective depth
d_f	Distance between flange centroids
d_o	Outside diameter of a circular hollow section, or
	Overall depth of section
e	Axial extension, or
	Eccentricity, or
	End distance in a plate
e_{st}	Axial extension at strain-hardening
e_Y	Axial extension at yield
f	Normal stress
f^*	Design normal stress range
f_{ac}, f_{at}	Stresses due to axial compression and tension
f_b	Bending stress in a web
f_{bcx}	Compression stress due to bending about x axis
f_{bg}^*	Design bending stress for gross section
f_{bn}^*	Design bending stress for net section
f_{btx}, f_{bty}	Tension stresses due to bending about x, y axes
f_{cw}, f_{tw}	Maximum compressive and tensile stresses in a web
f_e	Fatigue endurance limit
f_f^*	Mean stress in the smaller flange
f_i^*	ith design normal stress range
f_k	Specified or characteristic resistance
f_L	Limiting major axis stress in a crooked and twisted beam at first yield

f_{max}	Maximum stress
f_{min}	Minimum stress
f_{obl}	Elastic local buckling stress in bending
f_{oblo}	Elastic local buckling stress in bending alone
f_{ol}	Elastic local buckling stress in compression
f_{op}	Elastic buckling stress in bearing
f_{opo}	Elastic buckling stress in bearing alone
f_{ov}	Elastic buckling stress in shear
f_{ovo}	Elastic buckling stress in shear alone
f_p	Bearing stress
f_{tf}	Shear stress resisted by tension field
f_u	Minimum tensile strength
f_{uf}	Minimum tensile strength of fastener
f_{uw}	Tensile strength of weld
f_v^*	Maximum design shear stress in a web
f_w	Warping normal stress, or
	Normal stress on fillet weld throat
f_w^*	Design warping normal stress
f_x^*, f_y^*	Design normal stresses in x, y directions
f_y	Yield stress
h	Column height
h_s	Storey height, or
	Distance between flange shear centres
i	Integer
k	Deflection coefficient, or
	Modulus of foundation reaction
k_b	Plate buckling coefficient
k_{bs}	Plate bearing coefficient for bolt hole shape
k_e	Effective length factor
k_m	Member effective length factor
k_r	Factor for long shear connections
k_t	Axial stiffness of connector
k_v	Shear stiffness of connector
k_1, k_2	Stiffness factors
m	Torque per unit length
m, n	Integers
m_{LT}	Lateral buckling equivalent uniform moment factor
m_x, m_y	Flexural buckling equivalent uniform moment factors
n	Equivalent slenderness factor for tapered members, or Axial force ratio
n_{sc}	Number of load cycles
n_i^*	Number of cycles in the ith stress range
n_{im}	Fatigue life for ith stress range
p_b	Lateral buckling bending strength

p_{bb}	Bearing strength of a bolt
p_{bs}	Bearing strength of a plate
p_{cs}	Compressive strength of a slender section
p_{cx}, p_{cy}	Compressive strengths
p_E	Elastic buckling strength
p_F	Probability of failure
p_L	Compression member first yield strength
p_s	Design shear strength of a bolt
p_t	Compression member tangent modulus buckling strength, or Design tensile strength of a bolt
p_v	Design shear strength of steel
p_w	Design strength of a fillet weld
p_y	Design strength of steel
p_{yf}, p_{yw}	Flange and web design strengths
p_{yr}	Reduced design strength
p_{ys}	Design strength of a stiffener
$p(z)$	Particular integral
q	Intensity of uniformly distributed transverse load
q^*	Design distributed transverse load
q_{cr}	Critical shear strength of a web
q_D	Intensity of uniformly distributed dead load
q_e	Critical shear buckling parameter of a stiffened web
q_I	Intensity of uniformly distributed imposed load
q_i	Initial load
q_{ob}	Value of q at elastic buckling
q_u	Ultimate distributed load, or Value of q at plastic collapse
q_w	Shear buckling strength of a web
r	Radius of gyration, or Radius, or Radius of a fillet
r_{eff}	Effective radius of gyration
r_i	Distance from centre of rotation to ith connector
r_x, r_y	Radii of gyration about x, y axes
r_0	$= \sqrt{(r_x^2 + r_y^2)}$
r_1	$= \sqrt{(r_0^2 + x_0^2 + y_0^2)}$, or Stress ratio
s	Distance around thin-walled section
s_f, s_w	Distances along flange and web
s_g	Transverse gauge distance between lines of holes
s_p	Staggered pitch of holes
s_{pm}	Minimum staggered pitch for no reduction in effective area
s_1, s_2	Side widths of a fillet weld

t	Thickness of thin-walled section, or
	Web thickness
t_f, t_w	Flange and web thicknesses
t_s	Stiffener thickness
t_m	Maximum thickness
t_n	Thickness of nth rectangular element
t_p	Thickness of plate
t_1	Limiting brittle fracture thickness
u	Lateral buckling parameter
u, v	Deflections of shear centre in x, y directions
u_f	Flange deflection
u_0, v_0	Initial deflections
v	Lateral buckling slenderness factor
v_{AB}	Settlement of B relative to A
v_c	Mid-span deflection
w	Warping deflection in z direction, or
	Dimensionless warping section constant
x	Lateral buckling torsional index
x, y	Principal axes of section, or
	Principal axes of connector group
x_i, y_i	Coordinates of ith connector, or
	Distances from initial origin
x_{ic}, y_{ic}	Coordinates of centroid from initial origin
x_n, y_n	Coordinates of centroid of nth element
x_p, y_p	Distances to plastic neutral axes
x_r, y_r	Coordinates of centre of rotation
x_0, y_0	Coordinates of shear centre
$\bar{y}$	Distance to centroid
y_B, y_T	Distances to extreme bottom and top fibres
y_c	Distance to buckling centre of rotation, or
	Distance to centroid
y_n	Distance below centroid to neutral axis
y_p	Distance to plastic neutral axis
y_Q	Distance below centroid to load
y_r	Distance below centroid to restraint
z	Longitudinal axis through centroid, or
	Axis normal to connection plane
z_m	Distance to point of maximum moment
z_1, z_2	Indices in biaxial bending equation
α	Unit warping (see equation 10.31), or
	Bearing area ratio, or
	Inclination to principal axis, or
	Rotational or translational restraint stiffness, or
	Coefficient used to determine effective width

α_1, α_2	Rotational restraint stiffnesses at ends of member
α_{bc}	Buckling coefficient for beam columns with unequal end moments
α_{bcI}	Inelastic moment modification factor for bending and compression
α_{bcu}	Value of α_{bc} for ultimate strength
α_L	Limiting value of α for second mode buckling
α_L, α_0	Indices in interaction equations for biaxial bending
α_m	Moment modification factor for beam lateral buckling
α_n	$= \frac{1}{A}\int_0^E \alpha t \, ds$
α_r	Stiffness factor for effect of axial load
α_r, α_t	Rotational and translational stiffnesses
α_{rz}	Stiffness of torsional end restraint
α_s	Fatigue life index
α_{st}	Buckling moment factor for stepped and tapered beams
β	Safety index, or
	End moment ratio $(= -\beta_m)$
β_e	Stiffness factor for far end restraint conditions
β_m	Ratio of end moments, or
	End moment factor
β_w	Lateral buckling section modulus ratio
β_x	Monosymmetry section constant for I-beam
γ_f	Load partial factor
γ_m	Index used in biaxial bending equations for member capacity, or
	Material partial factor
$\gamma_m, \gamma_n, \gamma_s$	Factors used in moment amplification
γ_1, γ_2	Relative end stiffnesses of a member in a frame
Δ	Sway deflection
Δ_s	Storey sway
δ	Deflection, or
	Central deflection, or
	Moment amplification factor
δ_m	Second order central deflection
δ_p	Plastic design amplification factor
δ_s	Moment amplification factor for a sway member
δ_0	Initial central deflection
ε	Strain, or
	$= (K/\pi)2y_Q/d_f$, or
	Yield strength parameter
ε_s	Yield strength parameter for a stiffener
ε_{st}	Strain at strain-hardening
ε_Y	Strain at yield
η	Crookedness or imperfection parameter, or
	Perry imperfection factor for flexural buckling

η_{LT}	Perry imperfection factor for lateral buckling
θ	Central twist, or
	Torsion stress function, or
	Slope change at plastic hinge, or
	Joint rotation, or
	Member end rotation, or
	Inclination of fillet weld throat, or
	Angle between force and weld throat
θ_n	Orientation of nth rectangular element
$\theta_x, \theta_y, \theta_z$	Rotations about x, y, z axes
θ_0	Initial central twist
λ	Slenderness ratio $= L_E/r$
λ_c	Frame load parameter at elastic buckling
λ_{cm}	Member estimate of λ_c
λ_{cr}	Elastic buckling load factor
λ_d	Design load factor
λ_e	Plate element slenderness
λ_{eff}	Reduced slenderness ratio of a slender compression member
λ_{ep}	Plate element plasticity slenderness limit
λ_i	In-plane load factor
λ_{Lo}	Limiting slenderness ratio (lateral buckling)
λ_{LT}	Equivalent slenderness ratio (lateral buckling)
$\lambda_{L1}, \lambda_{L2}, \lambda_{L3}$	Class limiting element slendernesses
λ_{ms}	Storey estimate of λ_c
λ_o	Limiting slenderness ratio (axial compression)
λ_p	Load factor at full plasticity
λ_r	Rafter buckling load factor, or
	Equivalent slenderness ratio for lateral and flexural buckling
λ_{ro}	Limiting value of λ_r
$\lambda_{sf}, \lambda_{sp}$	Sway buckling load factors for fixed and pinned base portal frames
λ_u	Load factor at failure, or
	Uniform torsion load factor
λ_w	Warping torsion load factor, or
	$= (p_v/q_e)^{0.5}$
λ_x, λ_y	Compression member slenderness ratios for buckling about the x, y axes
λ_z	Torsion load factor
μ	$= \sqrt{(N/EI)}$
μ_n	Coefficient of friction for nth interface
v	Poisson's ratio
ρ	Perpendicular distance from centroid, or
	Reduction factor for shear buckling
ρ_m	Monosymmetric section parameter $= I_{yc}/I_y$

ρ_0	Perpendicular distance from shear centre
τ	Shear stress
τ_h, τ_v	Shear stresses due to V_x, V_y
τ_{hc}, τ_{vc}	Shear stresses due to a circulating shear flow
τ_{ho}, τ_{vo}	Shear stresses in an open section
τ_m	Maximum shear stress
τ_u	Uniform torsion shear stress
τ_u^*	Design uniform torsion shear stress
τ_{ub}	Shear ultimate strength of bolt material
τ_w	Warping shear stress, or
	Shear stress on fillet weld throat
τ_w^*	Design warping torsion shear stress
τ_{xy}^*	Design shear stress
τ_{xz}, τ_{yz}	Shear stresses in z direction
τ_y	Yield shear stress
τ_{yb}	Shear yield strength of bolt material
τ_{zx}, τ_{zy}	Transverse shear stresses
Φ	Cumulative frequency distribution of a standard normal variate, or
	Curvature
ϕ	Angle of twist rotation
ϕ_m	Maximum value of ϕ
ϕ_{um}	Uniform torsion value of ϕ_m
ϕ_{wm}	Warping torsion value of ϕ_m
ϕ_0	Initial angle of twist rotation
ψ	Monosymmetry index

1 Introduction

1.1 Steel structures

Engineering structures are required to support loads and resist forces, and to transfer these loads and forces to the foundations of the structures. The loads and forces may arise from the masses of the structure, or from man's use of the structures, or from the forces of nature. The uses of structures include the enclosure of space (buildings), the provision of access (bridges), the storage of materials (tanks and silos), transportation (vehicles), or the processing of materials (machines). Structures may be made from a number of different materials, including steel, concrete, wood, aluminium, stone and plastic, etc., or from combinations of these.

Structures are usually three-dimensional in their extent, but sometimes they are essentially two-dimensional (plates and shells), or even one-dimensional (lines and cables). Solid steel structures invariably include comparatively high volumes of high cost structural steel which are understressed and are uneconomic, except in very small scale components. Because of this, steel structures are usually formed from one-dimensional members (as in rectangular and triangulated frames), or from two-dimensional members (as in box girders), or from both (as in stressed skin industrial buildings). Three-dimensional steel structures are often arranged so that they act as if composed of a number of independent two-dimensional frames or one-dimensional members (Fig. 1.1).

Structural steel members may be one-dimensional as for beams and columns (whose lengths are much greater than their transverse dimensions), or two-dimensional as for plates (whose lengths and widths are much greater than their thicknesses), as shown in Fig. 1.2c. While one-dimensional steel members may be solid, they are usually thin-walled, in that their thicknesses are much less than their other transverse dimensions. Thin-walled steel members are rolled in a number of cross-sectional shapes [1] or are built up by connecting together a number of rolled sections or plates, as shown in Fig. 1.2b. Structural members can be classified as tension or compression members, beams, beam-columns, torsion members or plates (Fig. 1.3), according to the method by which they transmit the forces in the structure. The behaviour and design of these structural members are discussed in this book.

Structural members may be connected together in a number of ways, and by using a variety of connectors. These include pins, rivets, bolts and welds of various types. Steel plate gussets or angle cleats or other elements may also be used in the connections. The behaviour and design of these connectors and connections are also discussed in this book.

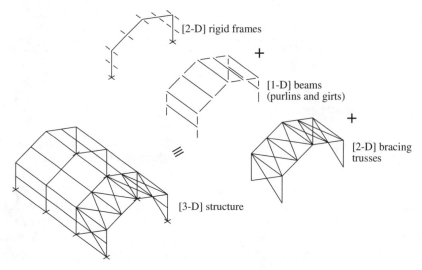

Fig. 1.1 Reduction of a [3-D] structure to simpler forms

This book deals chiefly with steel frame structures composed of one-dimensional members, but much of the information given is also relevant to plate structures. The members are generally assumed to be hot-rolled or fabricated from hot-rolled elements, while the frames considered are those used in buildings. However, much of the material presented is also relevant to bridge structures [2, 3], and to structural members cold-formed from light-gauge steel plates [4–7].

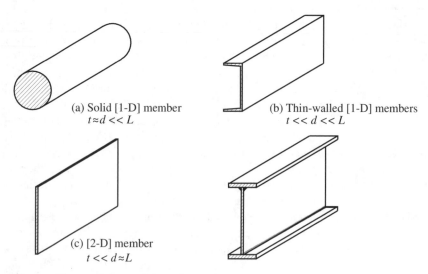

Fig. 1.2 Types of structural steel members

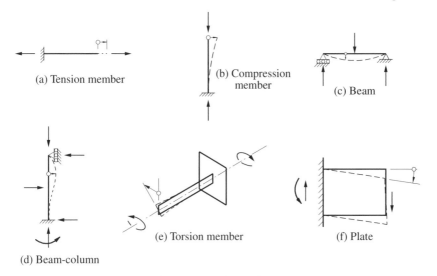

(d) Beam-column

Fig. 1.3 Load transmission by structural members

The purposes of this chapter are first, to consider the complete design process and the relationships between the behaviour and analysis of steel structures and their structural design, and second, to present information of a general nature (including information on material properties and structural loads) which is required for use in the later chapters. The nature of the design process is discussed first, and then brief summaries are made of the relevant material properties of structural steel and of the structural behaviour of members and frames. The loads acting on structures are considered, and the choice of appropriate methods of analysing steel structures is discussed. Finally, the considerations governing the synthesis of an understanding of structural behaviour with the results of analysis to form the design processes of BS5950 – Part 1 [8] are treated.

1.2 Design

1.2.1 DESIGN REQUIREMENTS

The principal design requirement of a structure is that it should be effective; that is, it should fulfil the objectives and satisfy the needs for which it was created. The structure may provide shelter and protection against the environment by enclosing space, as in buildings; or it may provide access for people and materials, as in bridges; or it may store materials, as in tanks and silos; or it may form part of a machine for transporting people or materials, as in vehicles, or for operating on materials. The design requirement of effectiveness is paramount, as there is little point in considering a structure which will not fulfil its purpose.

The satisfaction of the effectiveness requirement depends on whether the structure satisfies the structural and other requirements. The structural requirements relate to the way in which the structure resists and transfers the forces and loads acting on it. The primary structural requirement is that of safety, and the first consideration of the structural engineer is to produce a structure which will not fail in its design lifetime, or which has an acceptably low risk of failure. The other important structural requirement is usually concerned with the stiffness of the structure, which must be sufficient to ensure that the serviceability of the structure is not impaired by excessive deflections, vibrations, and the like.

The other design requirements include those of economy and of harmony. The cost of the structure, which includes both the initial cost and the cost of maintenance, is usually of great importance to the owner, and the requirement of economy usually has a significant influence on the design of the structure. The cost of the structure is affected not only by the type and quantity of the materials used, but also by the methods of fabricating and erecting it. The designer must therefore give careful consideration to the methods of construction as well as to the sizes of the members of the structure.

The requirements of harmony within the structure are affected by the relationships between the different systems of the structure, including the load resistance and transfer system (the structural system), the architectural system, the mechanical and electrical systems, and the functional systems required by the use of the structure. The serviceability of the structure is usually directly affected by the harmony, or lack of it, between the systems. The structure should also be in harmony with its environment, and should not react unfavourably with either the community or its physical surroundings.

1.2.2 THE DESIGN PROCESS

The overall purpose of design is to invent a structure which will satisfy the design requirements outlined in section 1.2.1. Thus the structural engineer seeks to invent a structural system which will resist and transfer the forces and loads acting on it with adequate safety, while making due allowance for the requirements of serviceability, economy, and harmony. The process by which this may be achieved is summarized in Fig. 1.4.

The first step is to define the overall problem by determining the effectiveness requirements and the constraints imposed by the social and physical environments and by the owner's time and money. The structural engineer will need to consult the owner; the architect, the site, construction, mechanical and electrical engineers; and any authorities from whom permissions and approvals must be obtained. A set of objectives can then be specified, which if met, will ensure the successful solution of the overall design problem.

The second step is to invent a number of alternative overall systems and their associated structural systems which appear to meet the objectives. In

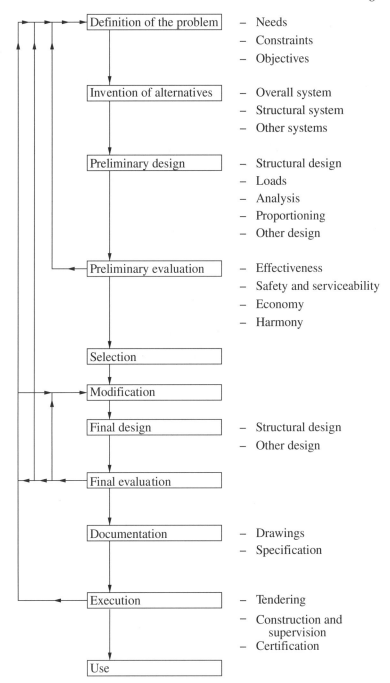

Fig. 1.4 The overall design process

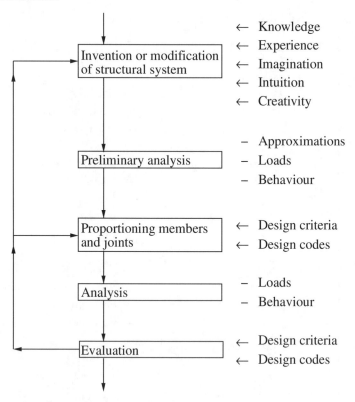

Fig. 1.5 The structural design process

doing so, the designer may use personal knowledge and experience or that which can be gathered from others [9–12]; or the designer may use his or her own imagination, intuition, and creativity [13], or a combination of all of these.

Following these first two steps of definition and invention comes a series of steps which include the structural design, evaluation, selection, and modification of the structural system. These may be repeated a number of times before the structural requirements are met and the structural design is finalized. A typical structural design process is summarized in Fig. 1.5.

After the structural system has been invented, it must be analysed to obtain information for determining the member sizes. First, the loads supported by and the forces acting on the structure must be determined. For this purpose, loading codes [14, 15] are usually consulted, but sometimes the designer determines the loading conditions or commissions experts to do this. A number of approximate assumptions are made about the behaviour of the structure, which is then analysed and the forces and moments acting on the members and joints of the structure are determined. These are used to proportion the structure so that it satisfies the structural requirements, usually by referring to a design code, such as BS5950 [8].

At this stage a preliminary design of the structure has been completed, but because of the approximate assumptions made about the structural behaviour, it is necessary to check the design. The first steps are to recalculate the loads and to reanalyse the structure designed, and these are carried out with more precision than was either possible or appropriate for the preliminary analysis. The performance of the structure is then evaluated in relation to the structural requirements, and any changes in the member and joint sizes are decided on. These changes may require a further reanalysis and reproportioning of the structure, and this cycle may be repeated until no further changes are required. Alternatively, it may be necessary to modify the original structural system and repeat the structural design process until a satisfactory structure is achieved.

The alternative overall systems are then evaluated in terms of their serviceability, economy, and harmony, and a final system is selected, as indicated in Fig. 1.4. This final overall system may be modified before the design is finalized. The detailed drawings and specifications can then be prepared, and tenders for the construction can be called for and let, and the structure can be constructed. Further modifications may have to be made as a consequence of the tenders submitted or due to unforeseen circumstances discovered during construction.

This book is concerned with the structural behaviour of steel structures, and the relationships between their behaviour and the methods of proportioning them, particularly in relation to the structural requirements of the UK steel structures code BS5950 [8]. This code consists of nine parts, with design using conventional members being treated in Part 1. Parts 2 and 7 of BS5950 deal with construction, while composite structures are covered by Part 3.1, cold-formed steel structures by Parts 4, 5, and 6, fire by Part 8, and stressed skin design by Part 9. These other eight parts of the code are also referenced within Part 1. Thus unless otherwise specified, all references to BS5950 should be taken to mean Part 1. Detailed discussions of the overall design process are therefore beyond the scope of this book, but further information is given in [13] on the definition of the design problem, the invention of solutions and their evaluation, and in [16–20] on the execution of design. Further, the conventional methods of structural analysis are adequately treated in many textbooks [21–23] and are discussed in only a few isolated cases in this book.

1.3 Material behaviour

1.3.1 MECHANICAL PROPERTIES UNDER STATIC LOAD

The important mechanical properties of most structural steels under static load are indicated in the idealized tensile stress–strain diagram shown in Fig. 1.6. Initially the steel has a linear stress–strain curve whose slope is the Young's modulus of elasticity E. The values of E vary in the range

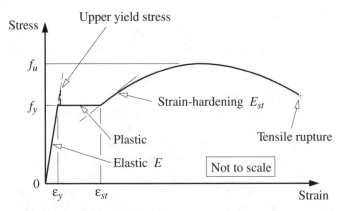

Fig. 1.6 Idealized stress–strain relationship for structural steel

200 000–210 000 N/mm², and the approximate value of 205 000 N/mm² is often assumed. The steel remains elastic while in this linear range, and recovers perfectly on unloading. The limit of the linear elastic behaviour is often closely approximated by the yield stress f_y and the corresponding yield strain $\varepsilon_y = f_y/E$. Beyond this limit the steel flows plastically without any increase in stress until the strain-hardening strain ε_{st} is reached. This plastic range is usually considerable, and accounts for the ductility of the steel. The stress increases above the yield stress f_y when the strain-hardening strain ε_{st} is exceeded, and this continues until the ultimate tensile stress f_u is reached. After this, large local reductions in the cross-section occur, and the load capacity decreases until tensile fracture takes place.

The yield stress f_y is perhaps the most important strength characteristic of a structural steel. This varies significantly with the chemical constituents of the steel, the most important of which are carbon and manganese, both of which increase the yield stress. The yield stress also varies with the heat treatment used and with the amount of working which occurs during the rolling process. Thus thinner plates which are more worked have higher yield stresses than thicker plates of the same constituency. The yield stress is also increased by cold working. The rate of straining affects the yield stress, and high rates of strain increase the upper or first yield stress (see the broken line in Fig. 1.6), as well as the lower yield stress f_y. The strain rates used in tests to determine the yield stress of a particular steel type are significantly higher than the nearly static rates often encountered in actual structures.

For design purposes, a 'minimum' yield stress is identified for each different steel classification. In the UK, these classifications are made on the basis of the chemical composition and the heat treatment, and so the yield stresses in each classification decrease as the greatest thickness of the rolled section or plate increases. The minimum yield stress of a particular steel is determined from the results of a number of standard tension tests. There is a significant scatter in these results because of small variations in the local composition, heat treatment, amount of working, thickness and rate of testing, and this scatter closely follows a normal distribution curve. Because of this, the minimum yield stress f_y quoted for a particular steel and used in design is usually a characteristic value which has a particular chance (often 95%) of being exceeded in any standard tension test. Consequently, it is likely that an isolated test result will be significantly higher than the quoted yield stress. This difference will, of course, be accentuated if the test is made for any but the thickest portion of the cross-section. In BS5950, the yield stress (or yield strength) used in design is taken as the lesser of f_y and $f_u/1.2$.

The yield stress f_y determined for uniaxial tension is usually accepted as being valid for uniaxial compression. However, the general state of stress at a point in a thin-walled member is one of biaxial tension and/or compression, and yielding under these conditions is not so simply determined. Perhaps the most generally accepted theory of two-dimensional yielding under biaxial stresses acting in the 1′2′ plane is the maximum distortion–energy theory (often associated with names of Huber, von Mises, or Hencky), and the stresses at yield according to this theory satisfy the condition

$$f_{1'}^2 - f_{1'}f_{2'} + f_{2'}^2 + 3f_{1'2'}^2 = f_y^2, \tag{1.1}$$

in which $f_{1'}$, $f_{2'}$ are the normal stresses and $f_{1'2'}$ is the shear stress at the point. For the case where 1′ and 2′ are the principal stress directions 1 and 2, equation 1.1 takes the form of the ellipse shown in Fig. 1.7, while for the case of pure shear ($f_{1'} = f_{2'} = 0$, so that $f_1 = -f_2 = f_{1'2'}$), equation 1.1 reduces to

$$f_{1'2'} = f_y/\sqrt{3} = \tau_y, \tag{1.2}$$

which defines the shear yield stress τ_y.

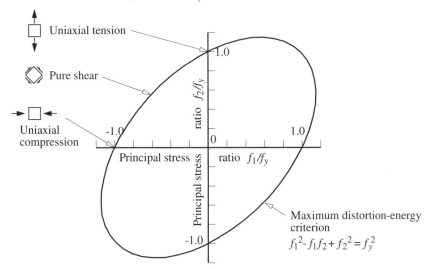

Fig. 1.7 Yielding under biaxial stresses

1.3.2 FATIGUE FAILURE UNDER REPEATED LOADS

Structural steel may fracture at low average tensile stresses after a large number of cycles of fluctuating load. This high-cycle fatigue failure is initiated by local damage caused by the repeated loads, which leads to the formation of a small local crack. The extent of the fatigue crack is gradually increased by the subsequent load repetitions, until finally the effective cross-section is so reduced that catastrophic failure may occur. High-cycle fatigue is only a design consideration when a large number of loading cycles involving tensile stresses is likely to occur during the design life of the structure (compressive stresses do not cause fatigue). This is often the case for bridges, cranes, and structures which support machinery; wind and wave loading may also lead to fatigue problems.

Factors which significantly influence the resistance to fatigue failure include the number of load cycles, the range of stress f^* during a load cycle, and the magnitudes of local stress concentrations. An indication of the effect of the number of load cycles is given in Fig. 1.8, which shows that the maximum tensile stress decreases from its ultimate static value f_u in an approximately linear fashion as the logarithm of the number of cycles n_{sc} increases. As the number of cycles increases further, the curve may flatten out and the maximum tensile stress may approach the endurance limit f_e.

The effects of the stress magnitude and stress ratio on the fatigue life are demonstrated in Fig. 1.9. It can be seen that the fatigue life n_{sc} decreases with increasing stress magnitude f_{max} and with decreasing stress ratio R.

The effect of stress concentration is to increase the stress locally, leading to local damage and crack initiation. Stress concentrations arise from sudden changes in the general geometry and loading of a member, and from local changes due to bolt and rivet holes and welds. Stress concentrations also occur at defects in the member, or its connectors and welds. These may be due to the

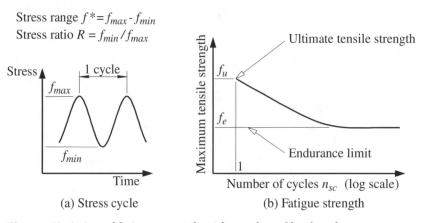

Stress range $f^* = f_{max} - f_{min}$
Stress ratio $R = f_{min} / f_{max}$

(a) Stress cycle   (b) Fatigue strength

Fig. 1.8 Variation of fatigue strength with number of load cycles

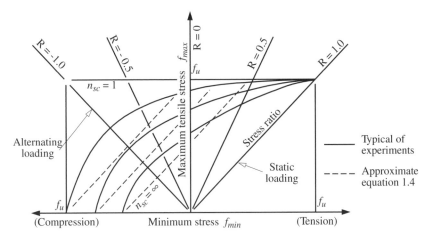

Fig. 1.9 Variation of fatigue life with stress magnitudes

original rolling of the steel, or due to subsequent fabrication processes, including punching, shearing, and welding, or due to damage such as that caused by stray arc fusions during welding.

It is generally accepted for design purposes that the fatigue life n_{sc} varies with the stress range

$$f^* = f_{max} - f_{min} \qquad (1.3)$$

according to equations of the type

$$n_{sc} = K(f^*)^{-\alpha_s} \qquad (1.4)$$

in which the constant K depends on the details of the fatigue site, and the constant α_s may increase with the number of cycles n_{sc}. This assumed dependence of fatigue life on the stress range produces the approximating straight lines shown in Fig. 1.9.

BS5950 does not provide a treatment of fatigue, since it is usually the case that either the stress range f^* or the number of high amplitude stress cycles n_{sc} is comparatively small. However, BS5950 requires structures supporting vibrating machinery and plant to be checked for fatigue. A comprehensive treatment of fatigue is given in BS7608 [24], which specifies the relationships shown in Fig. 1.10 between the fatigue life n_{sc} and the service stress range f^* for different detail categories. These relationships and detail categories are similar to the recommendations of the ECCS Code [25] for constant amplitude stress cycles.

Fatigue failure under variable amplitude stress cycles is assessed using Miner's rule [26]

$$\Sigma n_i^* / n_{im} \leqslant 1 \qquad (1.5)$$

in which n_i^* is the number of cycles of a particular stress range f_i^* and n_{im} the

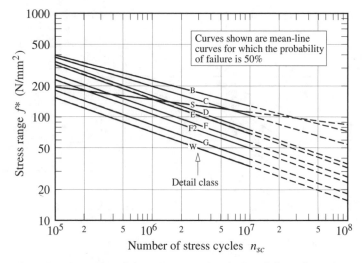

Fig. 1.10 Variation of the BS7608 mean fatigue life with stress range

constant amplitude fatigue life for that stress range. If any of the stress ranges exceeds the constant amplitude fatigue limit (at $n_{sc} = 5 \times 10^6$), then the effects of stress ranges below this limit are included in equation 1.5.

Designing against fatigue involves a consideration of joint arrangement as well as of permissible stress. Joints should generally be so arranged as to minimize stress concentrations and produce as smooth a 'stress flow' through the joint as is practicable. This may be done by giving proper consideration to the layout of a joint, by making gradual changes in section, and by increasing the amount of material used at points of concentrated load. Weld details should also be determined with this in mind, and unnecessary 'stress-raisers' should be avoided. It will also be advantageous to restrict, where practicable, the locations of joints to low stress regions such as at points of contraflexure or near the neutral axis. Further information and guidance on fatigue design are given in [27–30].

1.3.3 BRITTLE FRACTURE UNDER IMPACT LOAD

Structural steel does not always exhibit a ductile behaviour, and under some circumstances a sudden and catastrophic fracture may occur, even though the nominal tensile stresses are low. Brittle fracture is initiated by the existence or formation of a small crack in a region of high local stress. Once initiated, the crack may propagate in a ductile (or stable) fashion for which the external forces must supply the energy required to tear the steel. More serious are cracks which propagate at high speed in a brittle (or unstable) fashion, for which some of the internal elastic strain energy stored in steel is released and

used to fracture the steel. Such a crack is self-propagating while there is sufficient internal strain energy, and will continue until arrested by ductile elements
in its path which have sufficient deformation capacity to absorb the internal
energy released.

The resistance of a structure to brittle fracture depends on the magnitude of
local stress concentrations, on the ductility of the steel, and on the three-
dimensional geometrical constraints. High local stresses facilitate crack initiation, and so stress concentrations due to poor geometry and loading
arrangements (including impact loading) are dangerous. Also of great importance are flaws and defects in the material, which not only increase the local
stresses, but also provide potential sites for crack initiation.

The ductility of a structural steel depends on its composition, heat treatment, and thickness, and varies with temperature and strain rate. Figure 1.11
shows the increase with temperature of the capacity of the steel to absorb
energy during impact. At low temperatures the energy absorption is low and
initiation and propagation of brittle fractures are comparatively easy, while at
high temperatures the energy absorption is high because of ductile yielding,
and propagation of cracks can be arrested. Between these two extremes is a
transitional range in which crack initiation becomes increasingly difficult. The
likelihood of brittle fracture is also increased by high strain rates due to
dynamic loading, since the consequent increase in the yield stress reduces the
possibility of energy absorption by ductile yielding. The chemical composition
of a steel has a marked influence on its ductility: brittleness is increased by the
presence of excessive amounts of most non-metallic elements, while ductility
is increased by the presence of some metallic elements. A steel with large
grain size tends to be more brittle, and this is significantly influenced by heat

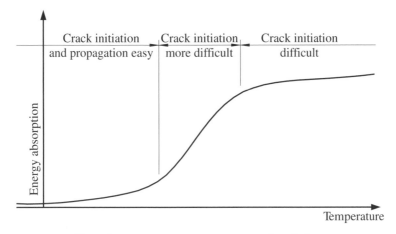

Fig. 1.11 Effect of temperature on resistance to brittle fracture

treatment of the steel, and by its thickness (the grain size tends to be larger in thicker sections). Tables 4 and 5 of BS5950 provide values of the maximum thickness t_1 for different steel grades and minimum service temperatures, while Table 3 provides values of the factor K by which t_1 must be multiplied for different detail types and stress levels.

Three-dimensional geometrical constraints, such as those occurring in thicker or more massive elements, also encourage brittleness, because of the higher local stresses, and because of the greater release of energy during cracking and consequent increase in the ease of propagation of the crack.

The risk of brittle fracture can be reduced by selecting steel types which have ductilities appropriate to the service temperatures, and by designing joints with a view to minimizing stress concentrations and geometrical constraints. Fabrication techniques should be such that they will avoid introducing potentially dangerous flaws or defects. Critical details in important structures may be subjected to inspection procedures aimed at detecting significant flaws. Of course the designer must give proper consideration to the extra cost of special steels, fabrication techniques, and inspection and correction procedures. Further information on brittle fracture is given in [28, 29, 31].

1.4 Member and structure behaviour

1.4.1 MEMBER BEHAVIOUR

Structural steel members are required to transmit axial and transverse forces and moments and torques as shown in Fig. 1.3. The response of a member to these actions can be described by the load-deformation characteristics shown in Fig. 1.12.

A member may have the linear response shown by curve 1 in Fig. 1.12, at least until the material reaches the yield stress. The magnitudes of the deformations depend on the elastic moduli E and G. Theoretically, a member can only behave linearly while the maximum stress does not exceed the yield stress f_y, and so the presence of residual stresses or stress concentrations will cause early non-linearity. However, the high ductility of steel causes a local redistribution after this premature yielding, and it can often be assumed without serious error that the member response remains linear until more general yielding occurs. The member behaviour then becomes non-linear (curve 2) and approaches the condition associated with full plasticity (curve 6). This condition depends on the yield stress f_y.

The member may also exhibit geometric non-linearity, in that the bending moments and torques acting at any section may be influenced by the deformations as well as by the applied forces. This non-linearity, which depends on the elastic moduli E and G, may cause the deformations to become very large (curve 3) as the condition of elastic buckling is approached (curve 4). This

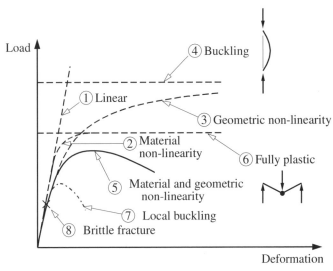

Fig. 1.12 Member behaviour

behaviour is modified when the material becomes non-linear after first yield, and the load may approach a maximum value and then decrease (curve 5).

The member may also behave in a brittle fashion because of local buckling in a thin plate element of the member (curve 7), or because of material fracture (curve 8).

The actual behaviour of an individual member will depend on the forces acting on it. Thus tension members, laterally supported beams, and torsion members remain linear until their material non-linearity becomes important, and then they approach the fully plastic condition. However, compression members and laterally unsupported beams show geometric non-linearity as they approach their buckling loads. Beam-columns are members which transmit both transverse and axial loads, and so they display both material and geometric non-linearities.

1.4.2 STRUCTURE BEHAVIOUR

The behaviour of a structure depends on the load-transferring action of its members and connections. This may be almost entirely by axial tension or compression, as in triangulated structures with joint loading as shown in Fig. 1.13a. Alternatively, the members may support transverse loads which are transferred by bending and shear actions. Usually the bending action dominates in structures composed of one-dimensional members, such as beams and many single-storey rigid frames (Fig. 1.13b), while shear becomes more important in two-dimensional plate structures (Fig. 1.13c). The members of

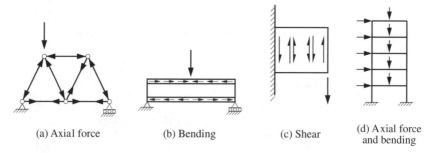

(a) Axial force   (b) Bending   (c) Shear   (d) Axial force and bending

Fig. 1.13 Structural load-transfer actions

many structures are subjected to both axial forces and transverse loads, such as those in multistorey buildings (Fig. 1.13d). The load-transferring action of the members of a structure depends on the arrangement of the structure, including the geometrical layout and the connection details, and on the loading arrangement.

In some structures, the loading and connections are such that the members are effectively independent. For example, in triangulated structures with joint loads, any flexural effects are secondary, and the members can be assumed to act as if pin-jointed, while in rectangular frames with simple flexible connections the moment transfers between beams and columns may be ignored. In such cases, the response of the structure is obtained directly from the individual member responses.

More generally, however, there will be interactions between the members, and the structure behaviour is not unlike the general behaviour of a member, as can be seen by comparing Figs 1.14 and 1.12. Thus, it has been traditional to assume that a steel structure behaves elastically under the service loads. This assumption ignores local premature yielding due to residual stresses and stress concentrations, but these are not usually serious. Purely flexural structures, and purely axial structures with lightly loaded compression members, behave as if linear (curve 1 in Fig. 1.14). However, structures with both flexural and axial actions behave non-linearly, even near the service loads (curve 3 in Fig. 1.14). This is a result of the geometrically non-linear behaviour of its members (see Fig. 1.12).

Most steel structures behave non-linearly near their ultimate loads, unless they fail prematurely due to brittle fracture, fatigue, or local buckling. This non-linear behaviour is due either to material yielding (curve 2 in Fig. 1.14), or member or frame buckling (curve 4), or both (curve 5). In axial structures, failure may involve yielding of some tension members, or buckling either of some compression members or of the frame, or both. In flexural structures, failure is associated with full plasticity occurring at a sufficient number of locations that the structure can form a collapse mechanism. In structures with both

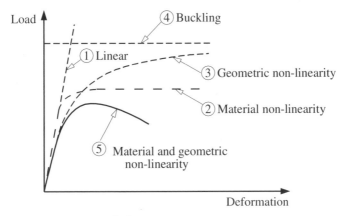

Fig. 1.14 Structure behaviour

axial and flexural actions, there is an interaction between yielding and buckling (curve 5 in Fig. 1.14), and the failure load is often difficult to determine. The transitions shown in Fig. 1.14 between the elastic and ultimate behaviour often take place in a series of non-linear steps as individual elements become fully plastic or buckle.

1.5 Loads

1.5.1 GENERAL

The loads acting on steel structures may be classified as dead loads, as imposed loads, including both gradually applied and dynamic loads, as wind loads, as earth or ground-water loads, or as indirect forces, including those due to temperature changes, foundation settlement, and the like. The structural engineer must evaluate the magnitudes of any of these loads which will act, and must determine which are the most severe combinations of loads for which the structure must be designed. These loads are discussed in the following subsections, both individually and in combinations.

1.5.2 DEAD LOADS

The dead loads acting on a structure arise from the weight of the structure including the finishes, and from any other permanent construction or equipment. The dead loads will vary during construction, but therefore will remain constant, unless significant modifications are made to the structure or its permanent equipment.

The dead load may be assessed from a knowledge of the dimensions and specific weights or from the total weights of all the permanent items which contribute to the total dead load. Guidance on specific weights is given in [14], the values in which are average values representative of the particular materials. The dimensions used to estimate dead loads should also be average and representative, in order that consistent estimates of the dead loads can be made. By making these assumptions, the statistical distribution of dead loads is often taken as being of a Weibull type [32]. The practice sometimes used of consistently overestimating dimensions and specific weights is often wasteful, and may also be dangerous in cases where the dead load component acts in the opposite sense to the resultant load.

1.5.3 IMPOSED LOADS

The imposed loads acting on a structure are gravity loads other than the dead loads, and arise from the weights of materials added to the structure as a result of its use, such as materials stored, people, and snow. Imposed loads usually vary both in space and time. Imposed loads may be sub-divided into two groups, depending on whether they are gradually applied, in which case static load equivalents can be used, or whether they are dynamic, including repeated loads and impact or impulsive loads.

Gradually applied imposed loads may be sustained over long periods of time, or may vary slowly with time [33]. The past practice, however, was to consider only the total imposed load, and so only extreme values (which occur rarely and may be regarded as lifetime maximum loads) were specified. The present imposed loads specified in loading codes [14] often represent peak loads which have 95% probability of not being exceeded over a 50 year period based on a Weibull type distribution [32].

It is usual to consider the most severe spatial distribution of the imposed loads, and this can only be determined by using both the maximum and minimum values of the imposed loads. In the absence of definite knowledge, it is often assumed that the minimum values are zero. When the distribution of imposed load over large areas is being considered, the maximum imposed loads specified, which represent rare events, are often reduced in order to make some allowance for the decreased probability that the maximum imposed loads will act on all areas at the same time.

Dynamic loads which act on structures include both repeated loads and impact and blast loads. Repeated loads are of significance in fatigue problems (see section 1.3.2), in which case the designer is concerned with both the magnitudes, ranges, and number of repetitions of loads which are very frequently applied. At the other extreme, impact loads (which are particularly important in the brittle fracture problems discussed in section 1.3.3) are usually specified by values of extreme magnitude which represent rare events. In structures for which the static loads dominate, it is common to replace the dynamic loads by static force equivalents [14]. However, such a procedure is likely to be inappropriate when the dynamic loads form a significant proportion of the total load, in which case a proper dynamic analysis [34, 35] of the structure and its response should be made.

1.5.4 WIND LOADS

The wind forces which act on structures have traditionally been allowed for by using static force equivalents. The first step is usually to determine a basic wind speed for the general region in which the structure is to be built by using information derived from meteorological studies. This basic wind speed may represent an extreme velocity measured at a height of 10 m and averaged over a period of 3 s which has a return period of 50 years (i.e. a velocity which will, on average, be reached or exceeded once in 50 years, or have a probability of being exceeded of 1/50). The basic wind speed may be adjusted to account for the topography of the site, for the ground roughness, structure size, and height above ground, and for the degree of safety required and the period of exposure. The resulting design wind speed may then be converted into the static pressure which will be exerted by the wind on a plane surface area (this is often referred to as the dynamic wind pressure because it is produced by decelerating the approaching wind velocity to zero at the surface area). The wind force acting on the structure may then be calculated by using pressure coefficients appropriate to each individual surface of the structure, or by using force coefficients appropriate to the total structure. Many values of these coefficients are tabulated in [15], but in special cases where these are inappropriate, the results of wind tunnel tests on model structures may be used.

In some cases it is not sufficient to treat wind loads as static forces. For example, when fatigue is a problem, both the magnitudes and the number of wind fluctuations must be estimated. In other cases, the dynamic response of a structure to wind loads may have to be evaluated (this is often the case with very flexible structures whose long natural periods of vibration are close to those of some of the wind gusts), and this may be done analytically [34, 35], or by specialists using wind tunnel tests. In these cases, special care must be taken to model correctly those properties of the structure which affect its response, including its mass, stiffness, and damping, as well as the wind characteristics and any interactions between wind and structure.

1.5.5 EARTH OR GROUND-WATER LOADS

Earth or ground-water loads act as pressure loads normal to the contact surface of the structure. Such loads are usually considered to be essentially static.

However, earthquake loads are dynamic in nature, and their effects on the structure must be allowed for. Very flexible structures with long natural periods of vibration respond in an equivalent static manner to the high frequencies of earthquake movements, and so can be designed as if loaded by static force equivalents. On the other hand, stiff structures with short natural periods of vibration respond significantly, and so in such a case a proper dynamic analysis [34, 35] should be made. The intensities of earthquake loads vary with the region in which the structure is to be built, but they are not considered to be significant in the UK.

1.5.6 INDIRECT FORCES

Indirect forces may be described as those forces which result from the straining of a structure or its components, and may be distinguished from the direct forces caused by the dead and applied loads and pressures. The straining may arise from temperature changes, from foundation settlement, from shrinkage, creep, or cracking of structural or other materials, and from the manufacturing process as in the case of residual stresses. The values of indirect forces are not usually specified, and so it is common for the designer to determine which of these forces should be allowed for, and what force magnitudes should be adopted.

1.5.7 COMBINATIONS OF LOADS

The different loads discussed in the preceding subsections do not occur alone, but in combinations, and so the designer must determine which combination is the most critical for the structure. However, if the individual loads, which have different probabilities of occurrence and degrees of variability, were combined directly, the resulting load combination would have a greatly reduced probability. Thus, it is logical to reduce the magnitudes of the various components of a combination according to their probabilities of occurrence. This is similar to the procedure used in reducing the live load intensities used over large areas.

The past design practice was to use the worst combination of dead load with live load and/or wind load, and to allow increased stresses whenever the wind load was included (which is equivalent to reducing the load magnitudes). These increases seem to be logical when live and wind loads act together because the probability that both of these loads will attain their maximum values simultaneously is greatly reduced. However, they are unjustified when applied in the case of dead and wind load, for which the probability of occurrence is virtually unchanged from that of the wind load.

A different and more logical method of combining loads is used in the BS5950 limit states design method [8], which is based on statistical analyses of the loads and the structure capacities (see section 1.7.3.4). Strength design is usually carried out for the most severe combination of

(1) $(1.4 \times \text{dead}) + (1.6 \times \text{imposed})$
(2) $(1.4 \times \text{dead}) + (1.4 \times \text{wind})$
(3) $-(1.0 \times \text{dead}) + (1.4 \times \text{wind})$
(4) $(1.2 \times \text{dead}) + (1.2 \times \text{imposed}) + (1.2 \times \text{wind})$

in which the negative sign indicates that the dead load acts in the opposite sense to the wind load. The different factors used for dead, imposed and wind loads reflect the different probabilities of overload associated with each individual load (for example, the probability of a 60% increase in the dead load is much less than that for an equal increase in the imposed load), and with each load combination. The use of load combinations in design according to BS5950 is illustrated in [36].

1.6 Analysis of steel structures

1.6.1 GENERAL

In the design process, the assessment of whether the structural design requirements will be met or not requires a knowledge of the stiffness and strength of the structure under load, and of its local stresses and deformations. The term structural analysis is used to denote the analytical process by which this knowledge of the response of the structure can be obtained. The basis for this process is a knowledge of the material behaviour, and this is used first to analyse the behaviour of the individual members and connections of the structure. The behaviour of the complete structure is then synthesized from these individual behaviours.

The methods of structural analysis are fully treated in many textbooks [e.g. 21–23], and so details of these are not within the scope of this book. However, some discussion of the concepts and assumptions of structural analysis is necessary so that the designer can make appropriate assumptions about the structure and make a suitable choice of the method of analysis.

In most methods of structural analysis, the distribution of forces and moments throughout the structure is determined by using the conditions of static equilibrium and of geometric compatibility between the members at the joints. The way in which this is done depends on whether a structure is statically determinate (in which case the complete distribution of forces and moments can be determined by statics alone), or is statically indeterminate (in which case the compatibility conditions for the deformed structure must also be used before the analysis can be completed).

An important feature of the methods of structural analysis is the constitutive relationships between the forces and moments acting on a member or connection and its deformations. These play the same role for the structural element as do the stress–strain relationships for an infinitesimal element of a structural material. The constitutive relationship may be linear (force proportional to deflection) and elastic (perfect recovery on unloading), or they may be non-linear because of material non-linearities such as yielding (inelastic), or because of geometrical non-linearities (elastic) such as when the deformations themselves induce additional moments, as in stability problems.

It is common for the designer to idealize the structure and its behaviour so as to simplify the analysis. A three-dimensional frame structure may be analysed as a number of independent two-dimensional frames, while individual members are usually considered as one-dimensional and the connections as points. The connections may be assumed to be frictionless hinges, or to be semi-rigid or rigid. In some cases the analysis may be replaced or supplemented by tests made on an idealized model which approximates part or all of the structure.

1.6.2 ANALYSIS OF STATICALLY DETERMINATE MEMBERS AND STRUCTURES

For an isolated statically determinate member, the forces and moments acting on the member are already known, and the structural analysis is only used to determine the stiffness and strength of the member. A linear elastic (or first-order elastic) analysis is usually made of the stiffness of the member when the material non-linearities are generally unimportant and the geometrical non-linearities are often small. The strength of the member, however, is not so easily determined, as one or both of the material and geometric non-linearities are most important. Instead, the designer usually relies on a design code or specification for this information. The strength of isolated statically determinate members is fully discussed in Chapters 2–7 and 10.

For a statically determinate structure, the principles of static equilibrium are used in the structural analysis to determine the member forces and moments, and the stiffness and strength of each member are then determined in the same way as for statically determinate members.

1.6.3 ANALYSIS OF STATICALLY INDETERMINATE STRUCTURES

A statically indeterminate structure can be approximately analysed if a sufficient number of assumptions are made about its behaviour to allow it to be treated as if determinate. One method of doing this is to guess the locations of points of zero bending moment and to assume there are frictionless hinges at a sufficient number of these locations that the member forces and moments can be determined by statics alone. Such a procedure is commonly used in the *preliminary analysis* of a structure, and followed at a later stage by a more precise analysis. However, a structure designed only on the basis of an approximate analysis can still be safe, provided the structure has sufficient ductility to redistribute any excess forces and moments. Indeed, the method is often conservative, and its economy increases with the accuracy of the estimated locations of the points of zero bending moment. More commonly, a preliminary analysis is made of the structure based on the linear elastic computer methods of analysis [37, 38], using approximate member stiffnesses.

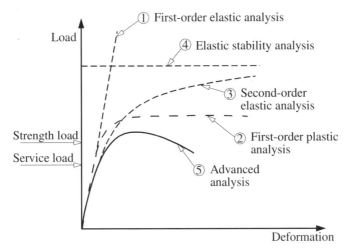

Fig. 1.15 Predictions of structural analyses

The accurate analysis of statically indeterminate structures is complicated by the interaction between members: the equilibrium and compatibility conditions and the constitutive relationships must all be used in determining the member forces and moments. There are a number of different types of analysis which might be made, and some indication of the relevance of these is given in Fig. 1.15 and in the following discussion. Many of these can only be used for two-dimensional frames.

For many structures, it is common to use a *first-order elastic analysis* which is based on linear elastic constitutive relationships and which ignores any geometrical non-linearities and associated instability problems. The deformations determined by such an analysis are proportional to the applied loads, and so the principle of superposition can be used to simplify the analysis. It is often assumed that axial and shear deformations can be ignored in structures whose action is predominantly flexural, and that flexural and shear deformations can be ignored in structures whose member forces are predominantly axial. These assumptions further simplify the analysis, which can then be carried out by any of the well-known methods [21–23], for which many computer programs are available [39, 40]. Some of these programs can be used for three-dimensional frames.

However, a first-order elastic analysis will underestimate the forces and moments in and the deformations of a structure when instability effects are present. Some estimate of the importance of these in the absence of flexural effects can be obtained by making an *elastic stability analysis*. A *second-order elastic analysis* accounts for both flexure and instability, but this is difficult to carry out, although computer programs are now generally available [39, 40].

BS5950 permits the use of the results of an elastic stability analysis in the amplification of the first-order moments as an alternative to second-order analysis.

The analysis of statically indeterminate structures near the ultimate load is further complicated by the decisive influence of the material and geometrical non-linearities. In structures without material non-linearities, an *elastic stability analysis* is appropriate when there are no flexural effects, but this is a rare occurrence. On the other hand, many flexural structures have very small axial forces and instability effects, in which case it is comparatively easy to use a *first-order plastic analysis*, according to which a sufficient number of plastic hinges must form to transform the structure into a collapse mechanism.

More generally, the effects of instability must be allowed for, and as a first approximation the nominal first yield load determined from a second-order elastic analysis may be used as a conservative estimate of the ultimate load. A much more accurate estimate may be obtained for structures where local and lateral buckling is prevented by using an *advanced analysis* [41] in which the actual behaviour is closely analysed by allowing for instability, yielding, residual stresses, and initial crookedness. However, this method is not yet in general use.

1.7 Design of steel structures

1.7.1 STRUCTURAL REQUIREMENTS AND DESIGN CRITERIA

The designer's task of assessing whether or not a structure will satisfy the structural requirements of serviceability and strength is complicated by the existence of errors and uncertainties in his or her analysis of the structural behaviour and estimation of the loads acting, and even in the structural requirements themselves. The designer usually simplifies this task by using a number of design criteria which allow him or her to relate the structural behaviour predicted by his or her analysis to the structural requirements. Thus the designer equates the satisfaction of these criteria by the predicted structural behaviour with satisfaction of the structural requirements by the actual structure.

In general, the various structural design requirements relate to corresponding *limit states*, and so the design of a structure to satisfy all the appropriate requirements is often referred to as a *limit states design*. The requirements are commonly presented in a deterministic fashion, by requiring that the structure shall not fail, or that its deflections shall not exceed prescribed limits. However, it is not possible to be completely certain about the structure and its loading, and so the structural requirements may also be presented in probabilistic forms, or in deterministic forms derived from probabilistic considerations. This may be done by defining an acceptably low risk of failure within the design life of the structure, after reaching some sort of balance between the initial cost of the structure and the economic and human losses resulting from failure. In many cases there will be a number of structural requirements which operate at different load levels, and it is not unusual to require a structure to suffer no damage at one load level, but to permit some minor damage to occur at a higher load level, provided there is no catastrophic failure.

The structural design criteria may be determined by the designer, or he or she may use those stated or implied in design codes. The stiffness design criteria adopted are usually related to the serviceability limit state of the structure under the service loads, and are concerned with ensuring that the structure has sufficient stiffness to prevent excessive deflections such as sagging, distortion, and settlement, and excessive motions under dynamic load, including sway, bounce, and vibration.

The strength limit state design criteria are related to the possible methods of failure of the structure under overload and understrength conditions, and so these design criteria are concerned with yielding, buckling, brittle fracture, and fatigue. Also of importance is the ductility of the structure at and near failure: ductile structures give warning of impending failure and often redistribute load effects away from the critical regions, while ductility provides a method of energy dissipation which will reduce damage due to earthquake and blast loading. On the other hand, a brittle failure is more serious, as it occurs with no warning of failure, and in a catastrophic fashion with a consequent release of stored energy and increase in damage. Other design criteria may also be adopted, such as those related to corrosion and fire.

1.7.2 ERRORS AND UNCERTAINTIES

In determining the limitations prescribed by design criteria, account must be taken of the deliberate and accidental errors made by the designer, and of the uncertainties in his or her knowledge of the structure and its loads. Deliberate errors include those resulting from the assumptions made to simplify the analysis of the loading and of the structural behaviour. These assumptions are often made so that any errors involved are on the safe side, but in many cases the nature of the errors involved is not precisely known, and some possibility of danger exists.

Accidental errors include those due to a general lack of precision, either in the estimation of the loads and the analysis of the structural behaviour, or in the manufacture and erection of the structure. The designer usually attempts to control the magnitudes of these, by limiting them to what he or she judges to be suitably small values. Other accidental errors include what are usually termed blunders. These may be of a gross magnitude leading to failure or to uneconomic structures, or they may be less important. Attempts are usually made to eliminate blunders by using checking procedures, but often these are unreliable, and the logic of such a process is open to debate.

As well as the errors described above, there exists a number of uncertainties about the structure itself and its loads. The material properties of steel fluctuate, especially the yield stress and the residual stresses. The practice of using a minimum or characteristic yield stress for design purposes usually leads to oversafe designs, especially for redundant structures of reasonable size, for which an average yield stress would be more appropriate because of the redistribution of load which takes place after early yielding. Variations in the residual stress levels are not often accounted for in design codes, but there is a growing tendency to adjust design criteria in accordance with the method of manufacture so as to make some allowance for gross variations in the residual stresses. This is undertaken to some extent in BS5950.

The cross-sectional dimensions of rolled steel sections vary, and the values given in section handbooks are only nominal, especially for the thicknesses of universal sections. The fabricated lengths of a structural member will vary slightly from the nominal length, but this is usually of little importance, except where the variation induces additional stresses because of lack-of-fit problems, or where there is a cumulative geometrical effect. Of some significance to members subject to instability problems are variations in their straightness which arise during manufacture, fabrication, and erection. Some allowances for these are usually made in design codes, while fabrication and erection tolerances are specified in BS5950 to prevent excessive crookedness.

The loads acting on a structure vary significantly. Uncertainty exists in the designer's estimate of the magnitude of the dead load because of variations in the densities of materials, and because of minor modifications to the structure during or subsequent to its erection. Usually these variations are not very significant and a common practice is to err on the safe side by making conservative assumptions. Imposed loadings fluctuate significantly during the design usage of the structure, and may change dramatically with changes in usage. These fluctuations are usually accounted for by specifying what appear to be extreme values in loading codes, but there is often a finite chance that these values will be exceeded. Wind loads vary greatly and the magnitudes specified in loading codes are usually obtained by probabilistic methods.

1.7.3 STRENGTH DESIGN

1.7.3.1 Load and capacity factors, and factors of safety

The errors and uncertainties involved in the estimation of the loads on and the behaviour of a structure may be allowed for in strength design by using load factors to increase the nominal loads and capacity factors to decrease the structural strength. In the previous codes that employed the traditional working stress design, this was achieved by using factors of safety to reduce the failure stresses to permissible working stress values. The purpose of using various factors is to ensure that the probability of failure under the most adverse conditions of structural overload and understrength, remains very small. The use of these factors is discussed in the following subsections.

1.7.3.2 Working stress design

The working stress methods of design given in previous codes and specifications required that the stresses calculated from the most adverse combination of loads must not exceed the specified permissible stresses. These specified stresses were obtained after making some allowances for the non-linear stability and material effects on the strength of isolated members, and in effect, were expressions of their ultimate strengths divided by the factors of safety SF. Thus

$$\text{Working stress} \leq \text{Permissible stress} \approx \frac{\text{Ultimate stress}}{\text{SF}}. \qquad (1.6)$$

It was traditional to use factors of safety of 1.7 approximately.

The working stress method of the previous steel design code [42] has been replaced by the limit states design method of BS5950. Detailed discussions of the working stress method are available in the first edition of this book [43].

1.7.3.3 Ultimate load design

The ultimate load methods of designing steel structures required that the calculated ultimate load carrying capacity of the complete structure must not exceed the most adverse combination of loads obtained by multiplying the working loads by the appropriate load factors LF. Thus

$$\Sigma(\text{Working load} \times \text{LF}) \leq \text{Ultimate load.} \qquad (1.7)$$

These load factors allowed some margins for any deliberate and accidental errors, and for the uncertainties in the structure and its loads, and also provided the structure with a reserve of strength. The values of the factors should depend on the load type and combination, and also on the risk of failure that could be expected and the consequences of failure. A simplified approach often employed (perhaps illogically) was to use a single load factor on the most adverse combination of the working loads.

The previous codes and specifications allowed the use of the plastic method of ultimate load design when stability effects were unimportant. These used load factors of 1.70 approximately. However, this ultimate load method has also been replaced in BS5950, and will not be discussed further.

1.7.3.4 Limit states design

It was pointed out in section 1.5.6 that different types of load have different probabilities of occurrence and different degrees of variability, and that the probabilities associated with these loads change in different ways as the degree of overload considered increases. Because of this, different load factors should be used for the different load types.

Thus for limit states design, the structure is deemed to be satisfactory if its *design load effect* does not exceed its *design resistance*. The design load effect S_d^* is an appropriate bending moment, torque, axial force or shear force, and is calculated from the sum of the effects of the specified (or characteristic) loads F_k multiplied by partial factors γ_f which allow for the variabilities of the loads and the structural behaviour. The design resistance f_k/γ_m is calculated from the specified (or characteristic) resistance f_k divided by the partial factor γ_m which allows for the variability of the material properties. Thus

$$\text{Design load effect} \leqslant \text{Design resistance} \tag{1.8a}$$

or

$$\Sigma\gamma_f \times (\text{effect of specified loads}) \leqslant (\text{specified resistance}/\gamma_m) \tag{1.8b}$$

Although the limit states design method is presented in deterministic form in equations 1.8, the partial factors involved are usually obtained by using probabilistic models based on statistical distributions of the loads and the capacities. Typical statistical distributions of the total load and the structural capacity are shown in Fig. 1.16. The probability of failure p_F is indicated by the region for which the load distribution exceeds that for the structural capacity.

In limit state codes, the probability of failure p_F is usually related to a parameter β, called the safety index, by the transformation

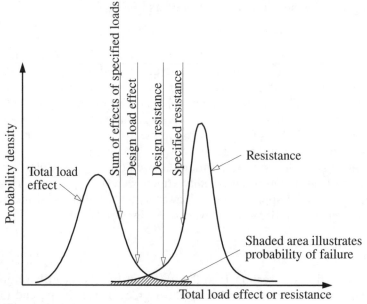

Fig. 1.16 Limit states design

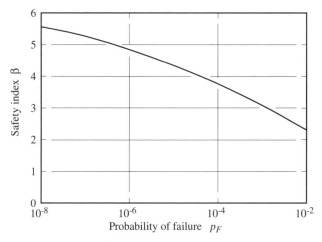

Fig. 1.17 Relationship between safety index and probability of failure

$$\Phi(-\beta) = p_F \tag{1.9}$$

where the function Φ is the cumulative frequency distribution of a standard normal variate [32]. The relationship between β and p_F shown in Fig. 1.17 indicates that an increase in β of 0.5 implies a decrease in the probability of failure by approximately an order of magnitude.

The concept of the safety index was used to derive the partial factors for BS5950. This was done with reference to the previous working stress code BS449 [42] to obtain comparable values of the probability of failure p_F, although much of the detailed calibration treated the load and resistance sides of equations 1.8 separately. Having decided upon values of 1.4 and 1.6 respectively for γ_f for the dead load and imposed loads (these figures having been set on the basis of the likely variabilities of these loads), a mean value of $\gamma_f = 1.5$ was used when considering γ_m. As a result of comparisons with the predictions of BS449 and with test data for selected components, a value of $\gamma_m = 1.0$ was chosen [44].

1.7.4 STIFFNESS DESIGN

In the stiffness design of steel structures, the designer seeks to make the structure sufficiently stiff so that its deflections under the most adverse working load conditions will not impair its strength or serviceability. These deflections are usually calculated by a first-order linear elastic analysis, although the effects of geometrical non-linearities should be included when these are significant, as in structures which are susceptible to instability problems. The design criteria used in the stiffness design relate principally to the serviceability of the structure, in that the flexibility of the structure should not lead to damage of any non-structural components, while the deflections should not be unsightly, and the structure should not vibrate excessively. It is usually left to the designer to choose limiting values for use in these criteria which are appropriate to the structure, although a few values are suggested in some design codes. The stiffness design criteria which relate to the strength of the structure itself are automatically satisfied when the appropriate strength design criteria are satisfied.

1.8 References

1. Steel Construction Institute (1997) *Steelwork Design Guide to BS5950: Part 1: 1990*, 5th edition, SCI, Ascot.
2. O'Connor, C. (1971) *Design of Bridge Superstructures*, John Wiley, New York.
3. Chatterjee, S. (1991) *The Design of Modern Steel Bridges*, BSP Professional Books, Oxford.
4. Rhodes, J. (editor) (1991) *Design of Cold Formed Members*, Elsevier Applied Science, London.
5. Rhodes, J. and Lawson, R.M. (1992) *Design of Steel Structures Using Cold Formed Steel Sections*, Steel Construction Institute, Ascot.
6. Yu, W.-W. (2000) *Cold-Formed Steel Design*, 3rd edition, John Wiley, New York.
7. Hancock, G.J. (1998) *Design of Cold-Formed Steel Structures*, 3rd edition, Australian Institute of Steel Construction, Sydney.
8. British Standards Institution (2000) *BS5950: The Structural Use of Steelwork in Building: Part 1: Code of practice for design: Rolled and welded sections*, BSI, London.
9. Nervi, P.L. (1956) *Structures*, McGraw-Hill, New York.
10. Salvadori, M.G. and Heller, R. (1963) *Structure in Architecture*, Prentice-Hall, Englewood Cliffs, New Jersey.
11. Torroja, E. (1958) *The Structures of Educardo Torroja*, F.W. Dodge Corp., New York.
12. Fraser, D.J. (1981) *Conceptual Design and Preliminary Analysis of Structures*, Pitman Publishing Inc., Marshfield, Massachusetts.
13. Krick, E.V. (1969) *An Introduction to Engineering and Engineering Design*, 2nd edition, John Wiley, New York.
14. British Standards Institution (1996) *BS6399: Loading for Buildings: Part 1: Code of practice for dead and imposed loads*, BSI, London.
15. British Standards Institution (1997) *BS6399: Loading for Buildings: Part 2: Code of practice for wind loads*, BSI, London.
16. Owens, G.W. and Knowles, P.R. (eds) (1992) *Steel Designers' Manual*, 5th edition, Blackwell Scientific Publications, Oxford.
17. CIMsteel (1997) *Design for Construction*, Steel Construction Institute, Ascot.
18. British Standards Institution (1990) *BS5950: The Structural Use of Steelwork in Building: Part 2: Specification for materials, fabrication, and erection: Rolled and welded sections*, BSI, London.

19. Antill, J.M. and Ryan, P.W.S. (1982) *Civil Engineering Construction*, 5th edition, McGraw-Hill, Sydney.
20. Australian Institute of Steel Construction (1991) *Economical Structural Steelwork*, AISC, Sydney.
21. Norris, C.H., Wilbur, J.B. and Utku, S. (1976) *Elementary Structural Analysis*, 3rd edition, McGraw-Hill, New York.
22. Coates, R.C., Coutie, M.G. and Kong, F.K. (1988) *Structural Analysis*, 3rd edition, Van Nostrand Reinhold (UK), Wokingham.
23. Ghali, A. and Neville A.M. (1989) *Structural Analysis – A Unified Classical and Matrix Approach*, Chapman and Hall, London.
24. British Standards Institution (1993) *BS7608: Code of Practice for Fatigue Design and Assessment of Steel Structures*, BSI, London.
25. European Committee for Standardization (1992) *Eurocode 3: Design of Steel Structures – Part 1.1 General Rules and Rules for Buildings*, ECS, Brussels.
26. Miner, M.A. (1945) Cumulative damage in fatigue, *Journal of Applied Mechanics, ASME*, **12**, No. 3, September, pp. A-159–A-164.
27. Gurney, T.R. (1979) *Fatigue of Welded Structures*, 2nd edition, Cambridge University Press.
28. Lay, M.G. (1982) *Structural Steel Fundamentals*, Australian Road Research Board, Melbourne.
29. Rolfe, S.T. and Barsoum, J.M. (1977) *Fracture and Fatigue Control in Steel Structures*, Prentice-Hall, Englewood Cliffs, New Jersey.
30. Grundy, P. (1985) Fatigue limit state for steel structures, *Civil Engineering Transactions*, Institution of Engineers, Australia, CE**27**, No. 1, February, pp. 143–8.
31. Navy Department Advisory Committee on Structural Steels (1970) *Brittle Fracture in Steel Structures* (ed. G.M. Boyd), Butterworth, London.
32. Walpole, R.E. and Myers, R.H. (1978) *Probability and Statistics for Engineers and Scientists*, Macmillan, New York.
33. Ravindra, M.K. and Galambos, T.V. (1978) Load and resistance factor design for steel, *Journal of the Structural Division, ASCE*, **104**, No. ST9, September, pp. 1337–54.
34. Clough, R.W. and Penzien, J. (1975) *Dynamics of Structures*, McGraw-Hill, New York.
35. Irvine, H.M. (1986) *Structural Dynamics for the Practising Engineer*, Allen and Unwin, London.
36. Steel Construction Institute (1991) *Steelwork Design Guide to BS5950: Part 1: 1990. Volume 2. Worked Examples*, 2nd edition, SCI, Ascot.
37. Harrison, H.B. (1973) *Computer Methods in Structural Analysis*, Prentice-Hall, Englewood Cliffs, New Jersey.
38. Harrison, H.B. (1990) *Structural Analysis and Design, Parts 1 and 2*, Pergamon Press, Oxford.
39. Computer Service Consultants (1999) *S-Frame Enterprise*, CSC (UK) Limited, Leeds.
40. Research Engineers Europe Limited (1999) *STAADpro-QSE Space Frame Analysis*, Bristol.
41. Clarke, M.J. (1994) Plastic-zone analysis of frames, Chapter 6 of *Advanced Analysis of Steel Frames: Theory, Software and Applications*, (eds W.F. Chen and S. Toma), CRC Press, Inc., Boca Raton, Florida, pp. 259–319.

42. British Standards Institution (1969) *BS449: Part 2: Specification for the Use of Structural Steel in Building*, BSI, London.
43. Trahair, N.S. (1977) *The Behaviour and Design of Steel Structures*, 1st edition, Chapman and Hall, London.
44. Building Research Establishment (undated) *A Deterministic Calibration of Draft BS5950*, BRE, Garston.

2 Tension members

2.1 Introduction

Concentrically loaded uniform tension members are perhaps the simplest structural elements, as they are nominally in a state of uniform axial stress. Because of this, their load–deformation behaviour very closely parallels the stress–strain behaviour of structural steel obtained from the results of tensile tests (see section 1.3.1). Thus a member remains sensibly linear and elastic until the general yield load is approached, even if it has residual stresses and initial crookedness.

However, in many cases a tension member is not loaded or connected concentrically or it has transverse loads acting, resulting in bending actions as well as an axial tension action. Simple design procedures are available which enable the bending actions in some members with eccentric connections to be ignored, but more generally special account must be taken of the bending action in design.

Tension members often have comparatively high average stresses, and in some cases the effects of local stress concentrations may be significant, especially when there is a possibility that the steel material may not act in a ductile fashion. In such cases, the causes of stress concentrations should be minimized, and the maximum local stresses should be estimated and accounted for.

In this chapter, the behaviour and design of steel tension members are discussed. The case of concentrically loaded members is dealt with first, and then a procedure which allows the simple design of some eccentrically connected tension members is presented. The design of tension members with eccentric or transverse loads is then considered, the effects of stress concentrations are discussed, and finally the design of tension members according to BS5950 is dealt with.

2.2 Concentrically loaded tension members

2.2.1 MEMBERS WITHOUT HOLES

The straight concentrically loaded steel tension member of length L and constant cross-sectional area A which is shown in Fig. 2.1a has no holes and is free from residual stress. The axial extension e of the member varies with the load N in the same way as does the average strain $\varepsilon = e/L$ with the average stress $f = N/A$, and so the load–extension relationship for the member shown in

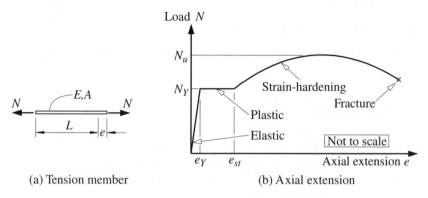

(a) Tension member   (b) Axial extension

Fig. 2.1 Load–extension behaviour of a perfect tension member

Fig. 2.1b is similar to the material stress–strain relationship shown in Fig. 1.6. Thus the extension at first increases linearly with the load and is equal to

$$e = \frac{NL}{EA},\tag{2.1}$$

where E is the Young's modulus of elasticity. This linear increase continues until the yield stress f_y of the steel is reached at the general yield load

$$N_Y = Af_y\tag{2.2}$$

when the extension increases with little or no increase in load until strain-hardening commences. After this, the load increases slowly until the maximum value

$$N_u = Af_u\tag{2.3}$$

is reached, in which f_u is the ultimate tensile strength of the steel. Beyond this, a local cross-section of the member necks down and the load N decreases until fracture occurs.

The behaviour of the tension member is described as ductile, in that it can reach and sustain the general yield load while significant extensions occur, before it fractures. The general yield load N_Y is often taken as the load capacity of the member.

If the tension member is not initially stress free, but has a set of residual stresses induced during its manufacture such as that shown in Fig. 2.2b, then local yielding commences before the general yield load N_Y is reached (Fig. 2.2c), and the range over which the load–extension behaviour is linear decreases. However, the general yield load N_Y at which the whole cross-section is yielded can still be reached because the early yielding causes a redistribution of the stresses. The residual stresses also cause local early strain-hardening, and while the plastic range is shortened (Fig. 2.2c), the member behaviour is still regarded as ductile.

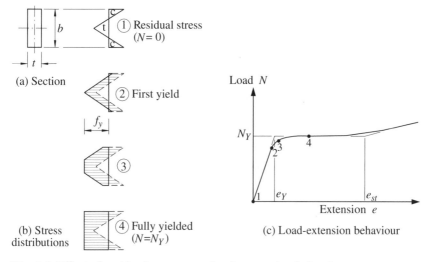

(a) Section

(b) Stress distributions

(c) Load-extension behaviour

Fig. 2.2 Effect of residual stresses on load–extension behaviour

If, however, the tension member has an initial crookedness v_0 (Fig. 2.3a), the axial load causes it to bend and deflect laterally v (Fig. 2.3b). These lateral deflections partially straighten the member so that the bending action in the central region is reduced. The bending induces additional axial stresses (see section 5.3), which cause local early yielding and strain-hardening, in much the same way as do residual stresses (see Fig. 2.2c). The resulting reductions in the linear and plastic ranges are comparatively small when the initial crookedness is small, as is normally the case, and the member behaviour is ductile.

2.2.2 MEMBERS WITH SMALL HOLES

The presence of small local holes in a tension member (such as small bolt holes used for the connections of the member) causes early yielding around the holes, so that the load–deflection behaviour becomes non-linear. When the holes are small, the member may reach the gross yield load

$$N_Y = A_g f_y \tag{2.4}$$

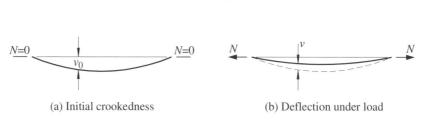

(a) Initial crookedness

(b) Deflection under load

Fig. 2.3 Tension member with initial crookedness

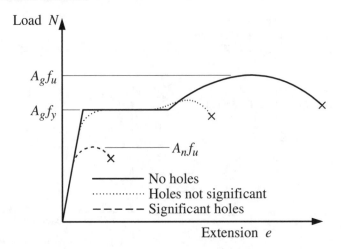

Fig. 2.4 Effect of holes on load–extension behaviour

calculated on the gross area A_g, as shown in Fig. 2.4, because of strain-hardening effects around the holes. In this case, the member behaviour is ductile, and the non-linear behaviour can be ignored because the axial extension of the member under load is not significantly increased, except when there are so many holes along the length of the member that the average cross-sectional area is significantly reduced. Thus the extension e can normally be calculated by using the gross cross-sectional area A_g in equation 2.1.

2.2.3 MEMBERS WITH SIGNIFICANT HOLES

When the holes are large, the member may fail before the gross yield load N_Y is reached by fracturing at a hole, as shown in Fig. 2.4. The local fracture load

$$N_u = A_n f_u \tag{2.5}$$

is calculated on the net area of the cross-section A_n measured perpendicular to the line of action of the load, and is given by

$$A_n = A_g - \Sigma\, dt, \tag{2.6}$$

where d is the diameter of a hole, t the thickness of the member at the hole, and the summation is carried out for all holes in the cross-section under consideration. The fracture load N_u is determined by the weakest cross-section, and therefore by the minimum net area A_n. A member which fails by fracture before the gross yield load can be reached is not ductile, and there is little warning of failure.

In many practical tension members with more than one row of holes, the

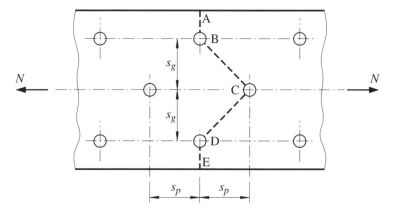

Fig. 2.5 Possible failure path with staggered holes

reduction in the cross-sectional area may be reduced by staggering the rows of holes (Fig. 2.5). In this case, the possibility must be considered of failure along a zig-zag path such as ABCDE in Fig. 2.5, instead of across the section perpendicular to the load. The minimum amount of stagger s_{pm} for which a hole no longer reduces the area of the member depends on the diameter d of the hole and the inclination s_g/s_p of the failure path, where s_g is the gauge distance between the rows of holes. An approximate expression for this minimum stagger is

$$s_{pm} \approx (4s_g d)^{1/2}. \tag{2.7}$$

When the actual stagger s_p is less than s_{pm}, some reduced part of the hole area A_h must be deducted from the gross area A_g, and this can be approximated by $A_h = dt(1 - s_p^2/s_{pm}^2)$, whence

$$A_h = dt(1 - s_p^2/4s_g d), \tag{2.8}$$

$$A_n = A_g - \Sigma dt + \Sigma s_p^2 t/4s_g, \tag{2.9}$$

where the summations are made for all the holes on the zig-zag path considered, and for all the staggers in the path. The use of equation 2.9 is allowed for in Clause 3.4.4.3 of BS5950, and is illustrated in section 2.7.1.

Holes in tension members also cause local stress increases at the hole boundaries, as well as the increased average stresses N/A_n discussed above. These local stress concentrations and their influence on member strength are discussed later in section 2.5.

2.3 Eccentrically connected tension members

In many cases the fabrication of tension members is simplified by making their end connections eccentric. Thus it is common to make connections to an angle section through one leg only, to a tee-section through the flange (or table), or to a channel section through the web (Fig. 2.6). The effect of these eccentric connections is to induce bending moments and bending stresses in the member, so that the stresses are increased. The increased stresses are local to the connection, and decrease away from the connection, and are reduced further by ductile stress redistribution after the onset of yielding.

  While tension members in bending can be designed rationally by using the procedure described in section 2.4, simpler methods [1, 2] also produce satisfactory results. In these simpler methods, the effects of bending are approximated by reducing the cross-sectional area of the member, and by designing it as if concentrically loaded. Thus the reduced area A_r, required by the BS5950 for a single angle bolted through one leg only (Fig 2.6a) is taken as

$$A_r = A_e - 0.5 \, a_2 \tag{2.10}$$

in which A_e is the effective area of the whole section and a_2 is the gross area of the unconnected leg, and for double angles bolted on either side of a plate (Fig. 2.6c) as

$$A_r = A_e - 0.25 \, a_2 \tag{2.11}$$

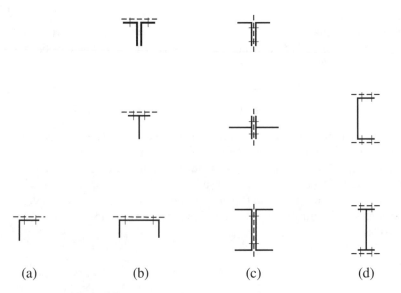

(a)   (b)   (c)   (d)

Fig. 2.6 Eccentrically connected tension members

BS5950 uses similar rules for welded connections, but with smaller reductions. An even simpler method has been suggested [3] in which the net area of an eccentrically connected tension member is reduced by multiplying it by a factor whose value depends on the member and connection types.

It should be noted that there may be significant local stress concentrations in members which are symmetrically connected through some but not all the elements of the cross-section, as in Fig. 2.6d. The effects of these stress concentrations are discussed in section 2.5.

2.4 Bending of tension members

Tension members often have bending actions caused by eccentric connections, eccentric loads, or transverse loads, including self-weight, as shown in Fig. 2.7. These bending actions, which interact with the tensile loads, reduce the ultimate strengths of tension members, and must therefore be accounted for in design.

The axial tension N and transverse deflections v of a tension member caused by any bending action induce restoring moments Nv which oppose the bending action. It is a common (and conservative) practice to ignore these restoring moments which are small when either the bending deflections v or the tensile load N are small. In this case the maximum stress f_{max} in the member can be safely approximated by

$$f_{max} = f_{at} + f_{btx} + f_{bty} \tag{2.12}$$

where $f_{at} = N/A$ is the average tensile stress and f_{btx} and f_{bty} are the maximum tensile bending stresses caused by the major and minor axis bending actions M_x and M_y alone (see section 5.3). The nominal first yield of the member therefore occurs when $f_{max} = f_y$, whence

$$f_{at} + f_{btx} + f_{bty} = f_y. \tag{2.13}$$

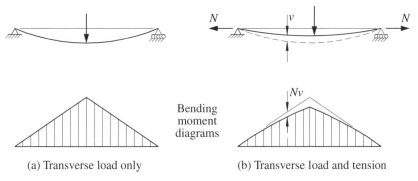

Bending moment diagrams

(a) Transverse load only   (b) Transverse load and tension

Fig. 2.7 Bending of a tension member

When either the tensile load N or the moments M_x and M_y are not small, the first yield prediction of equation 2.13 is inaccurate. A suggested interaction equation for failure is the linear equation

$$\frac{M_x}{M_{tx}} + \frac{M_y}{M_{ry}} = 1 \tag{2.14}$$

where M_{ry} is the cross-section strength for tension and bending about the minor principal (y) axis given by

$$M_{ry} = M_{sy}(1 - N/N_t) \tag{2.15}$$

and M_{tx} is the lesser of the cross-section strength M_{rx} for tension and bending about the major principal (x) axis given by

$$M_{rx} = M_{sx}(1 - N/N_t) \tag{2.16}$$

and the out-of-plane member buckling strength M_{btx} for tension and bending about the major principal axis given by

$$M_{btx} = M_{bx}(1 + N/N_t) \leqslant M_{sx} \tag{2.17}$$

in which M_{bx} is the lateral buckling capacity (see Chapter 6) when $N = 0$.

In these equations, N_t is the tensile strength in the absence of bending (taken as the lesser of N_Y and N_u), while M_{sx} and M_{sy} are the section strengths for bending alone about the x and y axes (see sections 4.7.2 and 5.6.1.3). Equation 2.14 is similar to the first yield condition of equation 2.13, but includes a simple approximation for the possibility of lateral buckling under large values of M_x through the use of equation 2.17.

2.5 Stress concentrations

High local stress concentrations are not usually important in ductile materials under static loading situations, because the local yielding resulting from these concentrations causes a favourable redistribution of stress. However, in situations for which it is doubtful whether the steel will behave in a ductile manner, as when there is a possibility of brittle fracture of a tension member under dynamic loads (see section 1.3.3), or when repeated load applications may lead to fatigue failure (see section 1.3.2), stress concentrations become very significant. It is usual to try to avoid or minimize stress concentrations by providing suitable joint and member details, but this is not always possible, in which case some estimate of the magnitudes of the local stresses must be made.

Stress concentrations in tension members occur at holes in the member, and where there are changes or very local reductions in the cross-section, and at points where concentrated forces act. The effects of a hole in a tension member (of net width b_n and thickness t) are shown by the uppermost curve in Fig. 2.8, which is a plot of the variation of the stress concentration factor (the

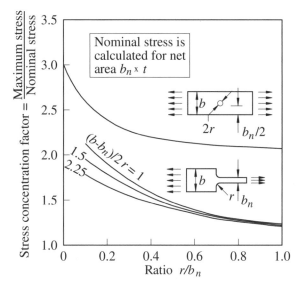

Fig. 2.8 Stress concentrations in tension members

ratio of the maximum tensile stress to the nominal stress averaged over the reduced cross-section $b_n t$) with the ratio of the hole radius r to the net width b_n. The maximum stress approaches three times the nominal stress when the plate width is very large ($r/b_n \to 0$).

This curve may also be used for plates with a series of holes spaced equally across the plate width, provided b_n is taken as the minimum width between adjacent holes. The effects of changes in the cross-section of a plate in tension are shown by the lower curves in Fig. 2.8, while the effects of some notches or very local reductions in the cross-section are given in [4]. Methods of analysing these and other stress concentrations are discussed in [5].

Large concentrated forces are usually transmitted to structural members by local bearing, and while this produces high local compressive stresses, any subsequent plastic yielding is not usually serious. Of more importance are the increased tensile stresses which result when only some of the plate elements of a tension member are loaded (see Fig. 2.6d). In this case conservative estimates can be made of these stresses by assuming that the resulting bending and axial actions are resisted solely by the loaded elements.

2.6 Design of tension members

2.6.1 GENERAL

In order to design or to check a tension member, the factored tensile force F_t^* in the member is obtained by a frame analysis (see Chapter 8) or by statics if the member is statically determinate, using the appropriate loads and partial load factors γ_f (see section 1.7). If there is substantial bending present, the factored moments M_x^* and M_y^* about the major and minor principal axes respectively are also obtained.

The process of checking a specified tension member or of designing an unknown member is summarised in Fig. 2.9 for the case where bending actions can be ignored. In this figure, the factored design tensile force F_t^* is distinguished by the addition of *, while the symbol p_y for the BS5950 design strength (usually taken as the specified minimum yield strength Y_s) replaces the symbol f_y used in earlier sections for the yield stress. If a specified member is to be checked, then both the strength limit states of gross yielding and net fracture are implicitly considered, so that both the design strength p_y and the effective net area factor K_e are required. When the member size is not known, an approximate target area A_g can be established. The trial member selected may be checked for the fracture limit state once its holes, connection eccentricities, and effective net area A_e have been established, and modified if necessary.

The following sub-sections describe each of the BS5950 check and design processes for a statically loaded tension member. Examples of their application are given in sections 2.7.1–2.7.5.

2.6.2 CONCENTRICALLY LOADED TENSION MEMBERS

The BS5950 method of strength design of tension members which are loaded concentrically follows the philosophy of section 2.2, with the two separate limit states of yield of the gross section and fracture of the net section represented by a single equation

$$P_t = p_y A_e \qquad (2.18)$$

for the tension capacity P_t.

The design strength p_y in this equation is usually determined from

$$p_y = Y_s \qquad (2.19)$$

in which Y_s is the specified minimum yield strength, except for steels with $U_s/Y_s < 1.2$, in which case the design strength is taken as $U_s/1.2$.

The effective area A_e in equation 2.18 is given by

$$A_e = A_g \qquad (2.20)$$

except when the net area A_n is less than A_g/K_e, in which case

$$A_e = K_e A_n \qquad (2.21)$$

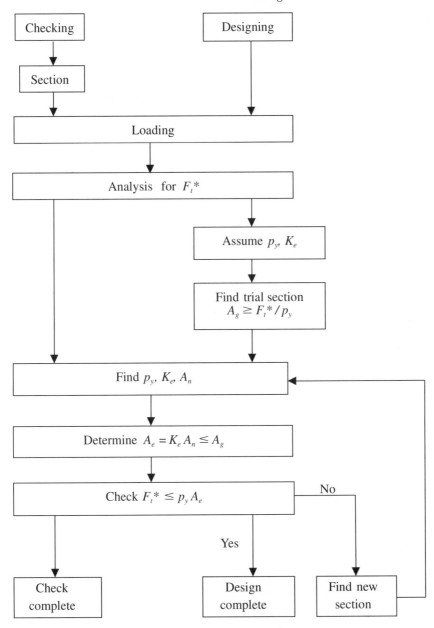

Fig. 2.9 Flow chart for the design of tension members

in which the effective net area coefficient K_e is equal to 1.2 for S275 steel, to 1.1 for S355 steel, to 1.0 for S460 steel, and to $U_s/(1.2\,p_y)$ for other steels. The net area of a cross-section A_n is obtained from the gross area A_g by deducting the area of all holes using equation 2.9.

When $A_e = A_g$, then equation 2.18 provides a capacity based on yielding of the gross section A_g at p_y, but when A_e is reduced to $K_e A_n$, then equation 2.18 can be considered as providing a check on fracture of the net section A_n at a reduced tensile strength given by $K_e p_y$.

Worked examples of concentrically loaded tension members are given in sections 2.7.1–2.7.4.

2.6.3 ECCENTRICALLY CONNECTED TENSION MEMBERS

The BS5950 method of strength design of tension members which are connected eccentrically as shown in Fig. 2.6 is similar to the method discussed in section 2.6.2 for concentrically loaded members, except that the effective area A_e used in equation 2.18 is replaced by a reduced area A_r calculated as in equations 2.10 or 2.11 for bolted connections, or by somewhat higher reduced areas for welded connections.

2.6.4 TENSION MEMBERS WITH BENDING

Tension members with factored bending moments M_x^* and M_y^* are required by BS5950 to satisfy the inequalities

$$\frac{F_t^*}{P_t} + \frac{M_x^*}{M_{cx}} + \frac{M_y^*}{M_{cy}} \leqslant 1 \tag{2.22}$$

and

$$M_x^* \leqslant M_b/m_{LT} \tag{2.23}$$

where M_{cx} and M_{cy} are the section moment capacities (sections 4.7.2 and 5.6.1.3) and M_b/m_{LT} is the lateral buckling moment capacity (section 6.4). These equations are similar in effect to equations 2.14–2.17, except that equation 2.23 conservatively omits the strengthening effect of tension on lateral buckling that is approximated by equation 2.17.

BS5950 also provides as an alternative to equation 2.22 the more accurate interaction equation

$$\left(\frac{M_x^*}{M_{rx}}\right)^{z_1} + \left(\frac{M_y^*}{M_{ry}}\right)^{z_2} \leqslant 1 \tag{2.24}$$

for Class 1 plastic or Class 2 compact doubly symmetric sections (see section 4.7.2), in which M_{rx} and M_{ry} are the full plastic moment capacities about the major (x) and minor (y) axes reduced for axial force (see section 7.2.4.1), and the values of z_1 and z_2 depend on the cross-section type ($z_1 = 2.0$ and $z_2 = 1.0$ for I-sections).

A worked example of a tension member with bending actions is given in section 2.7.5.

2.7 Worked examples

2.7.1 EXAMPLE 1 – NET AREA OF A BOLTED UNIVERSAL COLUMN SECTION MEMBER

Problem. Both flanges of a universal column section member have 22 mm diameter holes arranged as shown in Fig. 2.10a. If the gross area of the section is 201×10^2 mm^2 and the flange thickness is 25 mm, determine the net area A_n of the member which is effective in tension.

Solution. Using equation 2.7, the minimum stagger is $s_{pm} = \sqrt{(4 \times 60 \times 22)} = 72.7$ mm > 30 mm $= s_p$. The failure path through each flange is therefore staggered, and by inspection, it includes four holes and two staggers. The net area can therefore be calculated from equation 2.9 (or Clause 3.4.4.3) as

$$A_n = [201 \times 10^2] - [2 \times 4 \times (22 \times 25)] + [2 \times 2 \times (30^2 \times 25)/(4 \times 60)]$$
$$= 161 \times 10^2 \text{ mm}^2.$$

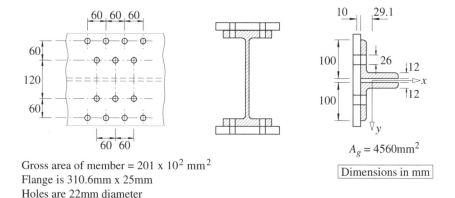

Gross area of member = 201×10^2 mm^2
Flange is 310.6mm x 25mm
Holes are 22mm diameter

$A_g = 4560$mm^2

Dimensions in mm

(a) Flange of a universal   (b) Universal beam   (c) Double angle
    column section   section   section

Fig. 2.10 Examples 1–6

2.7.2 EXAMPLE 2 – CHECKING A BOLTED UNIVERSAL COLUMN SECTION MEMBER

Problem. Determine the tension capacity of the tension member of example 1 if the design strength is $p_y = 345$ N/mm².

Solution.   From section 2.7.1, $A_n = 161 \times 10^2$ mm²
For S355 steel, $K_e = 1.1$ 3.4.3
$\qquad K_e A_n = 1.1 \times 161 \times 10^2 = 177.1 \times 10^2$ mm² $< 201 \times 10^2 = A_g$ 3.4.3
$\qquad A_e = 177.1 \times 10^2$ mm²   3.4.3
$\qquad P_t = 345 \times 177.1 \times 10^2$ N $= 6110$ kN   4.6.1

2.7.3 EXAMPLE 3 – CHECKING A BOLTED UNIVERSAL BEAM SECTION MEMBER

Problem.   A 610×229 UB125 tension member of S355 steel is connected through both flanges by 20 mm bolts in four lines, two in each flange as shown in Fig. 2.10b. Check the member for a factored tension force of $F_t^* = 4000$ kN.

Solution. $T = 19.6$ mm, $p_y = 345$ N/mm²   T9
$\qquad A_g = 15\ 900$ mm²
$\qquad A_n = 15\ 900 - 4 \times 22 \times 19.6 = 14\ 175$ mm²   3.4.4.2
For S355 steel, $K_e = 1.1$ 3.4.3
$\qquad K_e A_n = 1.1 \times 14\ 175 = 15\ 593$ mm² $< 15\ 900$ mm² $= A_g$ 3.4.3
$\qquad A_e = 15\ 593$ mm²   3.4.3
$\qquad P_t = 345 \times 15\ 593$ N $= 5380$ kN > 4000 kN $= F_t^*$ 4.6.1

and so the member is satisfactory.

2.7.4 EXAMPLE 4 – CHECKING AN ECCENTRICALLY CONNECTED DOUBLE ANGLE

Problem.   A tension member consists of two equal angles whose ends are connected to gusset plates as shown in Fig. 2.10c. Use the approximate method of section 2.3 to determine the tension capacity of the member if the design strength is $p_y = 355$ N/mm².

Solution. $t = 12$ mm, $p_y = 355$ N/mm²   T9
$\qquad A_n = 4560 - 2 \times 26 \times 12 = 3936$ mm²   3.4.4.2
For S355 steel, $K_e = 1.1$ 3.4.3
$\qquad K_e A_n = 1.1 \times 3936 = 4330$ mm² < 4560 mm² $= A_g$ 3.4.3
$\qquad A_e = 4330$ mm²   3.4.3
$\qquad a_1 = 2 \times 100 \times 12 = 2400$ mm²   4.6.3.1
$\qquad a_2 = 4560 - 2400 = 2160$ mm²   4.6.3.1
$\qquad P_t = 355 \times (4330 - 0.5 \times 2160)$ N $= 1154$ kN   4.6.3.1, 2

2.7.5 EXAMPLE 5 – CHECKING A MEMBER UNDER COMBINED TENSION AND BENDING

Problem. If the load eccentricity for the tension member of section 2.7.4 is 39.1 mm and the tension and bending capacities are 1537 kN and 31.3 kNm, determine the capacity of the member by treating it as a member under combined tension and bending.

Solution. Substituting into equation 2.22 leads to

$$F_t^*/1537 + F_t^* \times (39.1/1000)/31.3 \leqslant 1 \qquad\qquad 4.8.2.2$$

so that

$$F_t^* \leqslant 526 \text{ kN}.$$

Because the load eccentricity causes the member to bend about its minor axis, there is no need to check for lateral buckling, which only occurs when there is major axis bending.

It can be seen that the value of 526 kN obtained for the design capacity is substantially less than the approximate value of 1154 kN obtained as the solution of example 4. However, this does not necessarily indicate that the empirical method of section 2.3 is inadequate, since the method of section 2.4 neglects the beneficial effects of end restraints which substantially reduce the net end and central moments, and the deflections which reduce the central moment.

2.7.6 EXAMPLE 6 – ESTIMATING THE STRESS CONCENTRATION FACTOR

Problem. Estimate the maximum stress concentration factor for the tension member of section 2.7.1.

Solution. For the inner line of holes, the net width is

$$b_n = 60 + 30 - 22 = 68 \text{ mm},$$

and so

$$r/b_n = (22/2)/68 = 0.16,$$

and so using Fig. 2.8, the stress concentration factor is approximately 2.5.

However, the actual maximum stress is likely to be greater than 2.5 times the nominal average stress calculated from the effective area $A_n = 161 \times 10^2$ mm² determined in section 2.7.1, because the unconnected web is not completely effective. A safe estimate of the maximum stress can be determined from an effective area

$$A_n = [2 \times 310.6 \times 25] - [8 \times 22 \times 25] + [4 \times (30^2 \times 25)/(4 \times 60)]$$
$$= 115.1 \times 10^2 \text{ mm}^2,$$

calculated from the flange areas only.

2.8 Unworked examples

2.8.1 EXAMPLE 7 – BOLTING ARRANGEMENT

A channel section tension member has an overall depth of 381 mm, width of 101.6 mm, flange and web thicknesses of 16.3 and 10.4 mm, respectively, and a design strength of 345 N/mm². Determine an arrangement for bolting the web of the channel to a gusset plate so as to minimize the gusset plate length and maximize the tension member capacity.

2.8.2 EXAMPLE 8 – CHECKING A CHANNEL SECTION MEMBER CONNECTED BY BOLTS

Determine the axial load capacity of the tension member of section 2.8.1.

2.8.3 EXAMPLE 9 – CHECKING A CHANNEL SECTION MEMBER CONNECTED BY WELDS

If the tension member of section 2.8.1 is fillet welded to the gusset plate instead of bolted, determine its axial load capacity.

2.8.4 EXAMPLE 10 – CHECKING A MEMBER UNDER COMBINED TENSION AND BENDING

The beam of example 1 in section 6.13.1 has a concentric axial tensile load $5Q^*$ in addition to the central transverse load Q^*. Determine the maximum value of Q^*.

2.9 References

1. Regan, P.E. and Salter, P.R. (1984) Tests on welded-angle tension members, *Structural Engineer*, **62B**(2), pp. 25–30.
2. Nelson, H.M. (1953) *Angles in Tension*. Publication No. 7, British Constructional Steel Association, pp. 9–18.
3. Bennetts, I.D., Thomas, I.R. and Hogan, T.J. (1986) Design of statically loaded tension members, *Civil Engineering Transactions*, Institution of Engineers, Australia, CE**28**, No. 4, November, pp. 318–27.
4. Roark, R.J. (1965) *Formulas for Stress and Strain*, 4th edition, McGraw-Hill, New York.
5. Timoshenko, S.P. and Goodier, J.N. (1970) *Theory of Elasticity*, 3rd edition, McGraw-Hill, New York.

3 Compression members

3.1 Introduction

The compression member is the second type of axially loaded structural element, the first type being the tension member discussed in Chapter 2. Very stocky compression members behave in the same way as do tension members until after the material begins to flow plastically at the squash load $N_Y = Af_y$. However, the strength of a compression member decreases as its length increases, in contrast to the axially loaded tension member whose strength is independent of its length. Thus, the compressive strength of a very slender member may be much less than its tensile strength, as shown in Fig. 3.1.

This decrease in strength is caused by the action of the applied compressive load N which causes bending in a member with initial curvature (see Fig. 3.2a). In a tension member, the corresponding action decreases the initial curvature, and so this effect is usually ignored. However, the curvature and the lateral deflection of a compression member increase with the load, as shown in Fig. 3.2b. The compressive stresses on the concave side of the member also increase until the member fails due to excessive yielding. This bending action is accentuated by the slenderness of the member, and so the strength of a compression member decreases as its length increases.

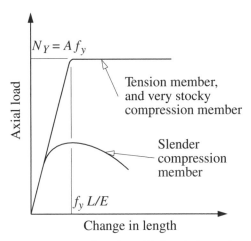

Fig. 3.1 Strengths of axially loaded members

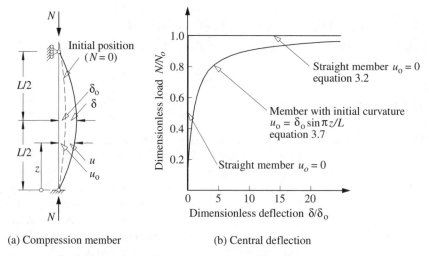

(a) Compression member   (b) Central deflection

Fig. 3.2 Elastic behaviour of a compression member

For the hypothetical limiting case of a perfectly straight elastic member, there is no bending until the applied load reaches the elastic buckling value N_o. At this load, the compression member begins to deflect laterally, as shown in Fig. 3.2b, and these deflections grow until failure occurs at the beginning of compressive yielding. This action of suddenly deflecting laterally is called flexural buckling.

The elastic buckling load N_o provides a measure of the slenderness of a compression member, while the squash load N_Y gives an indication of its resistance of yielding. In this chapter, the influences of the elastic buckling load and the squash load on the behaviour of concentrically loaded compression members are discussed and related to their design according to BS5950. Local buckling of thin plate elements in compression members is treated in Chapter 4, while the behaviour and design of eccentrically loaded compression members are discussed in Chapter 7, and compression members in frames in Chapter 8.

3.2 Elastic compression members

3.2.1 BUCKLING OF STRAIGHT MEMBERS

A perfectly straight member of a linear elastic material is shown in Fig. 3.3a. The member has a frictionless hinge at each end, its lower end being fixed in position while its upper end is free to move vertically but is prevented from deflecting horizontally. It is assumed that the deflections of the member remain small.

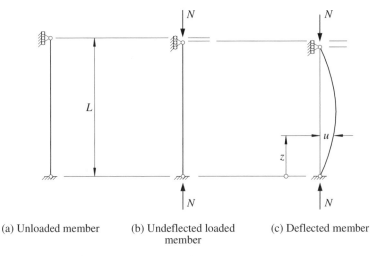

(a) Unloaded member   (b) Undeflected loaded   (c) Deflected member
member

Fig. 3.3 Straight compression member

The unloaded position of the member is shown in Fig. 3.3a. A concentric axial load N is applied to the upper end of the member, which remains straight (Fig. 3.3b). The member is then deflected laterally by a small amount u as shown in Fig. 3.3c, and is held in this position. If the original straight position of Fig. 3.3b is one of stable equilibrium, the member will return to it when released from the deflected position, while if the original position is one of unstable equilibrium, the member will collapse away from it when released. When the equilibrium of the original position is neither stable nor unstable, the member is in a condition described as one of neutral equilibrium. For this case, the deflected position is one of equilibrium, and the member will remain in this position when released. Thus, when the load N reaches the elastic buckling value N_o at which the original straight position of the member is one of neutral equilibrium, the member may deflect laterally without any change in the load, as shown in Fig. 3.2c.

The load N_o at which a straight compression member buckles laterally can be determined by finding a deflected position which is one of equilibrium. It is shown in section 3.8.1 that this position is given by

$$u = \delta \sin \pi z/L \qquad (3.1)$$

in which δ is the undetermined magnitude of the central deflection, and that the elastic buckling load is

$$N_o = \pi^2 EI/L^2 \qquad (3.2)$$

in which EI is the flexural rigidity of the compression member.

The elastic buckling N_o and the elastic buckling stress

$$p_E = N_o/A \qquad (3.3)$$

can be expressed in terms of the slenderness ratio L/r by

$$N_o = p_E A = \frac{\pi^2 EA}{(L/r)^2} \qquad (3.4)$$

in which $r = \sqrt{(I/A)}$ is the radius of gyration (which can be determined for a number of sections by using Fig. 5.6). The buckling load varies inversely as the square of the slenderness ratio L/r, as shown in Fig. 3.4, in which the dimensionless buckling load N_o/N_Y is plotted against a modified slenderness ratio

$$\sqrt{\left(\frac{N_Y}{N_o}\right)} = \sqrt{\left(\frac{f_y}{p_E}\right)} = \frac{L}{r}\sqrt{\left(\frac{f_y}{\pi^2 E}\right)} \qquad (3.5)$$

in which $N_Y = Af_y$ is the squash load. If the material ceases to be linear elastic at the yield stress f_y, then the above analysis is only valid for $\sqrt{(N_Y/N_o)} = \sqrt{(f_y/p_E)} \geq 1$. This limit is equivalent to a slenderness ratio L/r of approximately 85 for a material with a yield stress f_y of 275 N/mm^2.

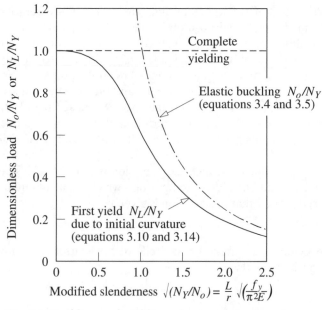

Fig. 3.4 Buckling and yielding of compression members

3.2.2 BENDING OF MEMBERS WITH INITIAL CURVATURE

Real structural members are not perfectly straight, but have small initial curvatures, as shown in Fig. 3.2a. The buckling behaviour of the hypothetical straight members discussed in section 3.2.1 must therefore be interpreted as the limiting behaviour of real members with infinitesimally small initial curvatures. The initial curvature of the real member causes it to bend from the commencement of application of the axial load, and this increases the maximum stress in the member.

If the initial curvature is such that

$$u_0 = \delta_0 \sin \pi z/L, \tag{3.6}$$

then the deflection of member is given by

$$u = \delta \sin \pi z/L, \tag{3.7}$$

where

$$\frac{\delta}{\delta_0} = \frac{N/N_o}{1 - N/N_o}, \tag{3.8}$$

as shown in section 3.8.2. The variation of the dimensionless central deflection δ/δ_0 is shown in Fig. 3.2b, and it can be seen that deflection begins at the commencement of loading and increases rapidly as the elastic buckling load N_o is approached.

The simple load–deflection relationship of equation 3.8 is the basis of the Southwell plot technique for extrapolating the elastic buckling load from experimental measurements. If equation 3.8 is rearranged as

$$\frac{\delta}{N} = \frac{1}{N_o}\delta + \frac{\delta_0}{N_o}, \tag{3.9}$$

then the linear relation between δ/N and δ shown in Fig. 3.5 is obtained. Thus, if a straight line is drawn which best fits the points determined from experimental measurements of N and δ, the reciprocal of the slope of this line gives an experimental estimate of the buckling load N_o. An estimate of the magnitude δ_0 of the initial crookedness can also be determined from the intercept on the horizontal axis.

As the deflections u increase with the load N, so also do the bending moments and the stresses. It is shown in section 3.8.2 that the limiting axial load N_L at which the compression member first yields is given by

$$\frac{N_L}{N_Y} = \left[\frac{1 + (1 + \eta)N_o/N_Y}{2}\right] - \left\{\left[\frac{1 + (1 + \eta)N_o/N_Y}{2}\right]^2 - \frac{N_o}{N_Y}\right\}^{1/2}, \tag{3.10}$$

where

$$\eta = \frac{\delta_0 b}{2r^2}, \tag{3.11}$$

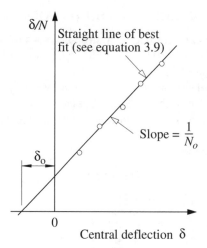

Fig. 3.5 Southwell plot

and b is the width of the member. Alternatively, the limiting compressive stress $p_L = N_L/A$ at which the member first yields is given by

$$p_L = \frac{p_E f_y}{\phi + \sqrt{(\phi^2 - p_E f_y)}} \qquad (3.12)$$

in which

$$\phi = \frac{f_y + (1 + \eta)p_E}{2} \qquad (3.13)$$

The variation of the dimensionless limiting axial load $N_L/N_Y = p_L/f_y$ with the modified slenderness ratio $\sqrt{(N_Y/N_o)} = \sqrt{(f_y/p_E)}$ is shown in Fig. 3.4 for the case when

$$\eta = \frac{1}{4} \frac{N_Y}{N_o} = \frac{1}{4} \frac{f_y}{p_E}. \qquad (3.14)$$

For stocky members, the limiting load N_L approaches the squash load N_Y, while for slender members the limiting load approaches the elastic buckling load N_o. Equation 3.12 is the basis for the member compression strength of BS5950.

3.3 Inelastic compression members

3.3.1 TANGENT MODULUS THEORY OF BUCKLING

The analysis of a perfectly straight elastic compression member given in section 3.2.1 applies only to a material whose stress–strain relationship remains linear. However, the buckling of an elastic member of a non-linear material,

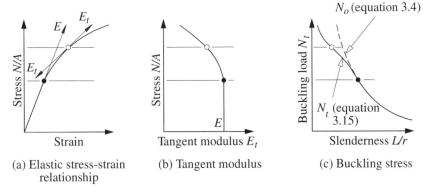

Fig. 3.6 Tangent modulus theory of buckling

such as that whose stress–strain relationship is shown in Fig. 3.6a, can be analysed by a simple modification of the linear elastic treatment. It is only necessary to note that the small bending stresses and strains, which occur during buckling, are related by the tangent modulus of elasticity E_t corresponding to the average compressive stress N/A (Fig. 3.6a and b), instead of the initial modulus E. Thus the flexural rigidity is reduced from EI to E_tI, and the tangent modulus buckling load N_t is obtained from equation 3.3 by substituting E_t for E, whence

$$N_t = \frac{\pi^2 E_t A}{(L/r)^2}.$$ (3.15)

The variation of this tangent modulus buckling load N_t is shown in Fig. 3.6c.

3.3.2 REDUCED MODULUS THEORY OF BUCKLING

The tangent modulus theory of buckling is only valid for elastic materials. For inelastic nonlinear materials, the changes in the stresses and strains are related by the initial modulus E when the total strain is decreasing, and the tangent modulus E_t only applies when the total strain is increasing, as shown in Fig. 3.7. The flexural rigidity E_rI of an inelastic member during buckling therefore depends on both E and E_t. As a direct consequence of this, the effective (or reduced) modulus of the section E_r depends on the geometry of the cross-section as well. It is shown in section 3.9.1 that the reduced modulus of a rectangular section is given by

$$E_r = \frac{4EE_t}{(\sqrt{E} + \sqrt{E_t})^2}.$$ (3.16)

The reduced modulus buckling load N_r can be obtained by substituting E_r for E in equation 3.3, whence

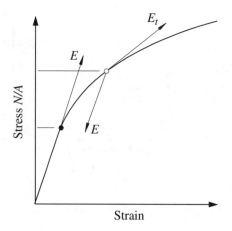

Fig. 3.7 Inelastic stress–strain relationship

$$N_r = \frac{\pi^2 E_r A}{(L/r)^2} . \tag{3.17}$$

Because the tangent modulus E_t is less than the initial modulus E, it follows that $E_t < E_r < E$, and so the reduced modulus buckling load N_r lies between the tangent modulus load N_t and the elastic buckling load N_o, as indicated in Fig. 3.8.

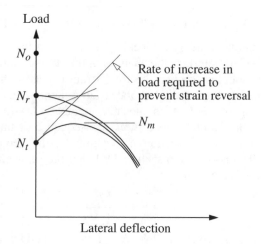

Fig. 3.8 Shanley's theory of inelastic buckling

3.3.3 SHANLEY'S THEORY OF INELASTIC BUCKLING

Although the tangent modulus theory appears to be invalid for inelastic materials, careful experiments have shown that it leads to more accurate predictions than the apparently rigorous reduced modulus theory. This paradox was resolved by Shanley [1], who reasoned that the tangent modulus theory is valid when buckling is accompanied by a simultaneous increase in the applied load (see Fig. 3.8) of sufficient magnitude to prevent strain reversal in the member. When this happens, all the bending stresses and strains are related by the tangent modulus of elasticity E_t, and so the buckling load is equal to the tangent modulus value N_t.

As the lateral deflection of the member increases as shown in Fig. 3.8, the tangent modulus E_t decreases (see Fig. 3.6b) because of the increased axial and bending strains, and the post-buckling curve approaches a maximum load N_m which defines the ultimate strength of the member. Also shown in Fig. 3.8 is a post-buckling curve which commences at the reduced modulus load N_r (at which buckling can take place without any increase in the load). The tangent modulus load N_t is the lowest load at which buckling can begin, and the reduced modulus load N_r is the highest load for which the member can remain straight. It is theoretically possible for buckling to begin at any load between N_t and N_r.

It can be seen that not only is the tangent modulus load more easily calculated, but it also provides a conservative estimate of the member strength, and is in closer agreement with experimental results than the reduced modulus load. For these reasons, the tangent modulus theory of inelastic buckling has gained wide acceptance.

3.3.4 BUCKLING OF MEMBERS WITH RESIDUAL STRESSES

The presence of residual stresses in an intermediate length steel compression member may cause a significant reduction in its buckling strength. Residual stresses are established during the cooling of a hot-rolled or welded steel member. The shrinking of the late-cooling regions of the member induces residual compressive stresses in the early-cooling regions, and these are balanced by equilibrating tensile stresses in the late-cooling regions. In hot-rolled I-section members, the flange–web junctions are least exposed to cooling influences, and so these are regions of residual tensile stress, as shown in Fig. 3.9, while the more exposed flange tips are regions of residual compressive stress. In a straight intermediate length compression member, the residual compressive stresses cause premature yielding under reduced axial loads, as shown in Fig. 3.10, and the member buckles inelastically at a load which is less than the elastic buckling load N_o.

In applying the tangent modulus concept of inelastic buckling to a steel which has the stress–strain relationships shown in Fig. 1.6 (see Chapter 1), the

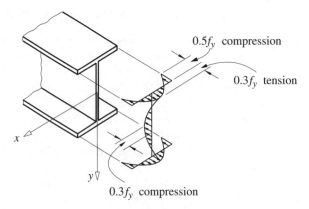

Fig. 3.9 Idealized residual stress pattern

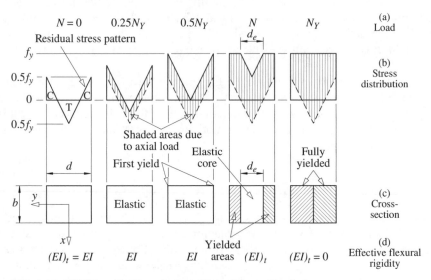

Fig. 3.10 Effective section of a member with residual stresses

strain-hardening modulus E_{st} is sometimes used in the yielded regions as well as in the strain-hardened regions. This use is based on the slip theory of dislocation, in which yielding is represented by a series of dynamic jumps instead of by a smooth quasi-static flow. Thus the material in the yielded region is either elastic or strain-hardened, and its subsequent behaviour may be estimated conservatively by using the strain-hardening modulus. However, the even more conservative assumption that the tangent modulus E_t is zero in both the yielded and strain-hardened regions is frequently used because of its simplicity. According to this assumption, these regions of the member are ineffective during buckling, and the moment of resistance is entirely due to the elastic core of the section.

Thus, for a rectangular section member which has the simplified residual stress distribution shown in Fig. 3.10, the effective flexural rigidity $(EI)_t$ about the y axis is (see section 3.9.2)

$$(EI)_t = EI[2(1 - N/N_Y)]^{1/2}, \tag{3.18}$$

when the axial load N is greater than $0.5N_Y$ at which first yield occurs, and the axial load at buckling N_t is given by

$$\frac{N_t}{N_Y} = \left(\frac{N_o}{N_Y}\right)^2 \{[1 + 2(N_Y/N_o)^2]^{1/2} - 1\}. \tag{3.19}$$

The variation of this dimensionless tangent modulus buckling load N_t/N_Y with the modified slenderness ratio $\surd(N_Y/N_o)$ is shown in Fig. 3.11. For stocky members, the buckling load N_t approaches the squash load N_Y (so that $p_t = N_t/A$ approaches f_y), while for intermediate length members it approaches the elastic buckling load N_o (so that p_t approaches p_E) as $\surd(N_Y/N_o) = \surd(f_y/p_E)$ approaches $\surd2$. For more slender members, premature yielding does not occur, and these members buckle at the elastic buckling load N_o.

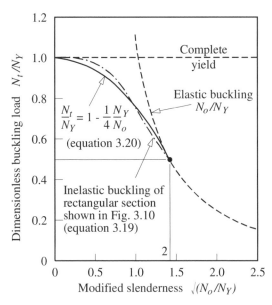

Fig. 3.11 Inelastic buckling of compression members with residual stresses

Also shown in Fig. 3.11 is the dimensionless tangent modulus buckling load N_t/N_Y given by

$$\frac{N_t}{N_Y} = 1 - \frac{1}{4}\frac{N_Y}{N_o},\tag{3.20}$$

which was developed as a compromise between major and minor axis buckling of hot-rolled I-section members. It can be seen that this simple compromise is very similar to the relationship given by equation 3.19 for the rectangular section member.

3.4 Real compression members

3.4.1 BEHAVIOUR OF REAL MEMBERS

The conditions under which real members act differ in many ways from the idealized conditions assumed in section 3.2.1 for the analysis of the elastic buckling of a perfect member. Real members are not perfectly straight, and their loads are applied eccentrically, while accidental transverse loads may act. The effects of small imperfections of these types are qualitatively the same as those of initial curvature, which were described in section 3.2.2. These imperfections can therefore be represented by an increased equivalent initial curvature which has a similar effect on the behaviour of the member as the combined effect of all of these imperfections. The resulting behaviour is shown by curve A in Fig. 3.12.

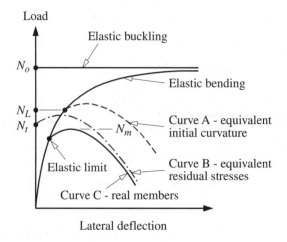

Fig. 3.12 Behaviour of real compression members

A real member also has residual stresses, and its elastic modulus E and yield stress f_y may vary throughout the member. The effects of these material variations are qualitatively the same as those of the residual stresses, which were described in section 3.3.4, and so they can be represented by an equivalent set of residual stresses. The resulting behaviour of the member is shown by curve B in Fig. 3.12.

Since real members have both kinds of imperfections, their behaviour is as shown by curve C in Fig. 3.12, which is a combination of curves A and B. Thus the real member behaves as a member with equivalent initial curvature (curve A) until the elastic limit is reached. It then follows a path which is similar to and approaches that of a member with equivalent residual stresses (curve B).

3.4.2 STRENGTHS OF REAL MEMBERS

Real compression members may be analysed using a model in which the initial crookedness and load eccentricity are specified. Residual stresses may be included, and the realistic stress–strain behaviour may be incorporated in the prediction of the load versus deflection relationship. Thus curve C in Fig. 3.12 may be generated, and the maximum load N_m ascertained.

Rational computer analyses based on the above modelling are rarely used except in research, and are inappropriate for the routine design of real compression members because of the uncertainties and variations that exist in the initial crookedness and residual stresses. Instead, simplified design predictions are used in BS5950 that have been developed from the results of computer analyses and correlations with available test data.

A close prediction of numerical solutions and test results may be obtained by using equation 3.12 which is based on the first yield of a geometrically imperfect member. This is achieved by writing the imperfection parameter η of equation 3.11 as

$$\eta = 0.001a(\lambda - \lambda_o) \geqslant 0 \tag{3.21}$$

in which $\lambda = L_E/r$ is the slenderness (the effective length L_E of a pin-ended member is equal to its actual length L, as discussed in section 3.5.1), λ_o is given by

$$\lambda_o = 0.2\sqrt{\left(\frac{\pi^2 E}{p_y}\right)} \tag{3.22}$$

in which p_y ($\equiv f_y$) is the design strength of the steel, and a is a constant which shifts the compression strength curve as shown in Fig. 3.13 for different cross-section types. The advantage of this approach is that the compressive strength of a particular group of sections can be determined by assigning an appropriate value of a to them. The compressive strength p_c is then defined by p_L in equation 3.12.

The dimensionless compressive strengths p_c/p_y for $a = 2.0, 3.5, 5.5,$ and 8.0,

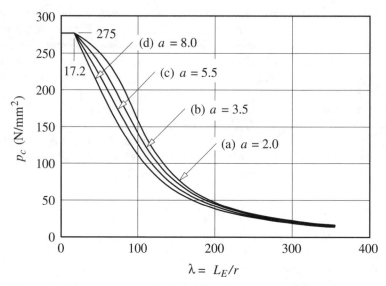

Fig. 3.13 Compression strengths of BS5950 ($p_y = 275\,\text{N/mm}^2$)

which represent the BS5950 strut curves (a), (b), (c), and (d) respectively, are shown in Fig. 3.13.

Members with higher initial crookednesses have lower strengths due to premature yielding, and these are associated with higher values of a. On the other hand, members with lower initial crookednesses are not as greatly affected by premature yielding, and have lower values of a assigned to them.

Compression members containing thin-plate elements are likely to be affected by local buckling of the cross-section (see Chapter 4). Local buckling reduces the strengths of short compression members below their squash loads N_Y and the strengths of longer members which fail by flexural buckling.

Local buckling of a short compression member is accounted for by using a reduced effective area A_{eff} instead of the gross area A, as discussed in section 4.7.1.

Local buckling of a longer compression member is accounted for by using A_{eff} instead of A and by reducing the slenderness to

$$\lambda = \left(\frac{L_E}{r}\right)\sqrt{\left(\frac{A_{\text{eff}}}{A}\right)} \tag{3.23}$$

but using the gross cross-sectional area A to determine the radius of gyration r.

A worked example of checking the resistance of a compression member is given in section 3.12.1, while worked examples of the design of compression members are given in sections 3.12.2 and 3.12.3.

3.5 Effective lengths of compression members

3.5.1 SIMPLE SUPPORTS AND RIGID RESTRAINTS

In the previous sections it was assumed that the compression member was supported only at its ends, as shown in Fig. 3.14a. If the member has an additional lateral support which prevents it from deflecting at its centre so that $(u)_{L/2} = 0$, as shown in Fig. 3.14b, then its buckled shape u is given by

$$u = \delta \sin 2\pi z/L, \tag{3.24}$$

and its elastic buckling load N_o is given by

$$N_o = \frac{4\pi^2 EI}{L^2}. \tag{3.25}$$

The end supports of a compression member may also differ from the simple supports shown in Fig. 3.14a which allow the member ends to rotate but prevent them from deflecting laterally. For example, one or both ends may be rigidly built-in so as to prevent end rotation (Fig. 3.14c and d), or one end may be completely free (Fig. 3.14e). In each case the elastic buckling load of the member may be obtained by finding the solution of the differential equilibrium equation which satisfies the boundary conditions. All of these buckling loads can be expressed in the form

$$N_o = \pi^2 EI/L_E^2 \tag{3.26}$$

in which L_E is the effective length. Expressions for L_E are shown in Fig. 3.14, and in each case it can be seen that the effective length of the member is equal to the distance between the inflexion points of its buckled shape.

The effects of the variations in the support and restraint conditions on the compression member resistance $P_c = Ap_c$ ($=N_L$ of equation 3.10) may be

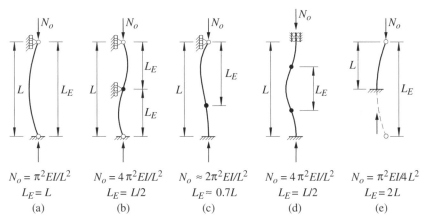

$$N_o = \pi^2 EI/L^2 \qquad N_o = 4\pi^2 EI/L^2 \qquad N_o \approx 2\pi^2 EI/L^2 \qquad N_o = 4\pi^2 EI/L^2 \qquad N_o = \pi^2 EI/4 L^2$$
$$L_E = L \qquad\qquad L_E = L/2 \qquad\qquad L_E \approx 0.7L \qquad\qquad L_E = L/2 \qquad\qquad L_E = 2L$$
$$\text{(a)} \qquad\qquad\qquad \text{(b)} \qquad\qquad\qquad \text{(c)} \qquad\qquad\qquad \text{(d)} \qquad\qquad\qquad \text{(e)}$$

Fig. 3.14 Effective lengths of columns

accounted for by using the effective length L_E instead of the actual length L in the calculation of the slenderness, and using this modified slenderness through-out the strength equations. It should be noted that it is often necessary to consider the member behaviour in each principal plane, since the effective lengths L_{Ex} and L_{Ey} may also differ, as well as the radii of gyration r_x and r_y.

3.5.2 INTERMEDIATE RESTRAINTS

In section 3.5.1 it was shown that the elastic buckling load of a simply supported member is increased by a factor of 4 to $N_o = 4\pi^2 EI/L^2$ when an additional lateral restraint is provided which prevents it from deflecting at its centre, as shown in Fig. 3.15b. This restraint need not be completely rigid, but may be elastic, provided its stiffness exceeds a certain minimum value. If the stiffness α of the restraint is defined by the force $\alpha\delta$ acting on the restraint which causes its length to change by δ, then the minimum stiffness α_L is determined in section 3.10.1 as

$$\alpha_L = 16\pi^2 EI/L^3. \tag{3.27}$$

The limiting stiffness α_L can be expressed in terms of the buckling load $N_o = 4\pi^2 EI/L^2$ as

$$\alpha_L = 4N_o/L. \tag{3.28}$$

Compression member restraints are generally required to be able to transmit 1% of the force in the member restrained. This is a little less than the value of 1.5% suggested recently [2] as leading to braces which are sufficiently stiff, and forms the basis of Clause 4.7.1.2 of BS5950.

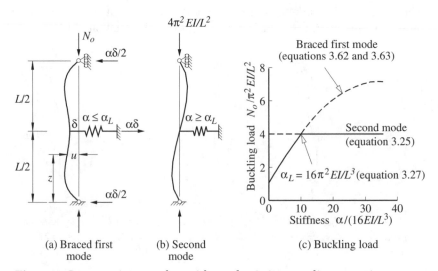

Fig. 3.15 Compression member with an elastic intermediate restraint

3.5.3 ELASTIC END RESTRAINTS

When the ends of a compression member are rigidly connected to adjacent elastic members, as shown in Fig. 3.16a, then at buckling the adjacent members exert total elastic restraining moments M_1, M_2 which oppose buckling and which are proportional to the end rotations θ_1, θ_2 of the compression member. Thus

$$\left.\begin{array}{l} M_1 = -\alpha_1\theta_1 \\ M_2 = -\alpha_2\theta_2 \end{array}\right\} \tag{3.29}$$

in which

$$\alpha_1 = \sum_1 \alpha, \tag{3.30}$$

where α is the stiffness of any adjacent member connected to the end 1 of the compression member, and α_2 is similarly defined. The stiffness α of an adjacent member depends not only on its length L and flexural rigidity EI but also on its support conditions and on the magnitude of any axial load transmitted by it.

The particular case of a braced restraining member, which acts as if simply supported at both ends as shown in Fig. 3.17a, and which provides equal and opposite end moments M and has an axial force N, is analysed in section 3.10.2, where it is shown that when the axial load N is compressive, the stiffness α is given by

$$\alpha = \frac{2EI}{L}\,\frac{\frac{\pi}{2}\sqrt{(N/N_{oL})}}{\tan\frac{\pi}{2}\sqrt{(N/N_{oL})}}, \tag{3.31}$$

where

$$N_{oL} = \pi^2 EI/L^2, \tag{3.32}$$

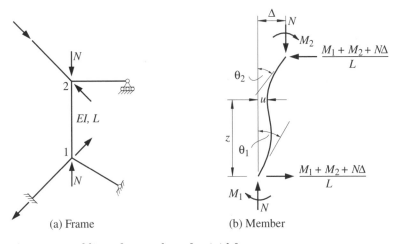

(a) Frame   (b) Member

Fig. 3.16 Buckling of a member of a rigid frame

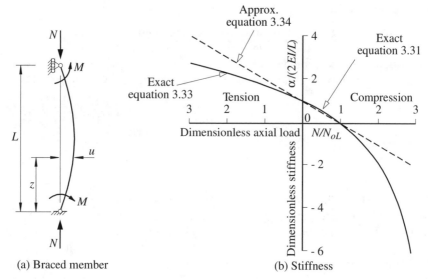

(a) Braced member   (b) Stiffness

Fig. 3.17 Stiffness of a braced member

and by

$$\alpha = \frac{2EI}{L} \frac{\frac{\pi}{2}\sqrt{(N/N_{oL})}}{\tanh \frac{\pi}{2}\sqrt{(N/N_{oL})}}, \tag{3.33}$$

when the axial load N is tensile. These relationships are shown in Fig. 3.17b, and it can be seen that the stiffness decreases almost linearly from $2EI/L$ to zero as the compressive axial load increases from zero to N_{oL}, and that the stiffness is negative when the axial load exceeds N_{oL}. In this case the adjacent member no longer restrains the buckling member, but disturbs it. When the axial load causes tension, the stiffness is increased above the value $2EI/L$.

Also shown in Fig. 3.17b is the simple approximation

$$\alpha = \frac{2EI}{L}\left(1 - \frac{N}{N_{oL}}\right). \tag{3.34}$$

The term $(1 - N/N_{oL})$ in this equation is the reciprocal of the amplification factor, which expresses the fact that the first-order rotations $ML/2EI$ associated with the end moments M are amplified by the compressive axial load N to $(ML/2EI)/(1 - N/N_{oL})$. The approximation provided by equation 3.34 is close and conservative in the range $0 < N/N_{oL} < 1$, but errs on the unsafe side and with increasing error as the axial load N increases away from this range.

Similar analyses may be made of braced restraining members under other end conditions, and some approximate solutions for the stiffnesses of these are summarized in Fig. 3.18. These approximations have accuracies comparable with that of equation 3.34.

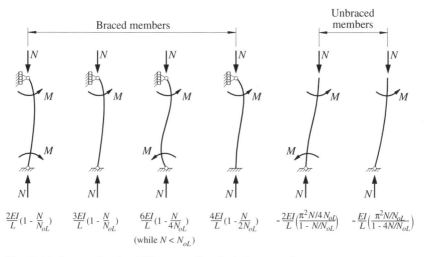

Fig. 3.18 Approximate stiffnesses of restraining members

When a restraining member is unbraced, its ends sway Δ as shown in Fig. 3.19a, and restoring end moments M are required to maintain equilibrium. Such a member which receives equal end moments is analysed in section 3.10.3, where it is shown that its stiffness is given by

$$\alpha = -\left(\frac{2EI}{L}\right)\frac{\pi}{2}\sqrt{\left(\frac{N}{N_{\mathrm{oL}}}\right)}\tan\frac{\pi}{2}\sqrt{\left(\frac{N}{N_{\mathrm{oL}}}\right)}, \tag{3.35}$$

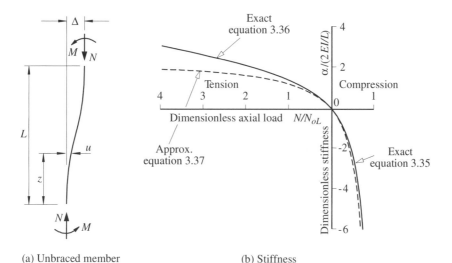

(a) Unbraced member   (b) Stiffness

Fig. 3.19 Stiffness of an unbraced member

when the axial load N is compressive, and by

$$\alpha = \left(\frac{2EI}{L}\right) \frac{\pi}{2} \sqrt{\left(\frac{N}{N_{oL}}\right)} \tanh \frac{\pi}{2} \sqrt{\left(\frac{N}{N_{oL}}\right)}, \qquad (3.36)$$

when the axial load N is tensile. These relationships are shown in Fig. 3.19b, and it can be seen that the stiffness α decreases as the tensile load N decreases, and becomes negative as the load changes to compressive. Also shown in Fig. 3.19b is the approximation

$$\alpha = -\frac{2EI}{L} \frac{\frac{\pi^2}{4}(N/N_{oL})}{(1 - N/N_{oL})}, \qquad (3.37)$$

which is close and conservative when the axial load N is compressive, and which errs on the safe side when the axial load N is tensile. It can be seen from Fig. 3.19b that the stiffness of the member is negative when it is subjected to compression, so that it disturbs the buckling member, rather than restraining it.

Similar analyses may be made of unbraced restraining members with other end conditions, and another approximate solution is given in Fig. 3.18. This has an accuracy comparable with that of equation 3.37.

3.5.4 BUCKLING OF BRACED MEMBERS WITH END RESTRAINTS

The buckling of an end-restrained compression member 1-2 is analysed in section 3.10.4, where it is shown that if the member is braced so that it cannot sway ($\Delta = 0$), then its elastic buckling load N_o can be expressed in the general form of equation 3.26 when the effective length ratio $k_e = L_E/L$ is the solution of

$$\frac{\gamma_1\gamma_2}{4} \left(\frac{\pi}{k_e}\right)^2 + \left(\frac{\gamma_1 + \gamma_2}{2}\right) \left(1 - \frac{\pi}{k_e} \cot \frac{\pi}{k_e}\right) + \frac{\tan \pi/2k_e}{\pi/2k_e} = 1, \qquad (3.38)$$

where the relative stiffness of the braced member at its end 1 is

$$\gamma_1 = \frac{(2EI/L)_{12}}{\sum_1 \alpha}, \qquad (3.39)$$

in which the summation $\sum_1 \alpha$ is for all the other members at end 1, and γ_2 is similarly defined. In BS5950, the relative stiffnesses γ_1, γ_2 ($0 \leqslant \gamma \leqslant \infty$) are replaced by

$$k_1 = \frac{(2EI/L)_{12}}{0.5\sum_1 \alpha + (2EI/L)_{12}} = \frac{2\gamma_1}{1 + 2\gamma_1} \qquad (3.40)$$

and a similar definition of k_2. Values of the effective length ratio L_E/L which satisfy equations 3.38 to 3.40 are presented in chart form in BS5950. This chart is reproduced in Fig. 3.20a.

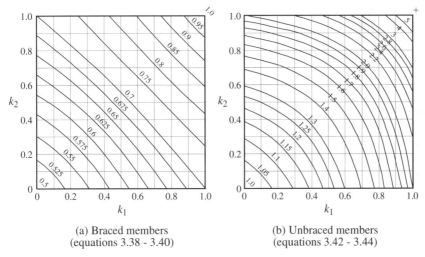

(a) Braced members
(equations 3.38 - 3.40)

(b) Unbraced members
(equations 3.42 - 3.44)

Fig. 3.20 Effective length ratios L_E/L for members in rigid-jointed frames

In using the chart of Fig. 3.20a, the stiffness factors k may be calculated from equation 3.40 by using the stiffness approximations of Fig. 3.18. Thus for restraining members of the type shown in Figs 3.17a and 3.18a, the factor k_1 can be approximated by

$$k_1 = \frac{(I/L)_{12}}{0.5\sum_{1}(I/L)(1 - N/N_{oL}) + (I/L)_{12}} \tag{3.41}$$

in which N is the compression force in the restraining member and N_{oL} its elastic buckling load.

3.5.5 BUCKLING OF UNBRACED MEMBERS WITH END RESTRAINTS

The buckling of an end-restrained compression member 1-2 is analysed in section 3.10.5, where it is shown that if the member is unbraced against side-sway, then its elastic buckling load N_o can be expressed in the general form of equation 3.26 when the effective length ratio $k_e = L_E/L$ is the solution of

$$\frac{\gamma_1\gamma_2(\pi/k_e)^2 - 36}{6(\gamma_1 + \gamma_2)} = \frac{\pi}{k_e}\cot\frac{\pi}{k_e}, \tag{3.42}$$

where the relative stiffness of the unbraced member at its end 1 is

$$\gamma_1 = \frac{(6EI/L)_{12}}{\sum_{1}\alpha} \tag{3.43}$$

in which the summation $\sum_1 \alpha$ is for all the other members at end 1, and γ_2 is similarly defined. In BS5950, the relative stiffnesses γ_1, γ_2 are replaced by

$$k_1 = \frac{(6EI/L)_{12}}{1.5\sum_1 \alpha + (6EI/L)_{12}} = \frac{\gamma_1}{1.5 + \gamma_1} \qquad (3.44)$$

and a similar definition of k_2. Values of the effective length ratio L_E/L which satisfy equation 3.42 are presented in chart form in BS5950. This chart is reproduced in Fig. 3.20b. In using the chart of Fig. 3.20b, the stiffness factors k may be calculated from equation 3.44 by using the stiffness approximations of Fig. 3.18. Thus

$$k_1 = \frac{(I/L)_{12}}{1.5\sum_1 (I/L)(1 - N/4N_{oL}) + (I/L)_{12}} \qquad (3.45)$$

3.5.6 BRACING STIFFNESS REQUIRED FOR A BRACED MEMBER

The elastic buckling load of an unbraced compression member may be increased substantially by providing a translational bracing system which effectively prevents sway. The bracing system need not be completely rigid, but may be elastic as shown in Fig. 3.21, provided its stiffness α exceeds a certain minimum value α_L. It is shown in section 3.10.6 that the minimum value for a pin-ended compression member is

$$\alpha_L = \pi^2 EI/L^3. \qquad (3.46)$$

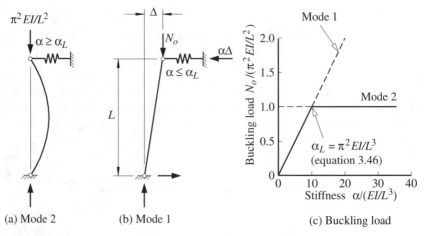

(a) Mode 2   (b) Mode 1   (c) Buckling load

Fig. 3.21 Compression member with an elastic sway brace

This conclusion can be extended to compression members with rotational end restraints, and it can be shown that if the sway bracing stiffness α is greater than

$$\alpha_L = N_o/L, \tag{3.47}$$

where N_o is the elastic buckling load for the braced mode, then the member is effectively braced against sway.

3.6 Design by buckling analysis

3.6.1 GENERAL

The effective length concept developed in section 3.5 is essentially a convenient method of expressing the elastic flexural buckling load N_o of a uniform member in uniform compression (equation 3.26). The use of this concept in the design of other compression members is an example of a more general approach to the analysis and design of compression members whose ultimate strengths are governed by the interaction between yielding and buckling. This more general buckling analysis approach originates from the dependence of the design compression resistance P_c ($\equiv N_L$ of equation 3.10) of a simply supported uniform compression member on its squash load N_Y and its elastic buckling load N_o, as shown in Fig. 3.22, which is adapted from Fig. 3.13. The generalization of this relationship to other members allows the compression resistance P_c of any member to be determined from its yield load N_Y and its elastic buckling load N_o by using Fig. 3.22.

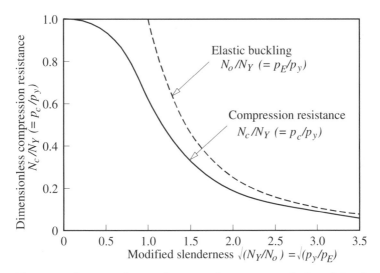

Fig. 3.22 Compression resistance of structures designed by buckling analysis

BS5950 allows the use of the method of design by buckling analysis in Annex E.6, in which it uses the elastic buckling load N_o of the member in equation 3.26 to determine the effective length L_E.

In the following subsections, this method of design by buckling analysis is extended to the in-plane behaviour of rigid-jointed frames with joint loading, and also to the design of compression members which either are non-uniform, or have intermediate loads, or twist during buckling. Similar and related methods may be used for the flexural–torsional buckling of beams (section 6.6).

3.6.2 RIGID-JOINTED FRAMES WITH JOINT LOADS ONLY

The application of the method of design by buckling analysis to rigid-jointed frames which only have joint loads is a simple extrapolation of the design method for simply supported compression members. For this extrapolation, it is assumed that the resistance P_c of a compression member in a frame is related to its squash load N_Y and to the axial force N_o carried by it when the frame buckles elastically, and that this relationship (Fig. 3.22) is the same as that used for simply supported compression members of the same type (Fig. 3.13).

This method of design by buckling analysis is virtually the same as the effective length method which uses the design charts of Fig. 3.20, because these charts were obtained from elastic buckling analyses of restrained members. The only difference is that the elastic buckling load N_o may be calculated directly for the method of design by buckling analysis, as well as by determining the approximate effective length from the design charts.

The elastic buckling load of the member may often be determined approximately by the effective length method discussed in section 3.5. In cases where this is not satisfactory, a more accurate method must be used. Many such methods have been developed, and some of these are discussed in sections 8.3.5.2 and 8.3.5.3 and in [3–12], while other methods are referred to in the literature [13–15]. The buckling loads of many frames have already been determined, and extensive tabulations of approximations for some of these are available [13, 15–19].

While the method of design by buckling analysis might also be applied to rigid frames whose loads act between joints, the bending actions present in those frames make this less rational. Because of this, consideration of the effects of buckling on the design of frames with bending actions will be deferred until Chapter 8.

3.6.3 NON-UNIFORM MEMBERS

It was assumed in the preceding discussions that the members are uniform, but some practical compression members are of variable cross-section. Non-uniform members may be stepped or tapered, but in either case the elastic buckling load N_o can be determined by solving a differential equilibrium equation similar to that governing the buckling of uniform members (see equation 3.63), but which has variable values of EI. In general, this can best be done using numerical techniques [3–5], and the tedium of these can be relieved by making use of a suitable computer program. Many particular cases have been solved, and tabulations and graphs of solutions are available [4, 5, 15, 20–22].

Once the elastic buckling load N_o of the member has been determined, the method of design by buckling analysis can be used. For this, the squash load N_Y is calculated for the most highly stressed cross-section, which is the section of minimum area. The compression resistance $P_c = Ap_c$ can then be obtained from Fig. 3.22.

3.6.4 MEMBERS WITH INTERMEDIATE AXIAL LOADS

The elastic buckling of members with intermediate as well as end loads is also best analysed by numerical methods [3–5], while solutions of many particular cases are available [4, 15, 20]. Once again, the method of design by buckling analysis can be used, and for this the squash load will be determined by the most heavily stressed section. If the elastic buckling load N_o is calculated for the same section, then the compression resistance P_c can be determined from Fig. 3.22.

3.6.5 FLEXURAL–TORSIONAL BUCKLING

In the previous sections, attention was confined to compression members which buckle by deflecting laterally, either perpendicular to the section minor axis at an elastic buckling load

$$N_{oy} = \pi^2 EI_y / L_{Ey}^2, \tag{3.48}$$

or perpendicular to the major axis at

$$N_{ox} = \pi^2 EI_x / L_{Ex}^2, \tag{3.49}$$

However, thin-walled open section compression members may also buckle by twisting about a longitudinal axis, as shown in Fig. 3.23 for a cruciform section, or by combined bending and twisting.

A compression member of doubly symmetric cross-section may buckle elastically by twisting at a torsional buckling load (see section 3.11.1) given by

$$N_{oz} = \frac{GJ}{r_1^2} \left(1 + \frac{\pi^2 EI_w}{GJL_{Ez}^2}\right) \tag{3.50}$$

in which GJ and EI_w are the torsional and warping rigidities (see Chapter 10), L_{Ez} is the distance between inflexion points of the twisted shape, and

$$r_1^2 = r_0^2 + x_0^2 + y_0^2 \tag{3.51}$$

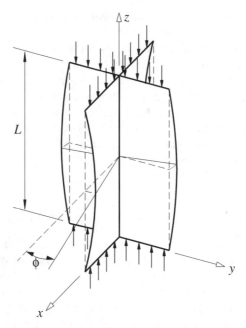

Fig. 3.23 Torsional buckling of a cruciform section

in which x_0, y_0 are the shear centre coordinates (which are zero for doubly symmetric sections, see section 5.4.3), and

$$r_0 = \sqrt{\{(I_x + I_y)/A\}} \tag{3.52}$$

is the polar radius of gyration. For most rolled steel sections, the minor axis buckling load N_{oy} is less than N_{oz}, and the possibility of torsional buckling can be ignored. However, short members which have low torsional and warping rigidities (such as thin-walled cruciforms) should be checked. Such members can be designed by using Fig. 3.22 with the value of N_{oz} substituted for the elastic buckling load N_o.

Monosymmetric and asymmetric section members (such as thin-walled tees and angles) may buckle in a combined mode by twisting and deflecting. This action takes place because the axis of twist through the shear centre does not coincide with the loading axis through the centroid, and any twisting which occurs causes the centroidal axis to deflect. For simply supported members, it is shown in [4, 23, 24] that the elastic buckling load N_o is the lowest root of the cubic equation

$$N_o^3\{r_1^2 - x_0^2 - y_0^2\} - N_o^2\{(N_{ox} + N_{oy} + N_{oz})r_1^2 - N_{oy}x_0^2 - N_{ox}y_0^2\} + N_o r_1^2\{N_{ox}N_{oy} + N_{oy}N_{oz} + N_{oz}N_{ox}\} - N_{ox}N_{oy}N_{oz}r_1^2 = 0. \tag{3.53}$$

For example, the elastic buckling load N_o for a pin-ended unequal angle is

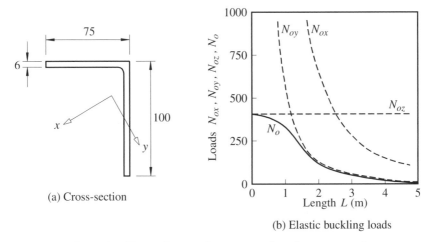

(a) Cross-section

(b) Elastic buckling loads

Fig. 3.24 Elastic buckling of a simply supported angle section column

shown in Fig. 3.24, where it can be seen that N_o is less than any of N_{ox}, N_{oy}, or N_{oz}.

These and other cases of flexural–torsional buckling are treated in a number of textbooks [3–5, 23, 24], and tabulations of solutions are also available [15, 25]. Once the buckling load N_o has been determined, the compression resistance P_c can be found from Fig. 3.22.

3.7 Design of compression members

3.7.1 GENERAL

For the design of a compression member, the axial force is determined by a rational frame analysis, as in Chapter 8, or by statics for a statically determinate structure. The factored loads F^* are for the strength limit state, and are determined by summing up the specified loads multiplied by the appropriate partial load factors γ_f (see section 1.5.6).

The procedure for checking a specified compression member is summarized in Fig. 3.25. In this figure, the factored design compression force F_c^* is distinguished by the addition of *, while the symbol p_y for the BS5950 design strength (usually taken as the specified minimum yield stress Y_s, except for slender sections) replaces the symbol f_y used in earlier sections for the yield stress. The cross-section is checked to determine if it is effective ($A_{eff} = A_g$, where A_g is the gross area) or slender (in which case the reduced effective area A_{eff} is used), and the compression resistance P_c is then found and compared with the design compression force F_c^*.

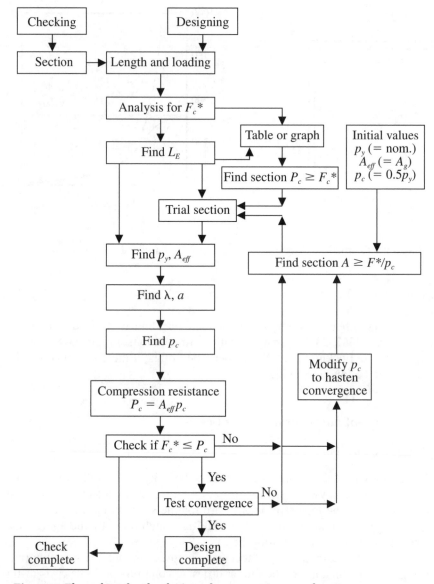

Fig. 3.25 Flow chart for the design of compression members

An iterative series of calculations is required when designing a compression member, as indicated in Fig. 3.25. An initial trial section is first selected, either by using tabulations or graphs of compression resistance P_c versus effective length L_E, or by making initial guesses for p_y, A_{eff}/A and p_c (say $0.5\,p_y$) and calculating a target area A. A trial section is then selected and checked. If the section is not satisfactory, then a new section is selected using the latest values of p_y, A_{eff}/A, and p_c, and the checking process is repeated. The iterations usually converge within a few cycles, but convergence can be hastened by using the mean of the previous and current values of p_c in the calculation of the target area A.

The following subsection describes the compression resistance check of BS5950.

3.7.2 COMPRESSION RESISTANCE CHECK

Compression members with insufficient bracing may buckle in a flexural or flexural–torsional mode at a load below that to cause failure of the cross-section (the squash load $A_g p_y$ or $A_{eff} p_y$). This is usually the case, and in BS5950, the reduced capacity is termed the compression resistance P_c. For the compression resistance check, the inequality

$$F_c^* \leqslant P_c \tag{3.54}$$

must be satisfied, where the compression resistance is

$$P_c = A_g p_c \tag{3.55}$$

if the cross-section is effective $(A_{eff} = A_g)$. If the cross-section is slender $(A_{eff} < A_g)$, then the compression resistance is

$$P_c = A_{eff} p_{cs} \tag{3.56}$$

where p_{cs} is obtained in the same way as p_c, but using the reduced slenderness of equation 3.23.

Values of p_c are given in Table 24 of BS5950 for selected values of a, p_y, and λ. Linear interpolation may be used between these values of p_c, and it is not necessary to undertake the solution of equation 3.10. In practice, designers often interpolate by eye, but for the worked examples of this book, linear interpolation is used.

Worked examples of checking the compression resistance are given in sections 3.12.1–3.12.3.

3.8 Appendix – elastic compression members

3.8.1 BUCKLING OF STRAIGHT MEMBERS

The elastic buckling load N_o of the compression member shown in Fig. 3.3 can be determined by finding a deflected position which is one of equilibrium. The differential equilibrium equation of bending of the member is

$$EI \frac{d^2u}{dz^2} = -N_o u. \tag{3.57}$$

This equation states that for equilibrium, the internal moment of resistance $EI(d^2u/dz^2)$ must exactly balance the external disturbing moment $-N_o u$ at any point along the length of the member. When this equation is satisfied at all points, the displaced position is one of equilibrium.

The solution of equation 3.57 which satisfies the boundary condition at the lower end that $(u)_0 = 0$ is

$$u = \delta \sin \frac{\pi z}{k_e L},$$

where

$$\frac{1}{k_e^2} = \frac{N_o}{\pi^2 EI/L^2},$$

and δ is an undetermined constant. The boundary condition at the upper end that $(u)_L = 0$ is satisfied when either

$$\left. \begin{array}{l} \delta = 0 \\ u = 0 \end{array} \right\} \tag{3.58}$$

or $k_e = 1/n$ in which n is an integer, so that

$$N_o = n^2 \pi^2 EI/L^2, \tag{3.59}$$

$$u = \delta \sin n\pi z/L. \tag{3.60}$$

The first solution (equations 3.58) defines the straight stable equilibrium position which is valid for all loads N less than the lowest value of N_o, as shown in Fig. 3.2b. The second solution (equations 3.59 and 3.60) defines the buckling loads N_o at which displaced equilibrium positions can exist. This solution does not determine the magnitude δ of the central deflection, as indicated in Fig. 3.2b. The lowest buckling load is the most important, and this occurs when $n = 1$, so that

$$N_o = \pi^2 EI/L^2, \tag{3.2}$$

$$u = \delta \sin \pi z/L. \tag{3.1}$$

3.8.2 BENDING OF MEMBERS WITH INITIAL CURVATURE

The bending of the compression member with initial curvature shown in Fig. 3.2a can be analysed by considering the differential equilibrium equation

$$EI \frac{d^2u}{dz^2} = -N(u + u_0), \tag{3.61}$$

which is obtained from equation 3.57 for a straight member by adding the additional bending moment $-Nu_0$ induced by the initial curvature.

If the initial curvature of the member is such that

$$u_0 = \delta_0 \sin \pi z/L, \tag{3.6}$$

then the solution of equation 3.61 which satisfies the boundary conditions $(u)_{0,L} = 0$ is the deflected shape

$$u = \delta \sin \pi z/L, \tag{3.7}$$

where

$$\frac{\delta}{\delta_0} = \frac{N/N_o}{1 - N/N_o}. \tag{3.8}$$

The maximum moment in the compression member is $N(\delta + \delta_0)$, and so the maximum bending stress is $N(\delta + \delta_0)/Z$, where Z is the elastic section modulus. Thus the maximum total stress is

$$f_{max} = \frac{N}{A} + \frac{N(\delta + \delta_0)}{Z}.$$

If the elastic limit is taken as the yield stress f_y, then the limiting axial load N_L for which the above elastic analysis is valid is given by

$$N_L = N_Y - \frac{N_L(\delta + \delta_0)A}{Z}, \tag{3.62}$$

where $N_Y = Af_y$ is the squash load. By writing $Z = 2I/b$, in which b is the member width, equation 3.62 becomes

$$N_L = N_Y - \frac{\delta_0 b}{2r^2} \frac{N_L}{(1 - N_L/N_o)},$$

which can be rearranged to give the dimensionless limiting load N_L/N_Y as

$$\frac{N_L}{N_Y} = \left[\frac{1 + (1 + \eta)N_o/N_Y}{2} \right] - \left\{ \left[\frac{1 + (1 + \eta)N_o/N_Y}{2} \right]^2 - \frac{N_o}{N_Y} \right\}^{1/2}, \tag{3.10}$$

where

$$\eta = \frac{\delta_0 b}{2r^2}. \tag{3.11}$$

Alternatively, equation 3.10 can be rearranged to give the limiting stress p_L as

$$p_L = \frac{p_E f_y}{\phi + \sqrt{(\phi^2 - p_E f_y)}} \tag{3.12}$$

in which

$$\phi = \frac{f_y + (1 + \eta)p_E}{2} \tag{3.13}$$

3.9 Appendix – inelastic compression members

3.9.1 REDUCED MODULUS THEORY OF BUCKLING

The reduced modulus buckling load N_r of a rectangular section compression member which buckles in the x direction (see Fig. 3.26a) can be determined from the bending strain and stress distributions, which are related to the curvature $\Phi(= -d^2u/dz^2)$ and the moduli E and E_t as shown in Fig. 3.26b and c. The position of the line of zero bending stress can be found by using the condition that the axial force remains constant during buckling, from which it follows that the force resultant of the bending stresses must be zero, so that

$$\tfrac{1}{2}db_cE_tb_c\Phi = \tfrac{1}{2}db_tEb_t\Phi,$$

or

$$\frac{b_c}{b} = \frac{\sqrt{E}}{\sqrt{E} + \sqrt{E_t}}.$$

The moment of resistance of the section $E_rI(d^2u/dz^2)$ is equal to the moment resultant of the bending stresses, so that

$$E_rI\frac{d^2u}{dz^2} = -\frac{db_c}{2}E_tb_c\Phi\frac{2b_c}{3} - \frac{db_t}{2}Eb_t\Phi\frac{2b_t}{3}$$

or

$$E_rI\frac{d^2u}{dz^2} = -\frac{4b_c^2E_t}{b^2}\frac{db^3}{12}\Phi.$$

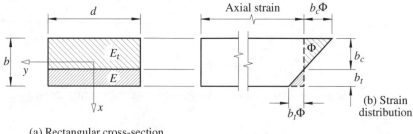

(a) Rectangular cross-section

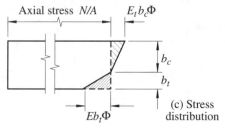

Fig. 3.26 Reduced modulus buckling of a rectangular section member

The reduced modulus of elasticity E_r is therefore given by

$$E_r = \frac{4EE_t}{(\sqrt{E} + \sqrt{E_t})^2}.$$ (3.16)

3.9.2 BUCKLING OF MEMBERS WITH RESIDUAL STRESSES

A rectangular section member with a simplified residual stress distribution is shown in Fig. 3.10. The section first yields at the edges $y = \pm d/2$ at a load $N = 0.5N_Y$, and yielding then spreads through the section as the load approaches the squash load N_Y. If the depth of the elastic core is d_e, then the flexural rigidity for bending about the y axis is

$$(EI)_t = EI\left(\frac{d_e}{d}\right),$$

where $I = b^3 d/12$, and the axial force is

$$N = N_Y\left(1 - \frac{1}{2}\frac{d_e^2}{d^2}\right).$$

By combining these two relationships, the effective flexural rigidity of the partially yielded section can be written as

$$(EI)_t = EI[2(1 - N/N_Y)]^{1/2}.$$ (3.18)

Thus the axial load at buckling N_t is given by

$$\frac{N_t}{N_o} = [2(1 - N_t/N_Y)]^{1/2},$$

which can be rearranged as

$$\frac{N_t}{N_Y} = \left(\frac{N_o}{N_Y}\right)^2 \{[1 + 2(N_Y/N_o)^2]^{1/2} - 1\}.$$ (3.19)

3.10 Appendix – effective lengths of compression members

3.10.1 INTERMEDIATE RESTRAINTS

A straight pin-ended compression member with a central elastic restraint is shown in Fig. 3.15a. It is assumed that when the buckling load N_o is applied to the member, it buckles symmetrically as shown in Fig. 3.15a with a central deflection δ, and the restraint exerts a restoring force $\alpha\delta$. The equilibrium equation for this buckled position is

$$EI\frac{d^2u}{dz^2} = -N_o u + \frac{\alpha\delta}{2}z$$

for $0 \leqslant z \leqslant L/2$.

The solution of this equation which satisfies the boundary conditions $(u)_0 = (du/dz)_{L/2} = 0$ is given by

$$u = \frac{\alpha \delta L}{2N_o} \left(\frac{z}{L} - \frac{\sin \pi z / k_e L}{2(\pi/2k_e) \cos \pi/2k_e} \right),$$

where $k_e = L_E/L$ and L_E is given by equation 3.26. Since $\delta = (u)_{L/2}$, it follows that

$$\frac{\left(\dfrac{\pi}{2k_e}\right)^3 \cot \dfrac{\pi}{2k_e}}{\left(\dfrac{\pi}{2k_e} \cot \dfrac{\pi}{2k_e} - 1\right)} = \frac{\alpha L^3}{16EI}. \tag{3.63}$$

The variation with the dimensionless restraint stiffness $\alpha L^3/16EI$ of the dimensionless buckling load

$$\frac{N_o}{\pi^2 EI/L^2} = \frac{4}{\pi^2} \left(\frac{\pi}{2k_e} \right)^2, \tag{3.64}$$

which satisfies equation 3.63 is shown in Fig. 3.15c. It can be seen that the buckling load for this symmetrical mode varies from $\pi^2 EI/L^2$ when the restraint is of zero stiffness to approximately $8\pi^2 EI/L^2$ when the restraint is rigid. When the restraint stiffness exceeds

$$\alpha_L = 16\pi^2 EI/L^3, \tag{3.27}$$

the buckling load obtained from equations 3.63 and 3.64 exceeds the value of $4\pi^2 EI/L^2$ for which the member buckles in the antisymmetrical second mode shown in Fig. 3.15b. Since buckling always takes place at the lowest possible load, it follows that the member buckles at $4\pi^2 EI/L^2$ in the second mode shown in Fig. 3.15b for all restraint stiffnesses α which exceed α_L.

3.10.2 STIFFNESS OF A BRACED MEMBER

When a structural member is braced so that its ends act as if simply supported as shown in Fig. 3.17a, then its response to equal and opposite disturbing end moments M can be obtained by considering the differential equilibrium equation

$$EI \frac{d^2u}{dz^2} = -Nu - M. \tag{3.65}$$

The solution of this which satisfies the boundary conditions $(u)_0 = (u)_L = 0$ when N is compressive is

$$u = \frac{M}{N} \left\{ \frac{1 - \cos \pi \sqrt{(N/N_{oL})}}{\sin \pi \sqrt{(N/N_{oL})}} \sin \left[\pi \sqrt{\left(\frac{N}{N_{oL}} \right)} \frac{z}{L} \right] \right.$$

$$\left. + \cos \left[\pi \sqrt{\left(\frac{N}{N_{oL}} \right)} \frac{z}{L} \right] - 1 \right\},$$

where

$$N_{oL} = \pi^2 EI/L^2. \tag{3.32}$$

The end rotation $\theta = (du/dz)_0$ is

$$\theta = \frac{2M}{NL} \frac{\pi}{2} \sqrt{\left(\frac{N}{N_{oL}}\right)} \tan \frac{\pi}{2} \sqrt{\left(\frac{N}{N_{oL}}\right)},$$

whence

$$\alpha = \frac{M}{\theta} = \frac{2EI}{L} \frac{\frac{\pi}{2}\sqrt{(N/N_{oL})}}{\tan \frac{\pi}{2}\sqrt{(N/N_{oL})}}. \tag{3.31}$$

When the axial load N is tensile, the solution of equation 3.65 is

$$u = \frac{M}{N} \left\{ \frac{1 - \cosh \pi\sqrt{(N/N_{oL})}}{\sinh \pi\sqrt{(N/N_{oL})}} \sinh \left[\pi\sqrt{\left(\frac{N}{N_{oL}}\right)} \frac{z}{L} \right] \right.$$

$$\left. + \cosh \left[\pi\sqrt{\left(\frac{N}{N_{oL}}\right)} \frac{z}{L} \right] - 1 \right\},$$

whence

$$\alpha = \frac{2EI}{L} \frac{\frac{\pi}{2}\sqrt{(N/N_{oL})}}{\tanh \frac{\pi}{2}\sqrt{(N/N_{oL})}}. \tag{3.33}$$

3.10.3 STIFFNESS OF AN UNBRACED MEMBER

When a structural member is unbraced so that its ends sway Δ as shown in Fig. 3.19a, restoring end moments M are required to maintain equilibrium. The differential equation for equilibrium of a member with equal end moments $M = N\Delta/2$ is

$$EI \frac{d^2u}{dz^2} = -Nu + M, \tag{3.66}$$

and its solution which satisfies the boundary conditions $(u)_0 = 0$, $(u)_L = \Delta$ is

$$u = \frac{M}{N} \left\{ \frac{1 + \cos \pi\sqrt{(N/N_{oL})}}{\sin \pi\sqrt{(N/N_{oL})}} \sin \left[\pi\sqrt{\left(\frac{N}{N_{oL}}\right)} \frac{z}{L} \right] \right.$$

$$\left. - \cos \left[\pi\sqrt{\left(\frac{N}{N_{oL}}\right)} \frac{z}{L} \right] + 1 \right\}.$$

The end rotation $\theta = (du/dz)_0$ is

$$\theta = \frac{2M}{NL} \frac{\pi}{2} \sqrt{\left(\frac{N}{N_{oL}}\right)} \cot \frac{\pi}{2} \sqrt{\left(\frac{N}{N_{oL}}\right)},$$

whence

$$\alpha = -\left(\frac{2EI}{L}\right)\frac{\pi}{2}\sqrt{\left(\frac{N}{N_{oL}}\right)}\tan\frac{\pi}{2}\sqrt{\left(\frac{N}{N_{oL}}\right)}, \tag{3.35}$$

where the negative sign occurs because the end moments M are restoring instead of disturbing moments.

When the axial load N is tensile, the end moments M are disturbing moments, and the solution of equation 3.66 is

$$u = \frac{M}{N}\left\{\frac{1 + \cosh \pi\sqrt{(N/N_{oL})}}{\sinh \pi\sqrt{(N/N_{oL})}}\sinh\left[\pi\sqrt{\left(\frac{N}{N_{oL}}\right)}\frac{z}{L}\right]\right.$$

$$\left. - \cosh\left[\pi\sqrt{\left(\frac{N}{N_{oL}}\right)}\frac{z}{L}\right] + 1\right\},$$

whence

$$\alpha = \frac{2EI}{L}\frac{\pi}{2}\sqrt{\left(\frac{N}{N_{oL}}\right)}\tanh\frac{\pi}{2}\sqrt{\left(\frac{N}{N_{oL}}\right)}. \tag{3.36}$$

3.10.4 BUCKLING OF A BRACED MEMBER

A typical compression member 1-2 in a rigid-jointed frame is shown in Fig. 3.16b. When the frame buckles, the member sways Δ, and its ends rotate by θ_1, θ_2 as shown. Because of these actions, there are end shears $(M_1 + M_2 + N\Delta)/L$ and end moments M_1, M_2. The equilibrium equation for the member is

$$EI\frac{d^2u}{dz^2} = -Nu - M_1 + (M_1 + M_2)\frac{z}{L} + N\Delta\frac{z}{L},$$

and the solution of this is

$$u = -\left(\frac{M_1 \cos \pi/k_e + M_2}{N_o \sin \pi/k_e}\right)\sin\frac{\pi z}{k_e L} + \frac{M_1}{N_o}\left(\cos\frac{\pi z}{k_e L} - 1\right)$$

$$+ \left(\frac{M_1 + M_2}{N_o}\right)\frac{z}{L} + \Delta\frac{z}{L},$$

where N_o is the elastic buckling load transmitted by the member (see equation 3.26) and $k_e = L_E/L$. By using this to obtain expressions for the end rotations θ_1, θ_2, and by substituting equations 3.29, the following equations can be obtained

$$\left.\begin{array}{l}M_1\left(\dfrac{N_o L}{\alpha_1} + 1 - \dfrac{\pi}{k_e}\cot\dfrac{\pi}{k_e}\right) + M_2\left(1 - \dfrac{\pi}{k_e}\csc\dfrac{\pi}{k_e}\right) + N_o\Delta = 0 \\[4mm] M_1\left(1 - \dfrac{\pi}{k_e}\csc\dfrac{\pi}{k_e}\right) + M_2\left(\dfrac{N_o L}{\alpha_2} + 1 - \dfrac{\pi}{k_e}\cot\dfrac{\pi}{k_e}\right) + N_o\Delta = 0\end{array}\right\} \tag{3.67}$$

If the compression member is braced so that joint translation is effectively prevented, the sway terms $N_o\Delta$ disappear from equations 3.67. The moments M_1, M_2 can then be eliminated, whence

$$\frac{(EI/L)^2}{\alpha_1\alpha_2}\left(\frac{\pi}{k_e}\right)^2 + \frac{EI}{L}\left(\frac{1}{\alpha_1} + \frac{1}{\alpha_2}\right)\left(1 - \frac{\pi}{k_e}\cot\frac{\pi}{k_e}\right) + \frac{\tan\pi/2k_e}{\pi/2k_e} = 1. \quad (3.68)$$

By writing the relative stiffness of the braced member at the end 1 as

$$\gamma_1 = \frac{2EI/L}{\alpha_1} = \frac{(2EI/L)_{12}}{\displaystyle\sum_1 \alpha} \quad (3.39)$$

with a similar definition for γ_2, equation 3.68 becomes

$$\frac{\gamma_1\gamma_2}{4}\left(\frac{\pi}{k_e}\right)^2 + \left(\frac{\gamma_1 + \gamma_2}{2}\right)\left(1 - \frac{\pi}{k_e}\cot\frac{\pi}{k_e}\right) + \frac{\tan\pi/2k_e}{\pi/2k_e} = 1. \quad (3.38)$$

3.10.5 BUCKLING OF AN UNBRACED MEMBER

If there are no reaction points or braces to supply the member end shears $(M_1 + M_2 + N\Delta)/L$, the member will sway as shown in Fig. 3.16b, and the sway moment $N\Delta$ will be completely resisted by the end moments M_1 and M_2, so that

$$N\Delta = -(M_1 + M_2).$$

If this is substituted into equation 3.67 and if the moments M_1 and M_2 are eliminated, then

$$\frac{(EI/L)^2}{\alpha_1\alpha_2}\left(\frac{\pi}{k_e}\right)^2 - 1 = \frac{EI}{L}\left(\frac{1}{\alpha_1} + \frac{1}{\alpha_2}\right)\frac{\pi}{k_e}\cot\frac{\pi}{k_e} \quad (3.69)$$

where $k_e = L_E/L$. By writing the relative stiffnesses of the unbraced member at end 1 as

$$\gamma_1 = \frac{6EI/L}{\alpha_1} = \frac{(6EI/L)_{12}}{\displaystyle\sum_1 \alpha} \quad (3.43)$$

with a similar definition of γ_2, equation 3.69 becomes

$$\frac{\gamma_1\gamma_2(\pi/k_e)^2 - 36}{6(\gamma_1 + \gamma_2)} = \frac{\pi}{k_e}\cot\frac{\pi}{k_e}. \quad (3.42)$$

3.10.6 STIFFNESS OF BRACING FOR A RIGID FRAME

If the simply supported member shown in Fig. 3.21 has a sufficiently stiff sway brace acting at one end, then it will buckle as if rigidly braced, as shown in

Fig. 3.21a, at a load $\pi^2 EI/L^2$. If the stiffness α of the brace is reduced below a minimum value α_L, the member will buckle in the rigid body sway mode shown in Fig. 3.21b. The elastic buckling load N_o for this mode can be obtained from the equilibrium condition that

$$N_o\Delta = \alpha\Delta L,$$

where $\alpha\Delta$ is the restraining force in the brace, so that

$$N_o = \alpha L.$$

The variation of N_o with α is compared in Fig. 3.21c with the braced mode buckling load of $\pi^2 EI/L^2$. It can be seen that when the brace stiffness α is less than

$$\alpha_L = \pi^2 EI/L^3, \tag{3.46}$$

the member buckles in the sway mode, and that when α is greater than α_L, it buckles in the braced mode.

3.11 Appendix – design by buckling analysis

3.11.1 TORSIONAL BUCKLING

Thin-walled open-section compression members may buckle by twisting, as shown in Fig. 3.23 for a cruciform section, or by combined bending and twisting. When this type of buckling takes place, the twisting of the member causes the axial compressive stresses N/A to exert a disturbing torque which is opposed by the torsional resistance of the section. For the typical longitudinal element of cross-sectional area δA shown in Fig. 3.27, which rotates $a_0(d\phi/dz)$ (where a_0 is the distance to the axis of twist) when the section rotates ϕ, the axial force $(N/A)\delta A$ has a component $(N/A)\delta A a_0(d\phi/dz)$ which exerts a torque $(N/A)\delta A a_0(d\phi/dz)a_0$ about the axis of twist (which passes through the shear

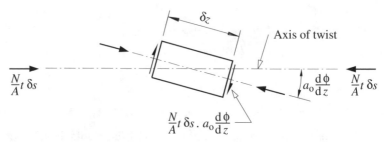

Fig. 3.27 Torque exerted by axial load during twisting

centre x_0, y_0 as shown in Chapter 5). The total disturbing torque T_P exerted is therefore

$$T_P = \frac{N}{A} \frac{d\phi}{dz} \int_A a_0^2 \, dA,$$

where

$$a_0^2 = (x - x_0)^2 + (y - y_0)^2.$$

This torque can also be written as

$$T_P = Nr_1^2 \frac{d\phi}{dz},$$

where

$$r_1^2 = r_0^2 + x_0^2 + y_0^2, \tag{3.51}$$

$$r_0 = \sqrt{\{(I_x + I_y)/A\}}. \tag{3.52}$$

For members of doubly symmetric cross-section ($x_0 = y_0 = 0$), a twisted equilibrium position is possible when the disturbing torque T_P exactly balances the internal resisting torque

$$M_z = GJ \frac{d\phi}{dz} - EI_w \frac{d^3\phi}{dz^3}$$

in which GJ and EI_w are the torsional and warping rigidities (see Chapter 10). Thus, at torsional buckling

$$\frac{N}{A} r_1^2 \frac{d\phi}{dz} = GJ \frac{d\phi}{dz} - EI_w \frac{d^3\phi}{dz^3}. \tag{3.70}$$

The solution of this which satisfies the boundary conditions of end twisting prevented (($\phi)_{0,L} = 0$) and ends free to warp (($d^2\phi/dz^2)_{0,L} = 0$) (see Chapter 10) is $\phi = (\phi)_{L/2} \sin \pi z/L$ in which ($\phi)_{L/2}$ is the undetermined magnitude of the angle of twist rotation at the centre of the member, and the buckling load N_{oz} is

$$N_{oz} = \frac{GJ}{r_1^2} \left(1 + \frac{\pi^2 EI_w}{GJL^2}\right).$$

This solution may be generalized for compression members with other end conditions by writing it in the form

$$N_{oz} = \frac{GJ}{r_1^2} \left(1 + \frac{\pi^2 EI_w}{GJL_{Ez}^2}\right) \tag{3.50}$$

in which the torsional buckling effective length L_{Ez} is the distance between inflexion points in the twisted shape.

3.12 **Worked examples**

3.12.1 EXAMPLE 1 – CHECKING A UB COMPRESSION MEMBER

Problem.   The 457×191 UB82 compression member of S275 steel of Fig. 3.28a is simply supported about both principal axes at each end ($L_{Ex} = 12.0$ m), and has a central brace which prevents lateral deflections in the minor principal plane ($L_{Ey} = 6.0$ m). Check the adequacy of the member for a factored axial compressive load corresponding to a nominal dead load of 150 kN and a nominal imposed load of 220 kN.

Factored axial load. $F_c^* = 1.4 \times 150 + 1.6 \times 220 = 562$ kN

Classifying the section.
For S275 steel with $T = 16$ mm, $p_y = 275$ N/mm^2 T9

$$\varepsilon = (275/275)^{0.5} = 1.0$$
$$b/(T\varepsilon) = (191.3/2)/(16.0 \times 1.0) = 5.98 < 15 \qquad \text{T11}$$
$$d/(t\varepsilon) = (460.2 - 2 \times 16.0 - 2 \times 10.2)/(9.9 \times 1.0) = 41.2 > 40 \qquad \text{T11}$$

and so the web is Class 4 slender.

Effective area.

$$d - d_{eff} = (460.2 - 2 \times 16.0 - 2 \times 10.2) - (40 \times 1.0 \times 9.9) = 11.8 \text{ mm}$$
                                                                                              3.6.2.2
$$A_{eff} = 105 \times 10^2 - 11.8 \times 9.9 = 10\ 383 \text{ mm}^2 \qquad\qquad 3.6.2.2$$

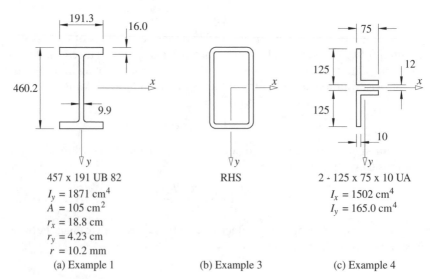

191.3 | 16.0

x

460.2

9.9

y

457 x 191 UB 82
$I_y = 1871$ cm^4
$A = 105$ cm^2
$r_x = 18.8$ cm
$r_y = 4.23$ cm
$r = 10.2$ mm
(a) Example 1

RHS
y
x
(b) Example 3

75

125

125

12

x

10

y

2 - 125 x 75 x 10 UA
$I_x = 1502$ cm^4
$I_y = 165.0$ cm^4
(c) Example 4

Fig. 3.28 Examples 1, 3, and 4

Compression resistance.

$$\lambda_x = 12\ 000/(18.8 \times 10) = 63.8 \qquad\qquad 4.7.2$$
$$\lambda_y = 6000/(4.23 \times 10) = 141.8 > 63.8 = \lambda_x \qquad 4.7.2$$
$$\lambda = \lambda_y = 141.8 \qquad\qquad 4.7.2$$
$$\lambda(A_{eff}/A_g)^{0.5} = 141.8 \times \{10\ 383/(105 \times 10^2)\}^{0.5} = 141.0 \qquad 4.7.4$$

For a rolled UB with $T = 16.0$ mm < 40 mm and y-axis buckling, use Strut Curve (b).   T23

$$p_{cs} = 84 - (84 - 79) \times (141.0 - 140)/(145 - 140) = 83.0\ \text{N/mm}^2 \quad \text{T24(4)}$$
$$P_c = 10\ 383 \times 83.0\ \text{N} = 861.8\ \text{kN} > 562\ \text{kN} = F^*_c \qquad 4.7.4$$

and so the member is satisfactory.

3.12.2 EXAMPLE 2 – DESIGNING A UC COMPRESSION MEMBER

Problem. Design a suitable UC of S355 steel to resist the loading of example 1 in section 3.12.1.

Design axial load. $F^*_c = 562$ kN, as in section 3.12.1.

Target area and first section choice.

Assume $p_y = 355$ N/mm^2 and $p_c = 0.5 \times 355 = 178$ N/mm^2
$A \geqslant 562 \times 10^3/178 = 3157$ mm^2 4.7.4
Try a 152×152 UC 30 with $A = 38.4$ cm^2, $r_x = 6.75$ cm, $r_y = 3.82$ cm, $T = 9.4$ mm.

$$\lambda_x = 12\ 000/(6.75 \times 10) = 177.8 \qquad\qquad 4.7.2$$
$$\lambda_y = 6000/(3.82 \times 10) = 157.1 < 177.8 = \lambda_x \qquad 4.7.2$$
$$\lambda = \lambda_x = 177.8 \qquad\qquad 4.7.2$$

For a rolled H-section with $T = 9.4$ mm < 40 mm and x-axis buckling, use Strut Curve (a).   T23

$$p_c \approx 60\ \text{N/mm}^2 \qquad\qquad \text{T24(2)}$$

which is much less than the guessed value of 178 N/mm^2.

Second section choice.

Guess $p_c = (178 + 60)/2 = 119$ N/mm^2

$$A \geqslant 562 \times 10^3/119 = 4723\ \text{mm}^2 \qquad\qquad 4.7.4$$

Try a 203×203 UC 46, with $A = 58.8$ cm^2, $r_x = 8.81$ cm, $T = 11.0$ mm.

$$p_y = 355\ \text{N/mm}^2 \qquad\qquad \text{T9}$$
$$\varepsilon = (275/355)^{0.5} = 0.880$$
$$b/(T\varepsilon) = (203.2/2)/(11.0 \times 0.880) = 10.5 < 15 \qquad \text{T11}$$
$$d/(t\varepsilon) = (203.2 - 2 \times 11.0 - 2 \times 10.2)/(7.3 \times 0.880) = 25.0 < 40 \qquad \text{T11}$$

and so the cross-section is fully effective.

$$\lambda_x = 12\,000/(8.81 \times 10) = 136.2 \qquad\qquad 4.7.2$$

For a rolled H-section with $T = 11.0 < 40$ mm and x-axis buckling, use Strut Curve (a).   T23

$$p_c = 101 - (101 - 94) \times (136.2 - 135)/(140 - 135) = 99.3 \text{ N/mm}^2 \quad \text{T24(2)}$$
$$P_c = 58.8 \times 10^2 \times 99.3 \text{ N} = 584.0 \text{ kN} > 562 \text{ kN} = F_c^* \qquad 4.7.4$$

and so the 203×203 UC 46 is satisfactory.

3.12.3 EXAMPLE 3 – DESIGNING AN RHS COMPRESSION MEMBER

Problem.   Design a suitable hot-finished RHS of S355 steel to resist the loading of example 1 in section 3.12.1.

Design axial load. $F_c^* = 562$ kN, as in section 3.12.1.

Solution.
Guess $p_c = 0.3 \times 355 = 107$ N/mm^2

$$A \geqslant 562 \times 10^3/107 = 5277 \text{ mm}^2 \qquad\qquad 4.7.4$$

Try a $250 \times 150 \times 8$ RHS, with $A = 61.1$ cm^2, $r_x = 9.19$ cm, $r_y = 6.16$ cm, $t = 8.0$ mm, $p_y = 355$ N/mm^2 T9

$$\varepsilon = (275/355)^{0.5} = 0.880$$
$$d/(t\varepsilon) = (250.0 - 2 \times 8.0 - 2 \times 4.2)/(8.0 \times 0.880) = 32.0 < 40 \qquad \text{T12}$$

and so the cross-section is fully effective.

$$\lambda_x = 12\,000/(9.19 \times 10) = 130.6 \qquad\qquad 4.7.2$$
$$\lambda_y = 6000/(6.16 \times 10) = 97.4 < 130.6 = \lambda_x \qquad\qquad 4.7.2$$
$$\lambda = \lambda_x = 130.6 \qquad\qquad 4.7.2$$

For a hot-finished RHS, use Strut Curve (a).   T23

$$p_c = 108 - (108 - 101) \times (130.6 - 130)/(135 - 130) = 107.2 \text{ N/mm}^2 \quad \text{T24(2)}$$
$$P_c = 61.1 \times 10^2 \times 107.2 \text{ N} = 654.9 \text{ kN} > 562 \text{ kN} = F_c^* \qquad 4.7.4$$

and so the $250 \times 150 \times 8$ RHS is satisfactory.

3.12.4 EXAMPLE 4 – BUCKLING OF DOUBLE ANGLES

Problem.   Two steel $125 \times 75 \times 10$ UA are connected together at 1.5 m intervals to form the long compression member whose properties are given in Fig. 3.28c. The minimum second moment of area of each angle is 50.2 cm^4. The member is simply supported about its major axis at 4.5 m intervals and about its minor axis at 1.5 m intervals. Determine the elastic buckling load of the member.

Member buckling about the major axis.

$$N_{ox} = \pi^2 \times 205\,000 \times 1502 \times 10^4/4500^2 \text{ N} = 1501 \text{ kN}.$$

Member buckling about the minor axis.

$$N_{oy} = \pi^2 \times 205\,000 \times 165 \times 10^4/1500^2 \text{ N} = 1484 \text{ kN}.$$

Single angle buckling.

$$N_{o\,min} = \pi^2 \times 205\,000 \times 50.2 \times 10^4/1500^2 \text{ N} = 451 \text{ kN}$$

and so for both angles $2N_{o\,min} = 2 \times 451 = 903 \text{ kN} < 1484 \text{ kN}$.

It can be seen that the lowest buckling load of 903 kN corresponds to the case where each unequal angle buckles about its own minimum axis.

3.12.5 EXAMPLE 5 – EFFECTIVE LENGTH FACTOR IN AN UNBRACED FRAME

Problem.   Determine the effective length ratio of member 1-2 of the unbraced frame shown in Fig. 3.29a.

Solution.   Using equation 3.44,

$$k_1 = \frac{6EI/L}{1.5 \times 0 + 6EI/L} = 1.0$$

$$k_2 = \frac{6EI/L}{1.5 \times 6(2EI)/(2L) + 6EI/L} = 0.4$$

Using Fig. 3.20b, $L_E/L = 2.3$

3.12.6 EXAMPLE 6 – EFFECTIVE LENGTH FACTOR IN A BRACED FRAME

Problem.   Determine the effective length ratio of member 1-2 of the frame shown in Fig. 3.29a, if bracing is provided which prevents sway buckling.

Solution.
Using equation 3.40,

$$k_1 = \frac{2EI/L}{0.5 \times 0 + 2EI/L} = 1.0$$

$$k_2 = \frac{2EI/L}{0.5 \times 2(2EI)/(2L) + 2EI/L} = 0.667$$

Using Fig. 3.20a, $L_E/L = 0.87$

3.12.7 EXAMPLE 7 – CHECKING A NON-UNIFORM MEMBER

Problem. The 5.0 m long simply supported 203×203 UC 60 member shown in Fig. 3.29b has two steel plates 250 mm $\times$ 12 mm $\times$ 2 m long welded to it, one to the central length of each flange. This increases the elastic buckling load about the minor axis by 70%. If the design strengths of the plates and the UC are 355 N/mm², determine the compression resistance.

Solution.

$$\varepsilon = (275/355)^{0.5} = 0.880$$

$b/(T\varepsilon) = (205.2/2)/(14.2 \times 0.880) = 8.21 < 15$	T11
$d/(t\varepsilon) = (209.6 - 2 \times 14.2 - 2 \times 10.2)/(9.3 \times 0.880) = 19.6 < 40$	T11

and so the cross-section is fully effective.

$\lambda_{cr}F_c^* = 1.7 \times \pi^2 \times 205\,000 \times 2047 \times 10^4/5000^2$ N $= 2816$ kN	
$L_E = \sqrt{\{\pi^2 \times 205\,000 \times 2047 \times 10^4/(2816 \times 10^3)\}} = 3835$ mm	E.6
$\lambda = 3835/(5.19 \times 10) = 73.9$	4.7.2

For a rolled H-section with welded flanges,

$$U/B = 205.2/250 = 0.821 > 0.8 \qquad\qquad \text{Fig. 14b}$$

and $T = 14.2 < 40$ mm and y-axis buckling, use Strut Curve (a).   T23

$p_c = 263 - (263 - 256) \times (73.9 - 72)/(74 - 72) = 256.4$ N/mm²	T24(1)
$P_c = 76 \times 10^2 \times 256.4$ N $= 1948$ kN	4.7.4

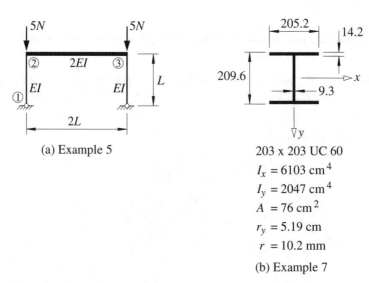

(a) Example 5

205.2

14.2

209.6

x

9.3

y

203 x 203 UC 60

$I_x = 6103$ cm⁴

$I_y = 2047$ cm⁴

$A = 76$ cm²

$r_y = 5.19$ cm

$r = 10.2$ mm

(b) Example 7

Fig. 3.29 Examples 5 and 7

3.12.8 EXAMPLE 8 – CHECKING A MEMBER WITH INTERMEDIATE AXIAL LOADS

Problem. A 5 m long simply supported 457×152 UB 82 compression member of S355 steel whose properties are given in Fig. 3.28a has concentric axial loads of F_c^* and $2F_c^*$ at its ends, and a concentric axial load F_c^* at its midpoint. At elastic buckling the maximum compression force in the member is 2000 kN. Determine the compression resistance.

Solution.
From section 3.12.1, the cross-section is Class 4 slender and $A_{eff} = 10\,383$ mm^2.

$$\lambda_{cr} F_c^* = 2000 \text{ kN}$$
$$L_E = \sqrt{\{\pi^2 \times 205\,000 \times 1871 \times 10^4/(2000 \times 10^3)\}} = 4351 \text{ mm} \quad \text{E.6}$$
$$\lambda (A_{eff}/A_g)^{0.5} = \{4351/(4.23 \times 10)\} \times \sqrt{\{10\,383/(105 \times 10^2)\}} = 102.3 \quad 4.7.4$$

For a rolled UB with $T = 16.0$ mm < 40 mm and y-axis buckling, use Strut Curve (b). T23
$$p_{cs} = 151 - (151 - 146) \times (102.3 - 102)/(104 - 102) = 150.3 \text{ N/mm}^2 \quad \text{T24(3)}$$
$$P_c = 10\,383 \times 150.3 \text{ N} = 1561 \text{ kN} \quad 4.7.4$$

3.13 Unworked examples

3.13.1 EXAMPLE 9 – CHECKING A WELDED COLUMN SECTION

The 14.0 m long welded column section compression member of S355 steel shown in Fig. 3.30a is simply supported about both principal axes at each end $(L_{Ex} = 14.0$ m), and has a central brace that prevents lateral deflections in the minor principal plane $(L_{Ey} = 7.0$ m). Check the adequacy of the member for an axial compressive force due to a nominal dead load of 400 kN and a nominal imposed load of 600 kN.

3.13.2 EXAMPLE 10 – DESIGNING A WELDED BEAM SECTION

Design a suitable UB of S355 steel to resist the loading of example 9 in section 3.13.1.

3.13.3 EXAMPLE 11 – CHECKING A COMPOUND SECTION

Determine the minimum elastic buckling load of the 2 laced 381×102 channels shown in Fig. 3.30b if the effective length about each axis is $L_E = 3.0$ m.

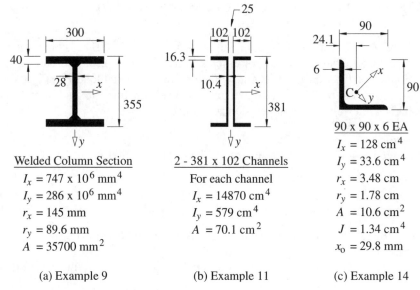

(a) Example 9

Welded Column Section
$I_x = 747 \times 10^6 \text{ mm}^4$
$I_y = 286 \times 10^6 \text{ mm}^4$
$r_x = 145 \text{ mm}$
$r_y = 89.6 \text{ mm}$
$A = 35700 \text{ mm}^2$

(b) Example 11

2 - 381 x 102 Channels
For each channel
$I_x = 14870 \text{ cm}^4$
$I_y = 579 \text{ cm}^4$
$A = 70.1 \text{ cm}^2$

(c) Example 14

90 x 90 x 6 EA
$I_x = 128 \text{ cm}^4$
$I_y = 33.6 \text{ cm}^4$
$r_x = 3.48 \text{ cm}$
$r_y = 1.78 \text{ cm}$
$A = 10.6 \text{ cm}^2$
$J = 1.34 \text{ cm}^4$
$x_o = 29.8 \text{ mm}$

Fig. 3.30 Examples 9, 11, and 14

3.13.4 EXAMPLE 12 – ELASTIC BUCKLING

Determine the elastic buckling load of the guyed column shown in Fig. 3.31. It should be assumed that the guys have negligible flexural stiffness, that $(EA)_{\text{guy}} = 2(EI)_{\text{column}}/L^2$, and that the column is built-in at its base.

3.13.5 EXAMPLE 13 – CHECKING A NON-UNIFORM COMPRESSION MEMBER

Determine the maximum design load F_c^* of a lightly welded box section cantilever compression member 1-2 made from plates of S355 steel which taper

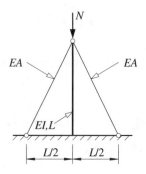

Fig. 3.31 Example 12

from 300 mm × 300 mm × 12 mm at end 1 (at the cantilever tip) to 500 mm × 500 mm × 12 mm at end 2 (at the cantilever support). The cantilever has a length of 10.0 m.

3.13.6 EXAMPLE 14 – FLEXURAL–TORSIONAL BUCKLING

A simply supported 90 × 90 × 6 EA compression member is shown in Fig. 3.30c. Determine the variation of the elastic buckling load with member length.

3.14 References

1. Shanley, F.R. (1947) Inelastic column theory, *Journal of the Aeronautical Sciences*, **14**, No. 5, May, pp. 261–7.
2. Trahair, N.S. (2000) Column bracing forces, *Australian Journal of Structural Engineering*, Institution of Engineers, Australia, **3**, Nos 2, 3, pp. 163–8.
3. Structural Stability Research Council (1988) *Guide to Stability Design Criteria for Metal Structures*, 4th edition (ed. T.V. Galambos), John Wiley, New York.
4. Timoshenko, S.P. and Gere, J.M. (1961) *Theory of Elastic Stability*, 2nd edition, McGraw-Hill, New York.
5. Bleich, F. (1952) *Buckling Strength of Metal Structures*, McGraw-Hill, New York.
6. Livesley, R.K. (1964) *Matrix Methods of Structural Analysis*, Pergamon Press, Oxford.
7. Horne, M.R. and Merchant, W. (1965) *The Stability of Frames*, Pergamon Press, Oxford.
8. Gregory, M.S. (1967) *Elastic Instability*, E. & F.N. Spon, London.
9. McMinn, S.J. (1961) The determination of the critical loads of plane frames, *The Structural Engineer*, **39**, No. 7, July, pp. 221–7.
10. Stevens, L.K. and Schmidt, L.C. (1963) Determination of elastic critical loads, *Journal of the Structural Division, ASCE*, **89**, No. ST6, pp. 137–58.
11. Harrison, H.B. (1967) The analysis of triangulated plane and space structures accounting for temperature and geometrical changes, *Space Structures*, (ed. R.M. Davies), Blackwell Scientific Publications, Oxford, pp. 231–43.
12. Harrison, H.B. (1973) *Computer Methods in Structural Analysis*, Prentice-Hall, Englewood Cliffs, New Jersey.
13. Lu, L.W. (1962) A survey of literature on the stability of frames, *Welding Research Council Bulletin*, No. 81, pp. 1–11.
14. Archer, J.S. *et al.* (1963) Bibliography on the use of digital computers in structural engineering, *Journal of the Structural Division, ASCE*, **89**, No. ST6, pp. 461–91.
15. Column Research Committee of Japan (1971) *Handbook of Structural Stability*, Corona, Tokyo.
16. Brotton, D.M. (1960) Elastic critical loads of multibay pitched roof portal frames with rigid external stanchions, *The Structural Engineer*, **38**, No. 3, March, pp. 88–99.
17. Switzky, H. and Wang, P.C. (1969) Design and analysis of frames for stability, *Journal of the Structural Division, ASCE*, **95**, No. ST4, pp. 695–713.
18. Davies, J.M. (1990) In-plane stability of portal frames, *The Structural Engineer*, **68**, No. 8, April, pp. 141–7.

19. Davies, J.M. (1991) The stability of multi-bay portal frames, *The Structural Engineer*, **69**, No. 12, June, pp. 223–9.
20. Bresler, B., Lin, T.Y. and Scalzi, J.B. (1968) *Design of Steel Structures*, 2nd edition, John Wiley, New York.
21. Gere, J.M. and Carter, W.O. (1962) Critical buckling loads for tapered columns, *Journal of the Structural Division, ASCE*, **88**, No. ST1, pp. 1–11.
22. Bradford, M.A. and Abdoli Yazdi, N. (1999) A Newmark-based method for the stability of columns, *Computers and Structures*, **71**, No. 6, pp. 689–700.
23. Trahair, N.S. (1993) *Flexural–Torsional Buckling of Structures*, E. & F.N. Spon, London.
24. Galambos, T.V. (1968) *Structural Members and Frames*, Prentice-Hall, Englewood Cliffs, New Jersey.
25. Chajes, A. and Winter, G. (1965) Torsional–flexural buckling of thin-walled members, *Journal of the Structural Division, ASCE*, **91**, No. ST4, pp. 103–24.

4 Local buckling of thin plate elements

4.1 Introduction

The behaviour of compression members was discussed in Chapter 3, where it was assumed that no local distortion of the cross-section took place, so that failure was only due to overall buckling and yielding. This treatment is appropriate for solid section members, and for members whose cross-sections are composed of comparatively thick plate elements, including many hot-rolled steel sections.

However, in some cases the member cross-section is composed of more slender plate elements, as for example in some built-up members and in most light-gauge, cold-formed members. These slender plate elements may buckle locally as shown in Fig. 4.1, and the member may fail prematurely, as indicated by the reduction from the full line to the dashed line in Fig. 4.2. A slender plate

Fig. 4.1 Local buckling of an I-section column

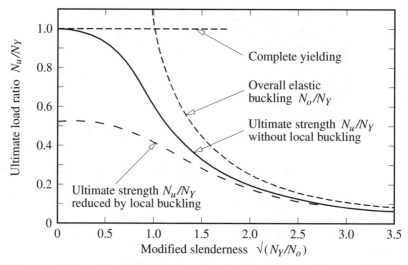

Fig. 4.2 Effect of local buckling on the strengths of compression members

element does not fail by elastic buckling, but exhibits significant post-buckling behaviour, as indicated in Fig. 4.3. Because of this, the plate's resistance to local failure depends not only on its slenderness, but also on its yield strength and residual stresses, as shown in Fig. 4.4. The strength of a plate element of intermediate slenderness is also influenced significantly by its geometrical

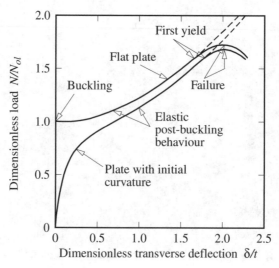

Fig. 4.3 Post-buckling behaviour of thin plates

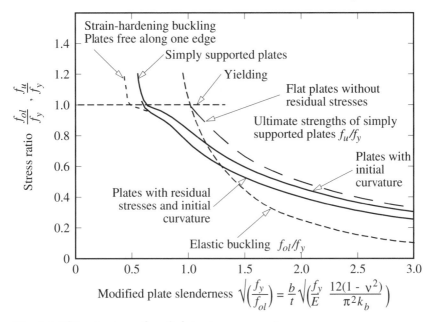

Fig. 4.4 Ultimate strengths of plates in compression

imperfections, while the strength of a stocky plate element depends primarily on its yield stress and strain-hardening moduli of elasticity, as indicated in Fig. 4.4.

In this chapter, the behaviour of thin rectangular plates subjected to in-plane compression, shear, bending or bearing is discussed. The behaviour under compression is applied to the design of plate elements in compression members and compression flanges in beams. The design of beam webs is also discussed, and the influence of the behaviour of thin plates on the design of plate girders to BS5950 is treated in detail.

4.2 Plate elements in compression

4.2.1 ELASTIC BUCKLING

4.2.1.1 Simply supported plates

The thin flat plate element of length L, width b, and thickness t shown in Fig. 4.5b is simply supported along all four edges. The applied compressive loads N are uniformly distributed over each end of the plate. When the applied loads are equal to the elastic buckling loads, the plate can buckle by deflecting u laterally out of its original plane into an adjacent position [1–4].

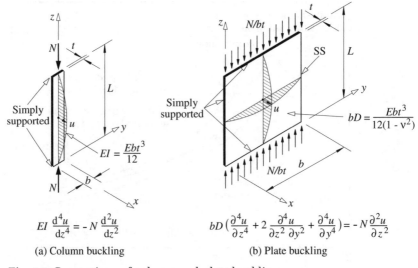

$$EI \frac{d^4 u}{dz^4} = -N \frac{d^2 u}{dz^2}$$

(a) Column buckling

$$bD\left(\frac{\partial^4 u}{\partial z^4} + 2 \frac{\partial^4 u}{\partial z^2 \partial y^2} + \frac{\partial^4 u}{\partial y^4}\right) = -N \frac{\partial^2 u}{\partial z^2}$$

(b) Plate buckling

Fig. 4.5 Comparison of column and plate buckling

It is shown in section 4.8.1 that this equilibrium position is given by

$$u = \delta \sin \frac{m\pi y}{b} \sin \frac{n\pi z}{L} \tag{4.1}$$

with $m = 1$, where δ is the undetermined magnitude of the central deflection. The elastic buckling load N_{ol} at which the plate buckles corresponds to the buckling stress

$$f_{ol} = N_{ol}/bt, \tag{4.2}$$

which is given by

$$f_{ol} = \frac{\pi^2 E}{12(1 - v^2)} \frac{k_b}{(b/t)^2}. \tag{4.3}$$

The lowest value of the buckling coefficient k_b (see Fig. 4.6) is

$$k_b = 4, \tag{4.4}$$

which is appropriate for the high aspect ratios L/b of most structural steel members (see Fig. 4.7).

The buckling stress f_{ol} varies inversely as the square of the plate slenderness or width–thickness ratio b/t, as shown in Fig. 4.4, in which the dimensionless buckling stress f_{ol}/f_y is plotted against a modified plate slenderness ratio

$$\sqrt{\left(\frac{f_y}{f_{ol}}\right)} = \frac{b}{t} \sqrt{\left(\frac{f_y}{E} \frac{12(1 - v^2)}{\pi^2 k_b}\right)}. \tag{4.5}$$

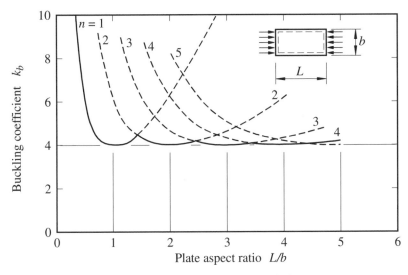

Fig. 4.6 Buckling coefficients of simply supported plates in compression

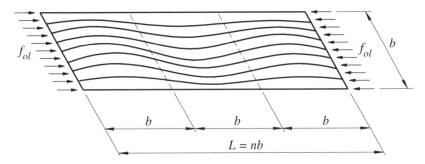

Fig. 4.7 Buckled pattern of a long simply supported plate in compression

If the material ceases to be linear elastic at the yield stress f_y, the above analysis is only valid for $\sqrt{(f_y/f_{ol})} \geqslant 1$. This limit is equivalent to a width–thickness ratio b/t given by

$$\frac{b}{t} \sqrt{\left(\frac{f_y}{275}\right)} = 51.9 \tag{4.6}$$

for a steel with $E = 205\,000$ N/mm² and $v = 0.3$. (The yield stress f_y in equation 4.6 and in all of the similar equations which appear later in this chapter must be expressed in N/mm².)

The values of the buckling coefficient k_b shown in Fig. 4.6 indicate that the use of intermediate transverse stiffeners to increase the elastic buckling stress of a plate in compression is not effective, except when their spacing is significantly less than the plate width. Because of this, it is more economical to use intermediate longitudinal stiffeners which cause the plate to buckle in a number of half waves across its width. When such stiffeners are used, equation 4.3 still holds, provided b is taken as the stiffener spacing. Longitudinal stiffeners have an additional advantage when they are able to transfer compressive stresses, in which case their cross-sectional areas may be added to that of the plate.

An intermediate longitudinal stiffener must be sufficiently stiff flexurally to prevent the plate from deflecting at the stiffener. An approximate value for the minimum second moment of area I_s of a longitudinal stiffener at the centre line of a simply supported plate is given by

$$I_s = 4.5bt^3 \left[1 + 2.3 \frac{A_s}{bt} \left(1 + 0.5 \frac{A_s}{bt} \right) \right] \tag{4.7}$$

in which b is now the half width of the plate and A_s is the area of the stiffener. This minimum increases with the stiffener area because the load transmitted by the stiffener also increases with its area, and the additional second moment of area is required to resist the buckling action of the stiffener load.

Stiffeners should also be proportioned to resist local buckling. A stiffener is usually fixed to one side of the plate rather than placed symmetrically about the mid-plane, and in this case its effective second moment of area is greater than the value calculated for its centroid. It is often suggested [2, 3] that the value of I_s can be approximated by the value calculated for the mid-plane of the plate, but this may provide an overestimate in some cases [4–6].

The behaviour of an edge stiffener is different from that of an intermediate stiffener, in that theoretically it must be of infinite stiffness before it can provide an effective simple support to the plate. However, if a minimum value of the second moment of area of an edge stiffener of

$$I_s = 2.25bt^3 \left[1 + 4.6 \frac{A_s}{bt} \left(1 + \frac{A_s}{bt} \right) \right] \tag{4.8}$$

is provided, the resulting reduction in the plate buckling stress is only a few percent [5]. Equation 4.8 is derived from equation 4.7 on the basis that an edge stiffener only has to support a plate on one side of the stiffener, while an intermediate stiffener has to support plates on both sides.

The effectiveness of longitudinal stiffeners decreases as their number increases, and the minimum stiffness required for them to provide effective lateral support increases. Some guidance on this is given in [3, 5].

4.2.1.2 Plates free along one longitudinal edge

The thin flat plate shown in Fig. 4.8 is simply supported along both transverse edges and one longitudinal edge, and is free along the other. The differential equation of equilibrium of the plate in a buckled position is the same as equation 4.97 (see section 4.8.1). The buckled shape which satisfies this equation

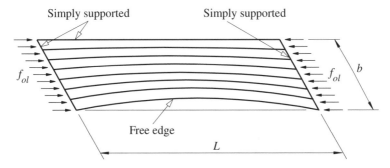

Simply supported   Simply supported

f_{ol} f_{ol} b

Free edge

L

Fig. 4.8 Buckled pattern of a plate free along one edge

differs, however, from the approximately square buckles of the simply supported plate shown in Fig. 4.7. The different boundary conditions along the free edge cause the plate to buckle with a single half wave along its length, as shown in Fig. 4.8. Despite this, the solution for the elastic buckling stress f_{ol} can still be expressed in the general form of equation 4.3 in which the buckling coefficient k_b is now approximated by

$$k_b = 0.425 + \left(\frac{b}{L}\right)^2,\tag{4.9}$$

as shown in Fig. 4.9.

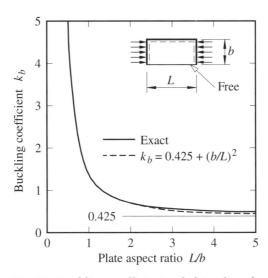

Buckling coefficient k_b

Plate aspect ratio L/b

Exact

$k_b = 0.425 + (b/L)^2$

Free

b

L

0.425

Fig. 4.9 Buckling coefficients of plates free along one edge

For the long plate elements which are used as flange outstands in many structural steel members, the buckling coefficient k_b is close to the minimum value of 0.425. In this case the elastic buckling stress (for a steel for which $E = 205\ 000$ N/mm^2 and $v = 0.3$) is equal to the yield stress f_y when

$$\frac{b}{t}\sqrt{\left(\frac{f_y}{275}\right)} = 16.9. \tag{4.10}$$

Once again it is more economical to use longitudinal stiffeners to increase the elastic buckling stress than transverse stiffeners. The stiffness requirements of intermediate and edge longitudinal stiffeners are discussed in section 4.2.1.1.

4.2.1.3 Plates with other support conditions

The edges of flat plate elements may be fixed or elastically restrained, instead of being simply supported or free. The elastic buckling loads of flat plates with various support conditions have been determined, and many values of the buckling coefficient k_b to be used in equation 4.3 are given in [2–6].

4.2.1.4 Plate assemblies

Many structural steel compression members are assemblies of flat plate elements which are rigidly connected together along their common boundaries. The local buckling of such an assembly can be analysed approximately by assuming that the plate elements are hinged along their common boundaries, so that each plate acts as if simply supported along its connected boundary or boundaries and free along any unconnected boundary. The buckling stress of each plate element can then be determined from equation 4.3 with $k_b = 4$ or 0.425 as appropriate, and the lowest of these can be used as an approximation for determining the buckling load of the member.

This approximation is conservative because the rigidity of the joints between the plate elements causes all plates to buckle simultaneously at a stress intermediate between the lowest and the highest of the buckling stresses of the individual plate elements. A number of analyses have been made of the stress at which simultaneous buckling takes place [4–6].

For example, values of the elastic buckling coefficient k_b for an I-section in uniform compression are shown in Fig. 4.10, and for a box section in uniform compression in Fig. 4.11. The buckling stress can be obtained from these figures by using equation 4.3 with the plate thickness t replaced by the flange thickness t_f. The use of these stresses and other results [4–6] will lead to more economic thin-walled compression members.

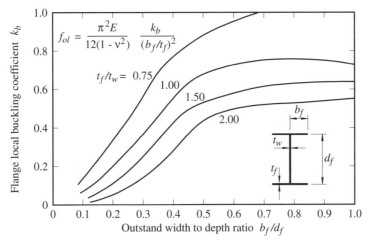

Fig. 4.10 Local buckling coefficients for I-section compression members

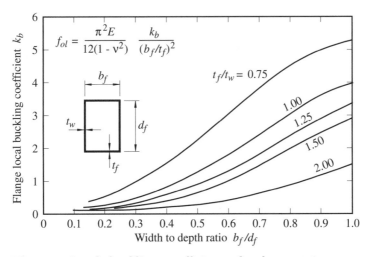

Fig. 4.11 Local buckling coefficients for box section compression members

4.2.2 ULTIMATE STRENGTH

4.2.2.1 *Inelastic buckling of thick plates*

The discussion given in section 4.2.1 on the elastic buckling of rectangular plates applies only to materials whose stress–strain relationships remain linear. Thus, for stocky steel plates for which the calculated elastic buckling stress exceeds the yield stress f_y, the elastic analysis must be modified accordingly.

One particularly simple modification which can be applied to strain-hardened steel plates is to use the strain-hardening modulus E_{st} and $\sqrt{(EE_{st})}$ instead of E in the terms of equation 4.97 (section 4.8.1) which represent the longitudinal bending and the twisting resistances to buckling, while still using E in the term for the lateral bending resistance. With these modifications, the strain-hardening buckling stress of a long simply supported plate is equal to the yield stress f_y when

$$\frac{b}{t} \sqrt{\left(\frac{f_y}{275}\right)} = 21.9 \qquad (4.11)$$

More accurate investigations [8] of the strain-hardening buckling of steel plates have shown that this simple approach is too conservative, and that the strain-hardening buckling stress is equal to the yield stress when

$$\frac{b}{t} \sqrt{\left(\frac{f_y}{275}\right)} = 29.7 \qquad (4.12)$$

for simply supported plates, and when

$$\frac{b}{t} \sqrt{\left(\frac{f_y}{275}\right)} = 7.6 \qquad (4.13)$$

for plates which are free along one longitudinal edge. If these are compared with equations 4.6 and 4.10, then it can be seen that these limits can be expressed in terms of the elastic buckling stress f_{ol} by

$$\sqrt{\left(\frac{f_y}{f_{ol}}\right)} = 0.58 \qquad (4.14)$$

for simply supported plates, and by

$$\sqrt{\left(\frac{f_y}{f_{ol}}\right)} = 0.46 \qquad (4.15)$$

for plates which are free along one longitudinal edge. These limits are shown in Fig. 4.4.

4.2.2.2 Post-buckling strength of thin plates

A thin elastic plate does not fail soon after it buckles, but can support loads significantly greater than its elastic buckling load without deflecting excessively. This is in contrast to the behaviour of an elastic compression member which can only carry very slightly increased loads before its deflections become excessive, as indicated in Fig. 4.12. This post-buckling behaviour of a thin plate is due to a number of causes, but the main reason is that the deflected shape of the buckled plate cannot be developed from the pre-buckled configuration without some redistribution of the in-plane stresses within the plate. This redistribution, which is ignored in the small deflection theory of elastic buckling,

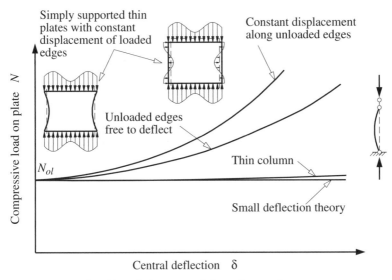

Fig. 4.12 Post-buckling behaviour of thin elastic plates

usually favours the less stiff portions of the plate, and causes an increase in the efficiency of the plate.

One of the most common causes of this redistribution is associated with the in-plane boundary conditions at the loaded edges of the plate. In long structural members, the continuity conditions along the transverse lines dividing consecutive buckled panels require that each of these boundary lines deflects a constant amount longitudinally. However, the longitudinal shortening of the panel due to its transverse deflections varies across the panel from a maximum at the centre to a minimum at the supported edges. This variation must, therefore, be compensated for by a corresponding variation in the longitudinal shortening due to axial strain, from a minimum at the centre to a maximum at the edges. The longitudinal stress distribution must be similar to the shortening due to strain, and so the stress at the centre of the panel is reduced below the average stress while the stress at the supported edges is increased above the average. This redistribution, which is equivalent to a transfer of stress from the more flexible central region of the panel to the regions near the supported edges, leads to a reduction in the transverse deflection, as indicated in Fig. 4.12.

A further redistribution of the in-plane stresses takes places when the longitudinal edges of the panel are supported by very stiff elements which ensure that these edges deflect a constant amount laterally, in the plane of the panel. In this case the variation along the panel of the lateral shortening due to the transverse deflections induces a self-equilibrating set of lateral in-plane stresses which are tensile at the centre of the panel and compressive at the loaded edges. The tensile stresses help support the less stiff central region of the panel, and lead to further reductions in the transverse deflections, as indicated in Fig. 4.12. However, this action is not usually fully developed, because the elements supporting the panels of most structural members are not very stiff.

The post-buckling effect is greater in plates supported along both longitudinal edges than it is in plates which are free along one longitudinal edge. This is because the deflected shapes of the latter have much less curvature than the former, and the redistributions of the in-plane stresses are not as pronounced. In addition, it is not possible to develop any lateral in-plane stresses along free edges, and so it is not uncommon to ignore any post-buckling reserves of slender flange outstands.

The redistribution of the in-plane stresses after buckling continues with increasing load until the yield stress f_y is reached at the supported edges. Yielding then spreads rapidly and the plate fails soon after, as indicated in Fig. 4.3. The occurrence of the first yield in an initially flat plate depends on its slenderness, and a thick plate yields before its elastic buckling stress f_{ol} is reached. As the slenderness increases and the elastic buckling stress decreases below the yield stress, the ratio of the ultimate stress f_u to the elastic buckling stress increases, as shown in Fig. 4.4.

Although the analytical determination of the ultimate strength of a thin flat plate is difficult, it has been found that the use of an effective width concept can lead to satisfactory approximations. According to this concept, the actual ultimate stress distribution in a simply supported plate (see Fig. 4.13) is replaced by a simplified distribution for which the central portion of the plate

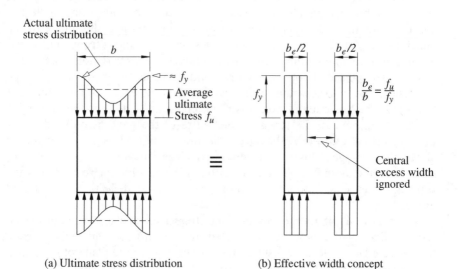

(a) Ultimate stress distribution   (b) Effective width concept

Fig. 4.13 Effective width concept for simply supported plates

is ignored and the remaining effective width b_e carries the yield stress f_y. It was proposed that this effective width should be approximated by

$$\frac{b_e}{b} = \sqrt{\left(\frac{f_{ol}}{f_y}\right)},$$

(4.16)

which is equivalent to supposing that the ultimate load carrying capacity of the plate $f_y b_e t$ is equal to the elastic buckling load of a plate of width b_e. Alternatively, this proposal can be regarded as determining an effective average ultimate stress f_u which acts on the full width b of the plate. This average ultimate stress, which is given by

$$\frac{f_u}{f_y} = \sqrt{\left(\frac{f_{ol}}{f_y}\right)},$$

(4.17)

is shown in Fig. 4.4.

Experiments on real plates with initial curvatures and residual stresses have confirmed the qualitative validity of this effective width approach, but suggest that the quantitative values of the effective width for hot-rolled and welded plates should be obtained from equations of the type

$$\frac{b_e}{b} = \alpha \sqrt{\left(\frac{f_{ol}}{f_y}\right)},$$

(4.18)

where α reflects the influence of the initial curvatures and residual stresses. For example, test results for the ultimate stresses $f_u (= f_y b_e/b)$ of hot-rolled simply supported plates with residual stresses and initial curvatures are shown in Fig. 4.14, and suggest a value of α equal to 0.65. Other values of α are given in [7]. Tests on cold-formed members also support the effective width concept, with quantitative values of the effective width being obtained from

$$\frac{b_e}{b} = \sqrt{\left(\frac{f_{ol}}{f_y}\right)}\left[1 - 0.22\sqrt{\left(\frac{f_{ol}}{f_y}\right)}\right].$$

(4.19)

4.2.2.3 Effects of initial curvature and residual stresses

Real plates are not perfectly flat, but have small initial curvatures similar to those in real columns (see section 3.2.2) and real beams (see section 6.2.2). The initial curvature of a plate causes it to deflect transversely as soon as it is loaded, as shown in Fig. 4.3. These deflections increase rapidly as the elastic buckling stress is approached, but slow down beyond the buckling stress and approach those of an initially flat plate.

In a thin plate with initial curvature, the first yield and failure occur only slightly before they do in a flat plate, as indicated in Fig. 4.3, while the initial curvature has little effect on the strength of a thick plate (see Fig. 4.4). It is only in a plate of intermediate slenderness that the initial curvature causes a significant reduction in the strength, as indicated in Fig. 4.4.

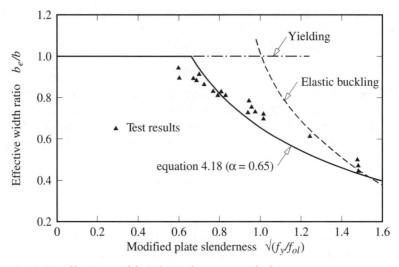

Fig. 4.14 Effective widths of simply supported plates

Real plates usually have residual stresses induced by uneven cooling after rolling or welding. These stresses are generally tensile at the junctions between plate elements, and compressive in the regions away from the junctions. Residual compressive stresses in the central region of a simply supported thin plate cause it to buckle prematurely and reduce its ultimate strength, as shown in Fig. 4.4. Residual stresses also cause premature yielding in plates of intermediate slenderness, as indicated in Fig. 4.4, but have a negligible effect on the strain-hardening buckling of stocky plates.

Some typical test results for thin supported plates with initial curvatures and residual stresses are shown in Fig. 4.14.

4.3 Plate elements in shear

4.3.1 ELASTIC BUCKLING

The thin flat plate of length, L, depth d, and thickness t shown in Fig. 4.15 is simply supported along all four edges. The plate is loaded by shear stresses distributed uniformly along its edges. When these stresses are equal to the elastic buckling value f_{ov}, then the plate can buckle by deflecting u laterally out of its original plane into an adjacent position. For this adjacent position to be one of equilibrium, the differential equilibrium equation [1–4]

$$\left(\frac{\partial^4 u}{\partial z^4} + \frac{2\partial^4 u}{\partial y^2 \partial z^2} + \frac{\partial^4 u}{\partial y^4} \right) = -\frac{2f_{ov}t}{D} \frac{\partial^2 u}{\partial y \partial z} \tag{4.20}$$

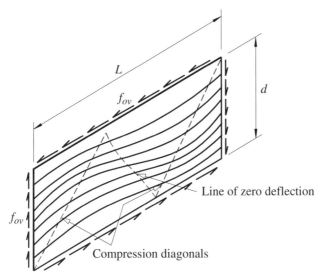

Fig. 4.15 Buckled pattern of a simply supported plate in shear

must be satisfied (this may be compared with the corresponding equation 4.97 of section 4.8.1, for a plate in compression).

Closed form solutions of this equation are not available, but numerical solutions have been obtained. These indicate that the plate tends to buckle along compression diagonals, as shown in Fig. 4.15. The shape of the buckle is influenced by the tensile forces acting along the other diagonal, while the number of buckles increases with the aspect ratio L/d. The numerical solutions for the elastic buckling stress f_{ov} can be expressed in the form

$$f_{ov} = \frac{\pi^2 E}{12(1 - v^2)} \frac{k_b}{(d/t)^2} \tag{4.21}$$

in which the buckling coefficient k_b is approximated by

$$k_b = 5.35 + 4\left(\frac{d}{L}\right)^2 \tag{4.22}$$

when $L \geq d$, and by

$$k_b = 5.35\left(\frac{d}{L}\right)^2 + 4 \tag{4.23}$$

when $L \leq d$, as shown in Fig. 4.16.

The shear stresses in many structural members are transmitted by unstiffened webs, for which the aspect ratio L/d is large. In this case the buckling

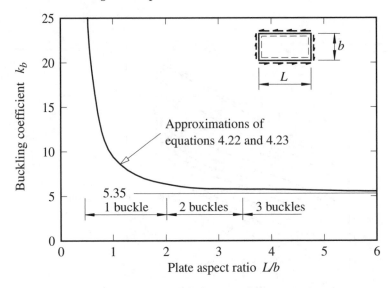

Fig. 4.16 Buckling coefficients of plates in shear

coefficient k_b approaches 5.35, and the buckling stress can be closely approximated by

$$f_{ov} = \frac{5.35\pi^2 E/12(1 - v^2)}{(d/t)^2}. \tag{4.24}$$

This elastic buckling stress is equal to the yield stress in shear $\tau_y = f_y/\sqrt{3}$ (see section 1.3.1) of a steel for which $E = 205\,000$ N/mm^2 and $v = 0.3$ when

$$\frac{d}{t}\sqrt{\left(\frac{f_y}{275}\right)} = 79.0. \tag{4.25}$$

The values of the buckling coefficient k_b shown in Fig. 4.16 and the form of equation 4.21 indicate that the elastic buckling stress may be significantly increased either by using intermediate transverse stiffeners to decrease the aspect ratio L/d and to increase the buckling coefficient k_b, or by using longitudinal stiffeners to decrease the depth–thickness ratio d/t. It is apparent from Fig. 4.16 that such stiffeners are likely to be most efficient when the stiffener spacing s is such that the aspect ratio of each panel lies between 0.5 and 2 so that only one buckle can form in each panel.

Any intermediate stiffeners used must be sufficiently stiff to ensure that the elastic buckling stress f_{ov} is increased to the value calculated for the stiffened panel. Some values of the required stiffener second moment of area I_s are given in [3, 5] for various panel aspect ratios a/d. These can be conservatively approximated by using

$$\frac{I_s}{at^3/12(1 - v^2)} = \frac{6}{a/d}, \tag{4.26}$$

when $a/d \geqslant 1$, and by using

$$\frac{I_s}{at^3/12(1 - v^2)} = \frac{6}{(a/d)^4}, \tag{4.27}$$

when $a/d \leqslant 1$.

4.3.2 ULTIMATE STRENGTH

4.3.2.1 Unstiffened plates

A stocky unstiffened web in an I-section beam in pure shear is shown in Fig. 4.17a. The elastic shear stress distribution in such a section is analysed in section 5.10.2. The web behaves elastically in shear until first yield occurs at $\tau_y = f_y/\sqrt{3}$, and then undergoes increasing plastification until the web is fully yielded in shear (Fig. 4.17b). Because the shear stress distribution at first yield is nearly uniform, the nominal first yield and fully plastic loads are nearly equal, and the shear shape factor is usually very close to 1.0. Stocky unstiffened webs in steel beams reach first yield before they buckle elastically, so that their strengths are determined by the shear stress τ_y, as indicated in Fig. 4.18.

Thus the resistance of a stocky web in a flanged section for which the shear shape factor is close to unity is closely approximated by the web plastic shear resistance

$$V_w = dt\tau_y. \tag{4.28}$$

When the depth to thickness ratio d/t of a long unstiffened steel web exceeds $79.0/\sqrt{(f_y/275)}$, its elastic buckling shear stress f_{ov} is less than the shear yield stress (see equation 4.25). The post-buckling reserve of shear strength of such an unstiffened web is not great, and its ultimate stress f_u can be approximated

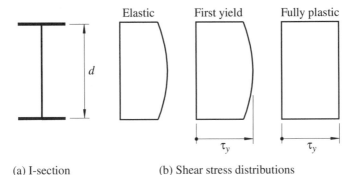

(a) I-section   (b) Shear stress distributions

Fig. 4.17 Plastification of an I-section web in shear

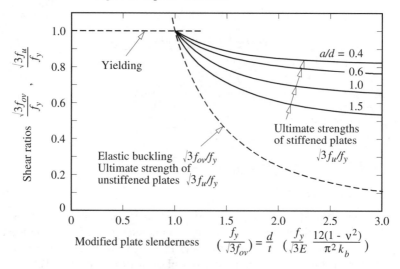

Fig. 4.18 Ultimate strengths of plates in shear

with reasonable accuracy by the elastic buckling stress f_{ov} given by equation 4.24 and shown by the dashed line in Fig. 4.18.

4.3.2.2 Stiffened plates

Stocky stiffened webs in steel beams yield in shear before they buckle, and so the stiffeners do not contribute to the ultimate strength. Because of this, stocky webs are usually unstiffened.

In a slender web with transverse stiffeners which buckles elastically before it yields, there is a significant reserve of strength after buckling. This is caused by a redistribution of stress, in which the diagonal tension stresses in the web panel continue to increase with the applied shear, while the diagonal compressive stresses remain substantially unchanged. The increased diagonal tension stresses form a tension field, which combines with the transverse stiffeners and the flanges to transfer the additional shear in a truss-type action [8, 9], as shown in Fig. 4.19. The ultimate shear stresses f_u of the web is reached soon after yield occurs in the tension field, and this can be approximated by

$$f_u = f_{ov} + f_{tf} \tag{4.29}$$

in which f_{ov} is the elastic buckling stress given by equations 4.21–4.23 with L equal to the stiffener spacing a, and f_{tf} is the tension field contribution at yield which can be approximated by

$$f_{tf} = \frac{f_y}{2} \frac{(1 - \sqrt{3f_{ov}/f_y})}{\sqrt{[1 + (a/d)^2]}}. \tag{4.30}$$

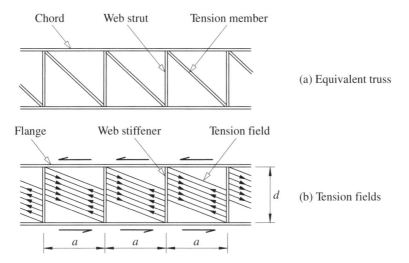

Fig. 4.19 Tension fields in a stiffened web

Thus the approximate ultimate strength can be obtained from

$$\frac{\sqrt{3}f_u}{f_y} = \frac{\sqrt{3}f_{ov}}{f_y} + \frac{\sqrt{3}}{2}\frac{(1 - \sqrt{3}f_{ov}/f_y)}{\sqrt{[1 + (a/d)^2]}},\qquad(4.31)$$

which is shown in Fig. 4.18. This equation is conservative, as it ignores any contributions made by the bending resistance of the flanges to the ultimate strength of the web [9, 10].

Not only must the intermediate transverse stiffeners be sufficiently stiff to ensure that the elastic buckling stress f_{ov} is reached, as explained in section 4.3.1, but they must also be strong enough to transmit the stiffener force

$$N_s = \frac{f_y d t}{2}\left(1 - \frac{\sqrt{3}f_{ov}}{f_y}\right)\left(\frac{a}{d} - \frac{(a/d)^2}{\sqrt{[1 + (a/d)^2]}}\right)\qquad(4.32)$$

required by the tension field action. Intermediate transverse stiffeners are often so stocky that their ultimate strengths can be approximated by their squash loads, in which case the required area A_s of a symmetrical pair of stiffeners is given by

$$A_s = N_s/f_y.\qquad(4.33)$$

However, the stability of very slender stiffeners should be checked, in which case the second moment of area I_s required to resist the stiffener force can be conservatively estimated from

$$I_s = \frac{N_s d^2}{\pi^2 E},\qquad(4.34)$$

which is based on the elastic buckling of a pin-ended column of length d.

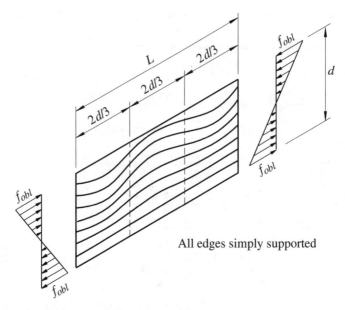

Fig. 4.20 Buckled pattern of a plate in bending

4.4 Plate elements in bending

4.4.1 ELASTIC BUCKLING

The thin flat plate of length L, width d, and thickness t shown in Fig. 4.20 is simply supported along all four edges. The plate is loaded by bending stress distributions which vary linearly across its width. When the maximum stress reaches the elastic buckling value f_{obl}, the plate can buckle out of its original plane as shown in Fig. 4.20. The elastic buckling stress can be expressed in the form

$$f_{obl} = \frac{\pi^2 E}{12(1 - v^2)} \frac{k_b}{(d/t)^2},\qquad(4.35)$$

where the buckling coefficient k_b depends on the aspect ratio L/d of the plate and the number of buckles along the plate. For long plates, the value of k_b is close to its minimum value of

$$k_b = 23.9 \qquad (4.36)$$

for which the length of each buckle is approximately $2d/3$. By using this value of k_b, it can be shown that the elastic buckling stress for a steel for which $E = 205\,000$ N/mm² and $v = 0.3$ is equal to the yield stress f_y when

$$\frac{d}{t}\sqrt{\left(\frac{f_y}{275}\right)} = 126.9. \tag{4.37}$$

The buckling coefficient k_b is not significantly greater than 23.9 except when the buckle length is reduced below $2d/3$. For this reason, transverse stiffeners are ineffective unless more closely spaced than $2d/3$. On the other hand, longitudinal stiffeners may be quite effective in changing the buckled shape, and therefore the value of k_b. Such a stiffener is most efficient when it is placed in the compression region at about $d/5$ from the compression edge. This is close to the position of the crests in the buckles of an unstiffened plate. Such a stiffener may increase the buckling coefficient k_b significantly, the maximum value of 129.4 being achieved when the stiffener acts as if rigid. The values given in [5, 6] indicate that a hypothetical stiffener of zero area A_s will produce this effect if its second moment of area I_s is equal to $4dt^3$. This value should be increased to allow for the compressive load in the real stiffener, and it is suggested that a suitable value might be given by

$$I_s = 4dt^3\left[1 + 4\frac{A_s}{dt}\left(1 + \frac{A_s}{dt}\right)\right], \tag{4.38}$$

which is of a similar form to that given by equation 4.7 for the longitudinal stiffeners of plates in uniform compression.

4.4.2 PLATE ASSEMBLIES

The local buckling of a plate assembly in bending, such as an I-beam bent about its major axis, can be analysed approximately by assuming that the plate elements are hinged along their common boundaries in much the same fashion as that described in section 4.2.1.4 for compression members. The elastic buckling moment can then be approximated by using $k_b = 0.425$ or 4 for the flanges, as appropriate, and $k_b = 23.9$ for the web, and determining the element which is closest to buckling. The results of more accurate analyses of the simultaneous buckling of the flange and web elements in I-beams and box section beams in bending are given in Figs. 4.21 and 4.22, respectively.

4.4.3 ULTIMATE STRENGTH

The ultimate strength of a thick plate in bending is governed by its yield stress f_y and by its plastic shape factor which is equal to 1.5 for a constant thickness plate, as discussed in section 5.5.2.

When the width to thickness ratio d/t exceeds $126.9/\sqrt{(f_y/275)}$, the elastic buckling stress f_{obl} of a simply supported plate is less than the yield stress f_y (see equation 4.37). A long slender plate such as this has a significant reserve of strength after buckling, because it is able to redistribute the compressive stresses from the buckled region to the area close to the supported

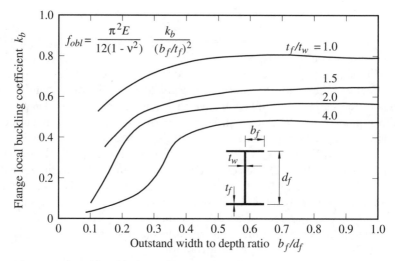

Fig. 4.21 Local buckling coefficients for I-beams

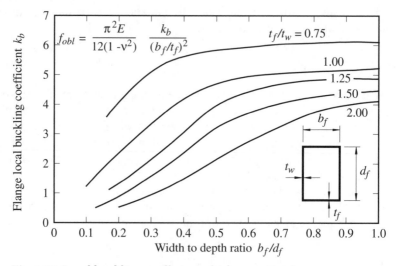

Fig. 4.22 Local buckling coefficients for box section beams

compression edge, in the same way as a plate in uniform compression (see section 4.2.2.2). Some solutions for the post-buckling reserve of strength of a plate in bending are given in [6, 11–13]. These show that an effective width treatment can be used, with the effective width being taken over part of the compressive portion of the plate. Unfortunately, the use of such a simplification is more difficult to implement than that for plates in uniform compression.

4.5 Plate elements in shear and bending

4.5.1 ELASTIC BUCKLING

The elastic buckling of the simply supported thin flat plate shown in Fig. 4.23 which is subjected to combined shear and bending can be predicted by using the approximate interaction equation [2, 4, 5, 13]

$$\left(\frac{f_{ov}}{f_{ovo}} \right)^2 + \left(\frac{f_{obl}}{f_{oblo}} \right)^2 = 1 \tag{4.39}$$

in which f_{ovo} is the elastic buckling stress when the plate is in pure shear (see equations 4.21–4.24) and f_{oblo} is the elastic buckling stress for pure bending (see equations 4.35 and 4.36). If the elastic buckling stresses f_{ov}, f_{obl} are used in the Hencky–Von Mises yield criterion (see section 1.3.1)

$$3f_{ov}^2 + f_{obl}^2 = f_y^2, \tag{4.40}$$

it can shown from equations 4.39 and 4.40 that the most severe loading condition for which elastic buckling and yielding occur simultaneously is that of pure shear. Thus, in an unstiffened web of a steel for which $E = 205\,000$ N/mm² and $v = 0.3$, yielding will occur before buckling while (see equation 4.25)

$$\frac{d}{t} \sqrt{\left(\frac{f_y}{275} \right)} \leqslant 79.0. \tag{4.41}$$

4.5.2 ULTIMATE STRENGTH

4.5.2.1 Unstiffened plates

Stocky unstiffened webs in steel beams yield before they buckle, and their design capacities can be estimated approximately by using the Hencky–Von

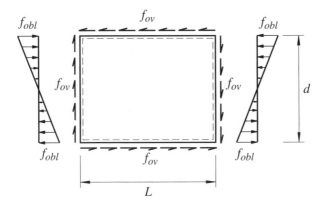

Fig. 4.23 Simply supported plate under shear and bending

Mises yield criterion (see section 1.3.1), so that the shear force V and moment M_w in the web satisfy

$$3\left(\frac{V}{dt}\right)^2 + \left(\frac{6M_w}{d^2t}\right)^2 = (f_y)^2. \qquad (4.42)$$

This approximation is conservative, as it assumes that the plastic shape factor for web bending is 1.0 instead of 1.5.

While the elastic buckling of slender unstiffened webs under combined shear and bending has been studied, the ultimate strengths of such webs have not been fully investigated. However, it seems possible that there is only a small reserve of strength after elastic buckling, so that the ultimate strength can be approximated conservatively by the elastic buckling stress combinations which satisfy equation 4.39.

4.5.2.2 Stiffened plates

The ultimate strength of a stiffened web under combined shear and bending can be discussed in terms of the interaction diagram shown in Fig. 4.24 for the ultimate shear force V and bending moment M acting on a plate girder. When there is no shear, the ultimate moment capacity is equal to the full plastic moment M_p, providing the flanges are plastic or compact (see section 4.7.2), while the ultimate shear capacity in the absence of bending moment is

$$V_u = dtf_u, \qquad (4.43)$$

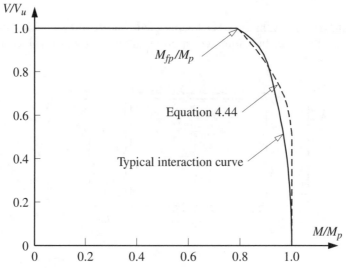

Fig. 4.24 Ultimate strengths of stiffened webs in shear and bending

where the ultimate stress f_u is given approximately by equation 4.31 (which ignores any contributions made by the bending resistance of the flanges). This shear capacity remains unchanged as the bending moment increases to the value M_{fp} which is sufficient to fully yield the flanges if they alone resist the moment. This fact forms the basis for the widely used proportioning procedure for which it is assumed that the web resists only the shear and that the flanges are required to resist the full moment. As the bending moment increases beyond M_{fp}, the ultimate shear capacity falls off rapidly as shown. A simple approximation for the reduced shear capacity is given by

$$\frac{V}{V_u} = 0.5\left\{1 + \sqrt{\frac{(1 - M/M_p)}{(1 - M_{fp}/M_p)}}\right\} \tag{4.44}$$

while $M_{fp}/M_p \leqslant M/M_p \leqslant 1$. This approximation ignores the influence of the bending stresses on the tension field and the bending resistance of the flanges. A more accurate method of analysis is discussed in [14].

4.6 Plate elements in bearing

4.6.1 ELASTIC BUCKLING

Plate elements are subjected to bearing stresses by concentrated or locally distributed edge loads. For example, a concentrated load applied to the top flange of a plate girder induces local bearing stresses in the web immediately beneath the load. In a slender girder with transverse web stiffeners, the load is resisted by vertical shear stresses acting at the stiffeners, as shown in Fig. 4.25a. Bearing

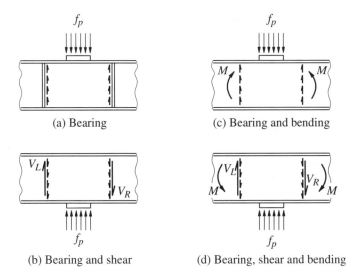

(a) Bearing   (c) Bearing and bending

(b) Bearing and shear   (d) Bearing, shear and bending

Fig. 4.25 Plate girder webs in bearing

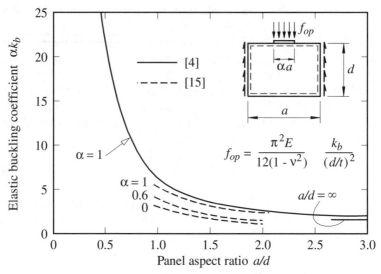

Fig. 4.26 Buckling coefficients for plates in bearing

commonly occurs in conjunction with shear (as at an end support, see Fig. 4.25b), bending (as at mid-span, see Fig. 4.25c), and with combined shear and bending (as at an interior support, see Fig. 4.25d).

In the case of a panel of a stiffened web, the edges of the panel may be regarded as simply supported. When the bearing load is distributed along the full length a of the panel and there is no shear or bending, the elastic bearing buckling stress f_{op} can be expressed as

$$f_{op} = \frac{\pi^2 E}{12(1 - v^2)} \frac{k_b}{(d/t)^2} \tag{4.45}$$

in which the buckling coefficient k_b varies with the panel aspect ratio a/d as shown in Fig. 4.26.

When a patch bearing load acts along a reduced length αs of the top edge of the panel, the value of αk_b is decreased. This effect has been investigated [15], and some values of αk_b are shown in Fig. 4.26. These values suggest a limiting value of $\alpha k_b = 0.8$ approximately for the case of unstiffened webs ($a/d \rightarrow \infty$) with concentrated loads ($\alpha \rightarrow 0$).

For the case of a panel in combined bearing, shear, and bending, it has been suggested [6, 15] that the elastic buckling stresses can be determined from the interaction equation

$$\frac{f_{op}}{f_{opo}} + \left(\frac{f_{ov}}{f_{ovo}}\right)^2 + \left(\frac{f_{obl}}{f_{oblo}}\right)^2 = 1 \tag{4.46}$$

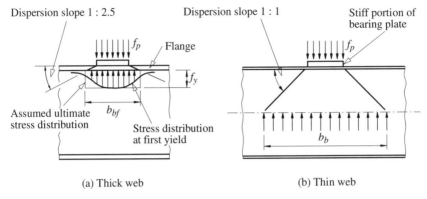

Fig. 4.27 Assumed dispersions of bearing stresses

in which the final subscripts $_o$ indicate the appropriate elastic buckling stress when only that type of loading is applied. Some specific interaction diagrams are given in [15].

4.6.2 ULTIMATE STRENGTH

The ultimate strength of a thick web in bearing depends chiefly on its yield stress f_y. Although yielding first occurs under the centre of the bearing plate as shown in Fig. 4.27a, general yielding does not take place until the applied load is large enough to yield a web area defined by a dispersion of the applied stress through the flange. Even at this load the web does not collapse catastrophically, and some further yielding and redistribution is possible. When the web is also subjected to shear and bending, general yielding in bearing can be approximated by using the Hencky–Von Mises yield criterion (see section 1.3.1)

$$f_p^2 + f_b^2 - f_p f_b + 3\tau^2 = f_y^2 \tag{4.47}$$

in which f_p is the bearing stress, f_b the bending stress and τ the shear stress.

Thin stiffened web panels in bearing have a reserve of strength after elastic buckling which is due to a redistribution of stress from the more flexible central region to the stiffeners. Studies of this effect suggest that the reserve of strength decreases as the bearing loads become more concentrated [15]. The collapse behaviour of stiffened panels is described in [16, 17].

4.7 Design against local buckling

4.7.1 COMPRESSION MEMBERS

Compression members must be designed so that the factored design compression force F_c^* does not exceed the compression resistance P_c, whence

$$F_c^* \leq P_c. \tag{4.48}$$

Section description	Hot-rolled UB, UC	Cold-formed CHS	Welded box	Cold-formed RHS
Section and element widths	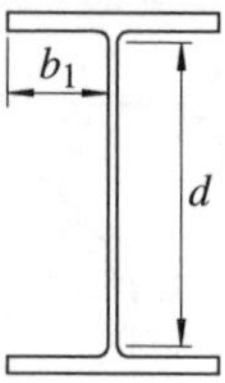 b_1 d	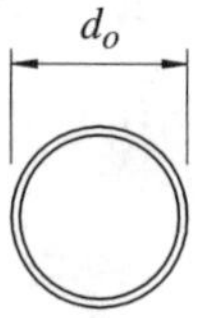 d_o	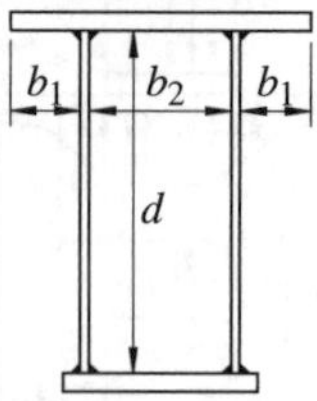 b_1 b_2 b_1 d	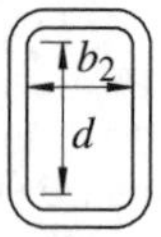 b_2 d
Flange outstand b_1	15ε	–	13ε	–
Flange b_2 supported along both edges	–	–	40ε	35ε
Web d supported along both edges	40ε	–	40ε	35ε
Diameter d_o	–	$80ε^2$	–	–

Fig. 4.28 Width-thickness yield limits $\lambda_{L3}ε$ for some compression in elements

The compression resistance P_c is given by

$$P_c = A_{eff} p_{cs} \qquad (4.49)$$

in which A_{eff} is the effective area of the cross-section and p_{cs} is the compressive strength.

For a member which has no local buckling effects, the effective area A_{eff} is equal to the gross area of the cross-section A_g, and the compressive strength p_{cs} depends only on the member slenderness $\lambda = L_E/r$ and the steel design strength p_y, as discussed in section 3.7.3. A member composed of flat plate elements has no local buckling effects when the width-thickness ratio b/t of every plate element of the member satisfies

$$b/(tε) \leq \lambda_{L3} \qquad (4.50)$$

in which λ_{L3} is the appropriate Class 3 semi-compact yield slenderness limit of Table 11 of BS5950 and $ε$ is a constant given by

$$ε = \sqrt{(275/p_y)} \qquad (4.51)$$

in which p_y is in N/mm^2 units. Some values of $\lambda_{L3}ε$ are given in Figs 4.28 and 4.29.

Flat plate elements which do not satisfy equation 4.50 are described by BS5950 as Class 4 slender elements. A member with one or more slender elements has its effective area reduced to

$$A_{eff} = \Sigma(b_{eff}t) \qquad (4.52)$$

in which the effective width of any slender plate element is determined from

$$b_{eff} = \lambda_{L3}tε \leq b \qquad (4.53)$$

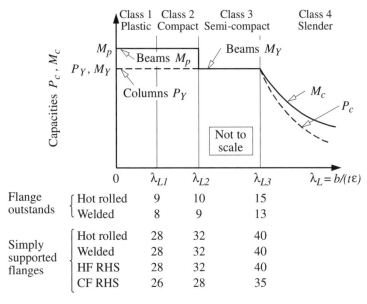

Fig. 4.29 BS5950 capacities of fully braced beams and columns

Flange outstands		λ_{L1}	λ_{L2}	λ_{L3}
Flange outstands	Hot rolled	9	10	15
	Welded	8	9	13
Simply supported flanges	Hot rolled	28	32	40
	Welded	28	32	40
	HF RHS	28	32	40
	CF RHS	26	28	35

A circular hollow section member has no local buckling effects when its diameter-thickness ratio D/t satisfies

$$D/(t\varepsilon^2) \leqslant 80 \tag{4.54}$$

A circular hollow section member which does not satisfy equation 4.54 is a Class 4 slender member. Its effective area is reduced to

$$A_{\text{eff}} = A_{\text{g}} \left\{ \frac{80}{D/(t\varepsilon^2)} \right\}^{0.5} \tag{4.55}$$

The compressive strength p_{cs} of a Class 4 slender member is increased by using a reduced slenderness

$$\lambda_{\text{eff}} = \lambda \sqrt{(A_{\text{eff}}/A_{\text{g}})} \tag{4.56}$$

Worked examples for checking the section capacity of a compression member are given in sections 4.9.1 and 3.12.1–3.12.3.

4.7.2 BEAM FLANGES AND WEBS IN COMPRESSION

Compression elements in a beam cross-section are classified in BS5950 as Class 1 plastic, Class 2 compact, Class 3 semi-compact, or Class 4 slender, depending on their local buckling strength.

Class 1 plastic elements are unaffected by local buckling, and are able to develop and maintain their fully plastic capacities (section 5.5) while inelastic moment redistribution takes place in the beam. Class 1 plastic elements satisfy

$$b/(t\varepsilon) \leq \lambda_{L1} \qquad (4.57)$$

in which λ_{L1} is the appropriate plasticity slenderness limit given in Table 11 of BS5950 (some values of λ_{L1} are shown in Figs 4.29 and 5.33). These limits are closely related to equations 4.12 and 4.13 for the strain-hardening buckling stresses of inelastic plates.

Class 2 compact elements are unaffected by local buckling in the development of their fully plastic capacities (section 5.5), but may be unable to maintain these capacities while inelastic moment redistribution takes place in the beam. Class 2 compact sections satisfy

$$\lambda_{L1} \leq b/(t\varepsilon) \leq \lambda_{L2} \qquad (4.58)$$

in which λ_{L2} is the appropriate Class 2 slenderness limit given in Table 11 of BS5950 (some values of λ_{L2} are shown in Figs 4.29 and 5.33). These limits are slightly greater than those implied in equations 4.12 and 4.13 for the strain-hardening buckling of inelastic plates.

Class 3 semi-compact elements are able to reach first yield, but buckle locally before they become fully plastic. Class 3 semi-compact elements satisfy

$$\lambda_{L2} \leq b/(t\varepsilon) \leq \lambda_{L3} \qquad (4.59)$$

in which λ_{L3} is the appropriate yield slenderness limit given in Table 11 of BS5950 (some values of λ_{L3} are shown in Figs 4.29 and 5.33). These limits are closely related to equations 4.18 used to define the effective widths of flange plates in uniform compression or modified from equation 4.37 for web plates in bending. Class 4 slender elements buckle locally before they reach first yield, and satisfy

$$\lambda_{L3} < b/(t\varepsilon) \qquad (4.60)$$

Beam cross-sections are classified as being Class 1 plastic, Class 2 compact, Class 3 semi-compact, or Class 4 slender, depending on the classification of their elements. A Class 1 plastic section has all elements plastic. A Class 2 compact section has no Class 3 semi-compact or Class 4 slender elements and has at least one Class 2 compact element, while a Class 3 semi-compact section has no Class 4 slender elements and at least one Class 3 semi-compact element. A Class 4 slender section has at least one Class 4 slender element.

A beam must be designed so that its factored design moment M^* does not exceed the section moment capacity, so that

$$M^* \leq M_c \qquad (4.61)$$

Beams must also be designed for shear (section 4.7.3) and bearing (sections 4.7.4 and 4.7.5) and against lateral buckling (section 6.9). The section capacity M_c is given by

$$M_c = p_y Z_{eff} \qquad (4.62)$$

in which Z_{eff} is the effective section modulus. For Class 1 plastic and Class 2 compact beams

$$Z_{eff} = S \leqslant 1.2Z, \qquad (4.63)$$

which allows many beams to be designed for the full plastic moment

$$M_p = p_y S \qquad (4.64)$$

as indicated in Fig. 4.29.

For Class 3 semi-compact beams, the effective section modulus is taken as the elastic section modulus Z. Alternatively, BS5950 presents equations that may be used to calculate a reduced plastic section modulus S_{eff} which is larger than the elastic section modulus Z_x.

For a beam with a slender compression flange supported along both edges, the effective section modulus Z_{eff} may be determined by calculating the elastic section modulus of an effective cross-section obtained by using a compression flange effective width b_{eff} obtained using equation 4.53. The calculation of Z_{eff} must incorporate the possibility that the effective section is not symmetric.

Worked examples of classifying the section and checking the section moment capacity are given in sections 4.9.2–4.9.4, 5.12.15 and 5.12.17.

4.7.3 LONGITUDINAL STIFFENERS

A logical basis for the design of a web in pure bending is to limit its proportions so that its maximum elastic bending strength can be used. When this is done, the section capacity M_c of the beam is governed by the slenderness of the flanges. Thus BS5950 requires a fully effective unstiffened web to satisfy

$$d/(t\varepsilon) \leqslant 120 \qquad (4.65)$$

This limit is close to the value of 126.9 at which the elastic buckling stress is equal to the yield stress (see equation 4.37). BS5950 places a general upper limit of

$$\frac{d}{t}\left(\frac{p_{yf}}{345}\right) \leqslant 250 \qquad (4.66)$$

on the slenderness of unstiffened webs, where p_{yf} is the design strength of the flange.

While the effectiveness of a web with transverse stiffeners may be increased by using longitudinal stiffeners to reduce the web panel depth, BS5950 provides no guidance for their design, but instead refers the designer to BS5400 [18]. The Australian standard AS4100 [19] gives a simple method of design in which a first longitudinal stiffener whose second moment of area is at least that of equation 4.38 is placed at one-fifth of the web depth from the compression flange when the overall depth to thickness ratio of the web exceeds the limit of

$$\frac{d}{t}\sqrt{\left(\frac{f_y}{250}\right)} = 200. \tag{4.67}$$

An additional stiffener whose second moment of area exceeds

$$I_s = dt^3 \tag{4.68}$$

is required at the neutral axis when $(d/t)\sqrt{(f_y/250)}$ exceeds 250.

4.7.4 BEAM WEBS IN SHEAR

4.7.4.1 Stocky webs

The design shear force F_v^* on a web for which the shear stress is approximately uniform must satisfy

$$F_v^* \leqslant P_v \tag{4.69}$$

in which P_v is the uniform shear capacity.

The uniform shear capacity of a stocky web of area A_v is given by

$$P_v = 0.6p_y A_v \tag{4.70}$$

which is very close to $dt\tau_y(\approx 0.58\,A_v p_y)$ given by equation 4.28 for the fully plastic shear capacity, where d is the depth of the web and t the web thickness.

An I-section web is classified by BS5950 as stocky when its depth to thickness ratio satisfies $d/(t\varepsilon) \leqslant 70$ for a hot-rolled beam (the webs of all British universal sections of S275 or S355 steel are stocky), and $d/(t\varepsilon) \leqslant 62$ for a welded plate girder, and as slender otherwise. These values are somewhat less than the value of 79 in equation 4.25 for the limiting slenderness at which the elastic buckling stress f_{ov} is equal to the shear yield stress τ_y.

The design force on a web for which the elastic shear stress varies significantly must be limited so that the maximum elastic shear stress does not exceed $0.7\,p_y$. This limit will usually apply when the ratio of the maximum and average elastic shear stresses in the web exceeds $0.7/0.6 = 1.17$.

4.7.4.2 Slender webs

While the shear resistance of a slender web in a welded plate girder decreases as the web slenderness $d/(t\varepsilon)$ increases, it may be increased by providing transverse stiffeners which increase the resistance to elastic buckling (section 4.3.1) and also permit the development of tension field action (section 4.3.2.2). The design shear force F_v^* on a slender web must satisfy

$$F_v^* \leqslant V_b = V_w + V_f \leqslant P_v \tag{4.71}$$

in which V_w is the simple shear buckling resistance arising from the increased elastic buckling resistance and the tension field action, and V_f is the flange-dependent shear buckling resistance which occurs only when the flanges are not fully stressed, and which may be neglected.

The simple shear buckling resistance V_w is given by

$$V_w = dtq_w \qquad (4.72)$$

in which the shear buckling strength q_w is given by

$$
\begin{aligned}
q_w &= p_v & \text{when } \lambda_w \leq 0.8 & \qquad (4.73a) \\
q_w &= [(13.48 - 5.6\,\lambda_w)/9]p_v & \text{when } 0.8 < \lambda_w < 1.25 & \qquad (4.73b) \\
q_w &= (0.9/\lambda_w)p_v & \text{when } 1.25 \leq \lambda_w & \qquad (4.73c)
\end{aligned}
$$

In these equations,

$$p_v = 0.6\, p_{yw} \qquad (4.74)$$
$$\lambda_w = (p_v/q_e)^{0.5} \qquad (4.75)$$

in which p_{yw} is the design strength of the web, and

$$q_e = \left[0.75 + \frac{1}{(a/d)^2}\right]\left[\frac{1000}{d/t}\right]^2 \qquad \text{when } a/d \leq 1 \qquad (4.76a)$$

$$q_e = \left[1 + \frac{0.75}{(a/d)^2}\right]\left[\frac{1000}{d/t}\right]^2 \qquad \text{when } 1 \leq a/d \qquad (4.76b)$$

in which a is the spacing of the stiffeners. Equations 4.76 are derived from the elastic buckling solutions of equations 4.21–4.23. Equation 4.73c provides a simple approximation for the elastic post-buckling strengths of stiffened webs, equation 4.37b provides an allowance for inelastic behaviour at intermediate web slendernesses, and equation 4.37a corresponds to the approximate shear yield strength of a stocky web. The dimensionless shear buckling strengths q_w/p_v given by these equations are shown in Fig. 4.30. These are similar to the shear ratios $\sqrt{3}f_{ov}/f_y$ shown in Fig. 4.18.

The flange-dependent shear buckling resistance V_f in equation 4.71 accounts for the formation of plastic hinges in understressed flanges during the development of the tension field in the web, which further increases the shear resistance. This additional resistance is approximated by

$$V_f = \frac{P_v(d/a)[1 - (f_f^*/p_{yf})^2]}{1 + 0.15(M_{pw}/M_{pf})} \qquad (4.77)$$

in which f_f^* is the design stress in the smaller flange and M_{pf} and M_{pw} are the plastic moments of the smaller flange and of the web, respectively.

Worked examples of checking the shear capacity of a slender web are given in sections 4.9.5 and 4.9.6.

4.7.4.3 Transverse stiffeners

Provisions are made in BS5950 for a minimum value for the second moment of area I_s of intermediate transverse stiffeners. These must be stiff enough to

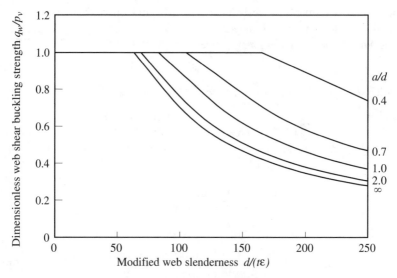

Fig. 4.30 Web shear buckling strengths of BS5950

ensure that the elastic buckling stress f_{ov} of a panel can be reached, and strong enough to transmit the tension field stiffener force. The stiffness requirement of BS5950 is that the second moment of area of a transverse stiffener must satisfy

$$I_s \geqslant 0.75 \, dt^3, \tag{4.78}$$

when $a/d \geqslant \sqrt{2}$, in which d is the depth of the web, and

$$I_s \geqslant 1.5 \, d^3 t^3/a^2, \tag{4.79}$$

when $a/d < \sqrt{2}$. These limits are shown in Fig. 4.31, and are close to those of equations 4.27 and 4.26.

The strength requirement of BS5950 is that the transverse stiffener must not buckle as a compression member when it transmits the tension field force. Thus

$$F_q^* \leqslant P_q \tag{4.80}$$

in which P_q is the compression buckling resistance (section 3.7.2) of the stiffener determined for an effective length L_E equal to 0.7 d, and F_q^* is the design compressive force in the stiffener given by

$$F_q^* = F_v^* - V_{cr} \tag{4.81}$$

in which V_{cr} is the critical shear buckling resistance given by

$$V_{cr} = (V_w/0.9)^2/P_v \leqslant (9V_w - 2P_v)/7 \tag{4.82}$$

This strength requirement is similar to that implied by equation 4.32.

In addition, the intermediate stiffeners must not themselves buckle locally

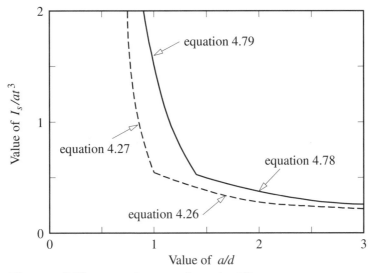

Fig. 4.31 Stiffness requirements for web stiffeners

before yielding, and so the effective width b_{effs} of a stiffener outstand of thickness t_s is limited to

$$b_{\text{effs}} = 13t_s\varepsilon_s \leq b_s \qquad (4.83)$$

A worked example of checking a pair of transverse stiffeners is given in section 4.9.8.

4.7.4.4 End panels

At the end of a plate girder, there is no adjacent panel to absorb the horizontal component of the tension field in the last panel, and so this load may have to be resisted by the end stiffener. Special treatment of this end stiffener may be avoided if the length of the last (anchor) panel is reduced so that the tension field contribution f_{tf} to the ultimate stress f_u is not required (see equation 4.29). This is achieved by designing the anchor panel so that its critical shear buckling resistance V_{cr} given by equation 4.82 is not less than the design shear force F_v^*. A worked example of checking an end panel is given in section 4.9.7.

Alternatively, an end post consisting of one or two load-bearing stiffeners (see section 4.7.6.2) may be used to anchor the tension field at the end of a plate girder. The end post must be designed to resist any reaction force and a longitudinal anchor force H_q^* obtained from

$$\frac{H_q^*}{0.5dtp_y\sqrt{(1 - V_{cr}/P_v)}} = \frac{F_v^* - V_{cr}}{V_w - V_{cr}} \leq 1 \qquad (4.83)$$

Single stiffener end posts are designed as beams to resist the moment effect of H_q^*, while twin stiffener end posts are designed as beams to resist the shear effect of H_q^*.

4.7.5 BEAM WEBS IN SHEAR AND BENDING

Where the design moment M^* and shear F_v^* are both high, the beam must be designed against combined shear and bending. BS5950 allows two methods of design, the proportioning method and the interaction method.

In the proportioning method, which is limited to beams whose flanges are not Class 4 slender, only the flanges are used to resist the moment, so that

$$M^* \leq M_f = A_{fm} d_f p_{yf} \tag{4.84}$$

in which A_{fm} is the lesser flange effective area and d_f the distance between flange centroids. This then allows the web to resist all of the design shear force, so that

$$F_v^* \leq V_b \tag{4.85}$$

The proportioning method is suitable for beams with slender webs, such as plate web girders.

The interaction method allows the web to make some contribution to the moment capacity, and is suitable for beams with less slender webs which have more than sufficient capacity to resist the design shear. In this method, BS5950 provides the interaction curve between the design shear force F_v^* and moment M^* shown in Fig. 4.32, for which

$$M^* \leq M_f + M_{cw}(1 - \rho) \tag{4.86}$$

in which M_f is the moment capacity of the flanges alone, M_{cw} is the additional moment capacity of the web in the absence of shear, and the factor ρ is given by

$$\rho = 0 \qquad \text{while } 0 \leq F_v^*/V_w \leq 0.6 \tag{4.87a}$$
$$\rho = (2F_v^*/V_w - 1)^2 \quad \text{while } 0.6 \leq F_v^*/V_w \leq 1 \tag{4.87b}$$

These equations are adapted from equation 4.44.

Worked examples of checking beams for combined shear and bending are given in sections 4.9.10 and 4.9.11.

4.7.6 BEAM WEBS IN BEARING

4.7.6.1 Unstiffened webs

For the design of thick webs in bearing according to BS5950, the bearing capacity P_{bw} based on yielding is taken as

$$P_{bw} = (b_1 + nT)t p_{yw} \tag{4.88}$$

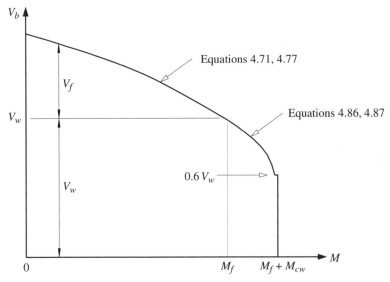

Fig. 4.32 BS5950 capacities of beams under bending and shear

in which b_1 is the stiff bearing length and nT is the additional length obtained by assuming a dispersion of the bearing at $1:2.5$ through the flange thickness (see Fig. 4.27a).

On the other hand, the buckling resistance P_x of a thin web is limited by the axial load capacity of the web acting as a compression member. In BS5950, this is taken as

$$P_x = \frac{25\varepsilon t P_{bw}}{\sqrt{\{(b_1 + nT)d\}}} \tag{4.89}$$

which is approximately equal to 0.6 times the geometric mean $\sqrt{(P_{bw}N_o)}$ of P_{bw} and the compression member elastic buckling load N_o (section 3.5) of a web of area $1.4dt$ and effective length $L_E = 0.7d$. While using this geometric mean may be optimistic, this will be at least partly compensated for by the factor 0.6 and when the area $1.4dt$ is conservative. Equation 4.89 applies to beams whose flanges are effectively restrained against rotation and relative lateral displacement and whose available web area is not less than $1.4dt$. For other beams, the web effective length and area must be modified.

4.7.6.2 Stiffened webs

When a web alone has insufficient bearing capacity, it may be strengthened by adding one or more pairs of load bearing stiffeners. These stiffeners increase the yield capacity to $P_{bw} + A_{snet}p_{yw}$ in which A_{snet} is the net area of the stiffeners.

Load bearing stiffeners also increase the buckling resistance, by increasing the effective section of the compression member to that of the stiffeners together with the web lengths $15t$ on either side of the stiffeners, if available. The effective length of the compression member is taken as the stiffener length L, or as $0.7L$ if flange restraints act to reduce the stiffener end rotations during buckling.

A worked example of checking load bearing stiffeners is given in section 4.9.9.

4.8 Appendix – elastic buckling of plate elements in compression

4.8.1 SIMPLY SUPPORTED PLATES

A simply supported rectangular plate element of length L, width b, and thickness t is shown in Fig. 4.5b. Applied compressive loads N are uniformly distributed over both edges b of the plate. The elastic buckling load N_{ol} can be determined by finding a deflected position such as that shown in Fig. 4.5b which is one of equilibrium. The differential equation for this equilibrium position is [1–4]

$$bD\left(\frac{\partial^4 u}{\partial z^4} + 2\frac{\partial^4 u}{\partial y^2 \partial z^2} + \frac{\partial^4 u}{\partial y^4}\right) = -N_{ol}\frac{\partial^2 u}{\partial z^2}, \tag{4.90}$$

where

$$bD = \frac{Ebt^3}{12(1 - v^2)} \tag{4.91}$$

is the flexural rigidity of the plate.

Equation 4.90 is compared in Fig. 4.5 with the corresponding differential equilibrium equation for a simply supported rectangular section column (obtained by differentiating equation 3.56 twice). It can be seen that bD for the plate corresponds to the flexural rigidity EI of the column, except for the $(1 - v^2)$ term which is due to the Poisson's ratio effect in wide plates. Thus the term $bD\partial^4 u/\partial z^4$ represents the resistance generated by longitudinal flexure of the plate to the disturbing effect $-N\partial^2 u/\partial z^2$ of the applied load. The additional terms $2bD\partial^4 u/\partial y^2\partial z^2$ and $bD\partial^4 u/\partial y^4$ in the plate equation represent the additional resistances generated by twisting and lateral bending of the plate.

A solution of equation 4.90 which satisfies the boundary conditions along the simply supported edges [1–4] is

$$u = \delta \sin\frac{m\pi y}{b} \sin\frac{n\pi z}{L}, \tag{4.1}$$

where δ is the undetermined magnitude of the deflected shape. When this is substituted into equation 4.90, an expression for the elastic buckling load N_{ol} is obtained as

$$N_{ol} = \frac{n^2\pi^2 bD}{L^2}\left[1 + 2\left(\frac{mL}{nb}\right)^2 + \left(\frac{mL}{nb}\right)^4\right],$$

which has its lowest values when $m = 1$ and the buckled shape has one half wave across the width of the plate b. Thus the elastic buckling stress

$$f_{ol} = \frac{N_{ol}}{bt} \qquad (4.2)$$

can be expressed as

$$f_{ol} = \frac{\pi^2 E}{12(1 - v^2)}\frac{k_b}{(b/t)^2} \qquad (4.3)$$

in which the buckling coefficient k_b is given by

$$k_b = \left[\left(\frac{nb}{L}\right)^2 + 2 + \left(\frac{L}{nb}\right)^2\right]. \qquad (4.92)$$

The variation of the buckling coefficient k_b with the aspect ratio L/b of the plate and the number of half waves n along the plate is shown in Fig. 4.6. It can be seen that the minimum value of k_b is 4, and that this occurs whenever the buckles are square ($nb/L = 1$), as shown in Fig. 4.7. In most structural steel members, the aspect ratio L/b of the plate elements is large so that the value of the buckling coefficient k_b is always close to the minimum value of 4. The elastic buckling stress can therefore be closely approximated by

$$f_{ol} = \frac{\pi^2 E/3(1 - v^2)}{(b/t)^2}. \qquad (4.93)$$

4.9 Worked examples

4.9.1. EXAMPLE 1 – COMPRESSION RESISTANCE OF A CLASS 4 SLENDER COMPRESSION MEMBER

Problem. Determine the compression resistance of the short welded I-section member shown in Fig. 4.33a if the slenderness is very low ($\lambda \to 0$).

Classifying the section plate elements.

$$\varepsilon = \sqrt{(275/355)} = 0.880 \qquad \text{T11}$$
$$b/(T\varepsilon) = [(400 - 10)/2]/(10 \times 0.880) = 22.16 > 13 \qquad \text{T11}$$

and so the flanges are Class 4 slender.

$$d/(t\varepsilon) = (420 - 2 \times 10)/(10 \times 0.880) = 45.45 > 40 \qquad \text{T11}$$

and so the web is Class 4 slender.

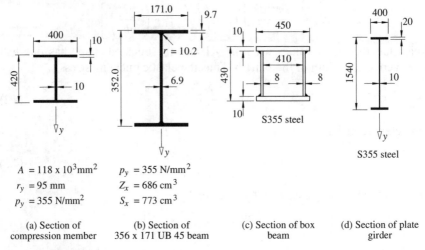

$A = 118 \times 10^3 \text{mm}^2$ $p_y = 355 \text{ N/mm}^2$
$r_y = 95 \text{ mm}$ $Z_x = 686 \text{ cm}^3$
$p_y = 355 \text{ N/mm}^2$ $S_x = 773 \text{ cm}^3$

(a) Section of compression member	(b) Section of 356 x 171 UB 45 beam	(c) Section of box beam	(d) Section of plate girder

Fig. 4.33 Worked examples

Effective area.

$b_{eff} = 13 \times 10 \times 0.880 = 114.4 \text{ mm}$ Fig. 8(a)
$d_{eff} = 40 \times 10 \times 0.880 = 350.0 \text{ mm}$ Fig. 8(a)
$A_{eff} = 2 \times (2 \times 114.4 + 10) \times 10 + (352.0 \times 10) = 8296 \text{ mm}^2$ 3.6.2.2

Compression resistance.

$\lambda(A_{eff}/A_g) \to 0$ since $\lambda \to 0$ 4.7.4
$p_{cs} = (355 - 20) = 335 \text{ N/mm}^2$ 4.7.5
$P_c = 8296 \times 335 \text{ N} = 2779 \text{ kN}$ 4.7.4

4.9.2 EXAMPLE 2 – SECTION MOMENT CAPACITY OF A CLASS 3 SEMI-COMPACT I-BEAM

Problem. Determine the section moment capacity and examine the suitability for plastic design of the 356×171 UB 45 of S355 steel shown in Fig. 4.33b.

Classifying the section plate elements.

$T = 9.7 \text{ mm}, p_y = 355 \text{ N/mm}^2$ T9
$\varepsilon = \sqrt{(275/355)} = 0.880$
$b/(T\varepsilon) = (171.0/2)/(9.7 \times 0.880) = 10.02 > 10$ T11

and so the flange is Class 3 semi-compact, but not quite Class 2 compact.

$d/(t\varepsilon) = (352.0 - 2 \times 9.7 - 2 \times 10.2)/(6.9 \times 0.880) = 51.4 < 80$ T11

and so the web is Class 1 plastic.

Because the flange is not Class 1 plastic, the section is not suitable for plastic design.

Section moment capacity.

$$M_c = 355 \times 686 \times 10^3 \text{ Nmm} = 243.5 \text{ kNm} \qquad 4.2.5.2$$

If the flange had been Class 2 compact instead of Class 3 semi-compact, then

$$M_c = 355 \times 773 \times 10^3 \text{ Nmm} = 274.4 \text{ kNm}.$$

Alternatively, using Clause 3.5.6.2,

$$S_{eff} = 686 + (773 - 686) \left(\frac{15/10.2 - 1}{15/10 - 1} \right) = 772.5 \text{ cm}^3$$

and $M_c = 355 \times 772.5 \times 10^3 \text{ Nmm} = 274.2 \text{ kNm}$.

4.9.3 EXAMPLE 3 – SECTION MOMENT CAPACITY OF A CLASS 4 SLENDER BOX BEAM

Problem. Determine the section moment capacity of the welded section beam of S355 steel shown in Fig. 4.33c.

Solution.

$$T = 12 \text{ mm}, p_y = 355 \text{ N/mm}^2 \qquad \text{T9}$$
$$\varepsilon = \surd(275/355) = 0.880$$
$$b/(T\varepsilon) = 410/(10 \times 0.880) = 46.6 > 40 \qquad \text{T11}$$

and so the flange is Class 4 slender.

$$d/(t\varepsilon) = (430 - 2 \times 10)/(8 \times 0.880) = 58.2 < 80 \qquad \text{T11}$$

and so the webs are Class 1 plastic.

$$\begin{aligned}
b_{eff} &= 40 \times 10 \times 0.880 = 352.0 \text{ mm} & \text{Fig. 8(c)} \\
A_{eff} &= (450 - 410 + 352.0) \times 10 + (450 \times 10) \\
&\quad + 2 \times (430 - 2 \times 10) \times 8 & 3.6.2.2 \\
&= 14\,980 \text{ mm}^2
\end{aligned}$$
$$\begin{aligned}
14\,980 \times y_c &= (450 - 410 + 352.0) \times 10 \times (430 - 10/2) \\
&\quad + 450 \times 10 \times 10/2 + 2 \times (430 - 2 \times 10) \times 8 \times 430/2 \\
y_c &= 206.87 \text{ mm} \\
I_{eff} &= (450 - 410 + 352.0) \times 10 \times (430 - 10/2 - 206.87)^2 \\
&\quad + 450 \times 10 \times (206.87 - 10/2)^2 + 2 \times (430 - 2 \times 10)^3 \times 8/12 \\
&\quad + 2 \times (430 - 2 \times 10) \times 8 \times (430/2 - 206.87)^2 \\
&= 462.2 \times 10^6 \text{ mm}^4 \\
Z_{eff} &= 462.2 \times 10^6/(430 - 206.87) = 2.0716 \times 10^6 \text{ mm}^3 & 3.6.2.3 \\
M_c &= 355 \times 2.0716 \times 10^6 \text{ Nmm} = 735.4 \text{ kNm} & 4.2.5.2
\end{aligned}$$

4.9.4 EXAMPLE 4 – SECTION MOMENT CAPACITY OF A SLENDER PLATE GIRDER

Problem.   Determine the section moment capacity of the welded plate girder of S355 steel shown in Fig. 4.33d.

Solution.

$$T = 20 \text{ mm}, p_y = 345 \text{ N/mm}^2 \qquad \text{T9}$$
$$\varepsilon = \sqrt{(275/345)} = 0.893$$
$$b/(T\varepsilon) = (410 - 10)/2/(20 \times 0.893) = 11.2, \ 9 < 11.2 < 13 \qquad \text{T11}$$

and so the flange is Class 3 semi-compact.

$$d/(t\varepsilon) = (1540 - 2 \times 20)/(10 \times 0.893) = 168.0 > 120 \qquad \text{T11}$$

and so the web is Class 4 slender.

A conservative approximation for the section moment capacity may be obtained by ignoring the web completely, so that

$$M_c = M_f = 345 \times (400 \times 20) \times (1540 - 20) \text{ Nmm} = 4195 \text{ kNm}$$
$$\text{4.4.4.2(b)}$$

A higher capacity may be obtained by using Clause 3.6.2.4 of BS5950 with $f_{cw} = f_{tw} = p_y = 345 \text{ N/mm}^2$ and $p_{yw} = 355 \text{ N/mm}^2$, whence

$$b_{eff} = \frac{(120 \times 0.893 \times 10)}{\{1 + (345 - 345)/355\}(1 + 345/345)} = 535.8 \text{ mm} \qquad 3.6.2.4$$

Thus $0.6b_{eff} = 321.5$ mm and $0.4b_{eff} = 214.3$ mm.

The position of the neutral axis is determined by the condition that the axial stress resultant on the effective web area must be zero. This condition can be expressed as

$$y_c/2 = (1500 - y_c)^2/(2y_c) - (1500 - y_c - 214.3)^2/(2y_c) + 321.5^2/(2y_c)$$

so that $y_c^2 + 428.6y_c - 700\,338 = 0$ and $y_c = 649.6$ mm.

Thus the ineffective width of the web is

$$1500 - 649.6 - 214.3 - 321.5 = 314.6 \text{ mm}$$

$$\begin{aligned}
I &= (400 \times 20) \times (10 + 1500 - 649.6)^2 + (400 \times 20) \times (649.6 + 10)^2 \\
&\quad + 1500^3 \times 10/12 + (1500 \times 10) \times (1500/2 - 649.6)^2 - 314.6^3 \times 10/12 \\
&\quad - (314.6 \times 10) \times (314.6/2 + 321.5)^2 \\
&= 11.619 \times 10^9 \text{ mm}^4
\end{aligned}$$

$$Z = 11.619 \times 10^9/(10 + 1500 - 649.6) = 13.505 \times 10^6 \text{ mm}^3$$
$$M_c = 345 \times 13.505 \times 10^6 \text{ Nmm} = 4659 \text{ kNm}.$$

4.9.5 EXAMPLE 5 – SHEAR BUCKLING RESISTANCE OF AN UNSTIFFENED PLATE GIRDER WEB

Problem. Determine the shear buckling resistance of the unstiffened plate girder web of S355 steel shown in Fig. 4.33d.

Solution.

$$t = 10 \text{ mm}, p_{yw} = 355 \text{ N/mm}^2 \qquad\qquad 4.4.2, \text{T9}$$
$$d/t = (1540 - 2 \times 20)/10 = 150$$
$$a/d = \infty/d = \infty$$
$$q_w = 88 \text{ N/mm}^2 \qquad\qquad \text{T21(3)}$$
$$V_b = V_w = (1540 - 2 \times 20) \times 10 \times 88 \text{ N} = 1320 \text{ kN} \qquad 4.4.5.2$$

4.9.6 EXAMPLE 6 – SHEAR BUCKLING RESISTANCE OF A STIFFENED PLATE GIRDER WEB

Problem. Determine the shear buckling resistance of the plate girder web of S355 steel shown in Fig. 4.33d if intermediate transverse stiffeners are spaced at 1800 mm.

Solution.

$$t = 10 \text{ mm}, p_{yw} = 355 \text{ N/mm}^2 \qquad\qquad 4.4.2, \text{T9}$$
$$d/t = (1540 - 2 \times 20)/10 = 150$$
$$a/d = 1800/(1540 - 2 \times 20) = 1.20$$
$$q_w = 108 \text{ N/mm}^2 \qquad\qquad \text{T21(3)}$$
$$V_b = V_w = (1540 - 2 \times 20) \times 10 \times 108 \text{ N} = 1620 \text{ kN} \qquad 4.4.5.2$$

4.9.7 EXAMPLE 7 – ANCHOR PANEL IN A STIFFENED PLATE GIRDER WEB

Problem. Determine the critical shear buckling capacity of the end anchor panel of the welded stiffened plate girder of section 4.9.6 if the width of the panel is 1800 mm.

Solution.

$$t = 10 \text{ mm}, p_{yw} = 355 \text{ N/mm}^2 \qquad\qquad 4.4.2, \text{T9}$$
$$d/t = (1540 - 2 \times 20)/10 = 150$$
$$a/d = 1800/(1540 - 2 \times 20) = 1.20$$
$$q_e = (1 + 0.75/1.20^2) (1000/150)^2 = 67.6 \text{ N/mm}^2 \qquad \text{H.1}$$
$$p_v = 0.6 \times 355 = 213.0 \text{ N/mm}^2 \qquad\qquad \text{H.1}$$
$$\lambda_w = (213.0/67.6)^{0.5} = 1.78 > 1.25 \qquad\qquad \text{H.1}$$
$$q_{cr} = 213.0/1.78^2 = 67.6 \text{ N/mm}^2 \qquad\qquad \text{H.2}$$
$$V_{cr} = (1540 - 2 \times 20) \times 10 \times 67.6 \text{ N} = 1014 \text{ kN} \qquad \text{H.2}$$

4.9.8 EXAMPLE 8 – INTERMEDIATE TRANSVERSE STIFFENER

Problem. Check the adequacy of a pair of intermediate transverse web stiffeners 100 × 16 of S460 steel for the plate girder of section 4.9.6.

Solution.

$$t_s = 16 \text{ mm}, p_{ys} = 460 \text{ N/mm}^2 \qquad \text{T9}$$
$$\varepsilon = \sqrt{(275/460)} = 0.773$$
$$13\varepsilon t_s = 13 \times 0.773 \times 16 = 161 \text{ mm} > 100 \text{ mm} \qquad 4.5.1.2$$
$$b_{eff} = 100 \text{ mm} \qquad 4.5.1.2$$
$$A_{eff} = 2 \times (15 \times 10) \times 10 + 2 \times 100 \times 16 = 6200 \text{ mm}^2 \qquad 4.5.3.3$$
$$I_{eff} = (2 \times 100 + 10)^3 \times 16/12 = 12.35 \times 10^6 \text{ mm}^4$$
$$r_{eff} = \sqrt{(12.35 \times 10^6/6200)} = 44.6 \text{ mm}$$
$$L_E = 0.7 \times 1500 = 1050 \text{ mm} \qquad 4.5.5$$
$$\lambda = 1050/44.6 = 23.5 \qquad 4.7.2$$
$$p_c = 442 - (442 - 428) \times 3.5/5 = 432.2 \text{ N/mm}^2 \qquad \text{T24(5)}$$
$$P_q = P_c = 6200 \times 432.2 \text{ N} = 2680 \text{ kN} \qquad 4.7.4, 4.4.6.6$$

Using Example 7, $V_{cr} = 1014$ kN.
  If $V^* = V_b = 1620$ kN using Example 6,

$$F_q^* = 1620 - 1014 = 606 \text{ kN} < 2680 \text{ kN} = P_q \quad \text{OK} \qquad 4.4.6.6$$
$$I_s = I_{eff} = 12.35 \times 10^6 \text{ mm}^4$$
$$a/d = 1800/1500 = 1.20 < \sqrt{2} \qquad 4.4.6.4$$
$$1.5(d/a)^2 dt_{min}^3 = 1.5 \times (1500/1800)^2 \times 1500 \times 10^3 \qquad 4.4.6.4$$
$$= 1.563 \times 10^6 \text{ mm}^4 < 12.35 \times 10^6 \text{ mm}^4 = I_s \quad \text{OK}$$

4.9.9 EXAMPLE 9 – LOAD-BEARING STIFFENER

Problem. Check the adequacy of a pair of load-bearing stiffeners 100 × 16 of S460 steel which are above the support of the plate girder of section 4.9.6. The flanges of the girder are not restrained by other structural elements against rotation. The girder is supported on a stiff bearing 300 mm long, the end panel width is 1000 mm, and the design reaction is 1400 kN.

Bearing check.

$$t_s = 16 \text{ mm}, p_{ys} = 460 \text{ N/mm}^2 \qquad \text{T9}$$
$$\varepsilon = \sqrt{(275/460)} = 0.773$$
$$b_1 = 300 \text{ mm}$$
$$nT = 2 \times 20 + 0.6 \times 0 = 40 \text{ mm} \qquad 4.5.2.1$$
$$P_{bw} = (300 + 40) \times 10 \times 460 \text{ N} = 1564 \text{ kN} \qquad 4.5.2.1$$
$$A_{s.net} = 2 \times 100 \times 16 = 3200 \text{ mm}^2 \qquad 4.5.2.2$$
$$P_s = 3200 \times 460 \text{ N} = 1472 \text{ kN} \qquad 4.5.2.2$$
$$P_{bw} + P_s = 1564 + 1472 = 3036 \text{ kN} > 1400 \text{ kN} = F_x^* \quad \text{OK} \qquad 4.5.2.2$$

Buckling check.

$$13\varepsilon t_s = 13 \times 0.773 \times 16 = 161 \text{ mm} > 100 \text{ mm} \qquad \text{4.5.1.2}$$
$$b_{\text{eff}} = 100 \text{ mm} \qquad \text{4.5.1.2}$$
$$A_{\text{eff}} = 2 \times (15 \times 10) \times 10 + 2 \times 100 \times 16 = 6200 \text{ mm}^2 \qquad \text{4.5.3.3}$$
$$I_{\text{eff}} = (2 \times 100 + 10)^3 \times 16/12 = 12.35 \times 10^6 \text{ mm}^4$$
$$r_{\text{eff}} = \sqrt{(12.35 \times 10^6/6200)} = 44.6 \text{ mm}$$
$$L_E = 1.0 \times 1500 = 1500 \text{ mm} \qquad \text{4.5.5}$$
$$\lambda = 1500/44.6 = 33.6 \qquad \text{4.7.2}$$
$$p_c = 413 - (413 - 397) \times 3.6/5 = 401.5 \text{ N/mm}^2 \qquad \text{T24(5)}$$
$$P_x = 6200 \times 401.5 \text{ N} = 2489 \text{ kN} > 1400 \text{ kN} = F^* \quad \text{OK} \qquad \text{4.5.3.3}$$

4.9.10 EXAMPLE 10 – SHEAR AND BENDING OF A CLASS 3 SEMI-COMPACT I-BEAM

Problem. Determine the design shear capacity of the 356 × 171 UB 45 of S355 steel shown in Fig. 4.33b at a point where the factored moment is $M_x^* = 230$ kNm.

Solution.
As in section 4.9.2, $p_y = 355$ N/mm², the flange is Class 3 semi-compact, the web is Class 1 plastic, the section is Class 3 semi-compact, and $S_{\text{eff}} = 772.5$ cm³.

$$d/t = (352.0 - 2 \times 9.7 - 2 \times 10.2)/6.9 = 45.2$$
$$q_w = 213 \text{ N/mm}^2 \qquad \text{T21(3)}$$
$$P_v = 213 \times 352.0 \times 6.9 \text{ N} = 517.3 \text{ kN} \qquad \text{4.2.3}$$
$$S_v = 352.0^2 \times 6.9/4 \text{ mm}^3 = 213.7 \text{ cm}^3 \qquad \text{4.2.5.3}$$

The shear capacity F_v corresponds to $M_x^* = 230$ kNm, and using $M_c = M_x^*$ in $M_c = p_y(S_{\text{eff}} - \rho S_v)$, leads to   4.2.5.3

$$\rho = \frac{(772.5 \times 10^3 - 230 \times 10^6/355)}{(213.7 \times 10^3)} = 0.583$$

Now $\rho = (2F_v/P_v - 1)^2$ which leads to   4.2.5.3

$$F_v/P_v = (\sqrt{(0.583)} + 1)/2 = 0.882$$

so that $F_v = 0.882 \times 517.3 = 456.1$ kN

4.9.11 EXAMPLE 11 – SHEAR AND BENDING OF A STIFFENED PLATE WEB GIRDER

Problem. Determine the shear resistance of the stiffened plate web girder of section 4.9.6 and shown in Fig. 4.33d at a point where the factored moment is $M_x^* = 4000$ kNm.

Solution.
As in section 4.9.4, p_y = 345 N/mm², the flange is Class 3 semi-compact and the web is Class 4 slender.

$$M_f = 345 \times (400 \times 20) \times (1540 - 20) \text{ Nmm} = 4195 \text{ kNm} > 4000 \text{ kNm} = M^*_x$$

and so the web shear resistance is not reduced.   4.4.4.2(b)
    Using section 4.9.6, V_b = 1620 kN.

4.10 Unworked examples

4.10.1 EXAMPLE 12 – ELASTIC LOCAL BUCKLING OF A BEAM

Determine the elastic local buckling moment for the beam shown in Fig. 4.34a.

4.10.2 EXAMPLE 13 – ELASTIC LOCAL BUCKLING OF A BEAM-COLUMN

Determine the elastic local buckling load for the beam-column shown in Fig. 4.34b when M/N = 20 mm.

4.10.3 EXAMPLE 14 – SECTION CAPACITY OF A WELDED BOX BEAM

Determine the nominal section moment capacity for the welded box beam of S275 steel shown in Fig. 4.33c.

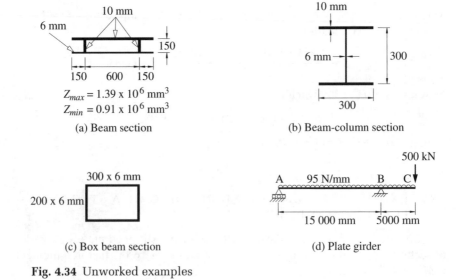

(a) Beam section

Z_{max} = 1.39 x 10⁶ mm³
Z_{min} = 0.91 x 10⁶ mm³

(b) Beam-column section

(c) Box beam section

(d) Plate girder

Fig. 4.34 Unworked examples

4.10.4 EXAMPLE 15 – DESIGNING A PLATE WEB GIRDER

The overall depth of the laterally supported plate girder of S275 steel shown in Fig. 4.34d must not exceed 1800 mm. Design a constant section girder for the factored loads shown, and determine:

(a) the flange proportions,
(b) the web thickness,
(c) the distribution of any intermediate stiffeners,
(d) the stiffener proportions,
(e) the proportions and arrangements of any load-bearing stiffeners.

4.11 References

1. Timoshenko, S.P. and Woinowsky-Krieger, S. (1959) *Theory of Plates and Shells*, 2nd edition, McGraw-Hill, New York.
2. Timoshenko, S.P. and Gere, J.M. (1961) *Theory of Elastic Stability*, 2nd edition, McGraw-Hill, New York.
3. Bleich, F. (1952) *Buckling Strength of Metal Structures*, McGraw-Hill, New York.
4. Bulson, P.S. (1970) *The Stability of Flat Plates*, Chatto and Windus, London.
5. Column Research Committee of Japan (1971) *Handbook of Structural Stability*, Corona, Tokyo.
6. Allen, H.G. and Bulson, P.S. (1980) *Background to Buckling*, McGraw-Hill (UK).
7. Bradford, M.A., Bridge, R.Q., Hancock, G.J., Rotter, J.M. and Trahair, N.S. (1987) Australian limit state design rules for the stability of steel structures. *Proceedings*, First Structural Engineering Conference, Institution of Engineers, Australia, Melbourne, pp. 209–16.
8. Basler, K. (1961) Strength of plate girders in shear, *Journal of the Structural Division, ASCE*, **87**, No. ST7, pp. 151–80.
9. Evans, H.R. (1983) Longitudinally and transversely reinforced plate girders, Chapter 1 in *Plated Structures: Stability and Strength* (ed. R. Narayanan), Applied Science Publishers, London, pp. 1–37.
10. Rockey, K.C. and Skaloud, M. (1972) The ultimate load behaviour of plate girders loaded in shear. *The Structural Engineer*, **50**, No. 1, pp. 29–47.
11. Usami, T. (1982) Postbuckling of plates in compression and bending, *Journal of the Structural Division, ASCE*, **108**, No. ST3, pp. 591–609.
12. Kalyanaraman, V. and Ramakrishna, P. (1984) Non-uniformly compressed stiffened elements, *Proceedings*, Seventh International Specialty Conference Cold-Formed Structures, St Louis, Department of Civil Engineering, University of Missouri-Rolla, pp. 75–92.
13. Merrison Committee of the Department of Environment (1973) *Inquiry into the Basis of Design and Method of Erection of Steel Box Girder Bridges*, Her Majesty's Stationery Office, London.
14. Rockey, K.C. (1971) An ultimate load method of design for plate girders, *Developments in Bridge Design and Construction*, (eds K.C. Rockey, J.L. Bannister, and H.R. Evans), Crosby Lockwood and Son, London, pp. 487–504.
15. Rockey, K.C., El-Gaaly, M.A. and Bagchi, D.K. (1972) Failure of thin-walled

members under patch loading, *Journal of the Structural Division, ASCE*, **98**, No. ST12, pp. 2739–52.

16. Roberts, T.M. (1981) Slender plate girders subjected to edge loading, *Proceedings*, Institution of Civil Engineers, **71**, Part 2, September, pp. 805–19.

17. Roberts, T.M. (1983) Patch loading on plate girders, Chapter 3 in *Plated Structures: Stability and Strength* (ed. R. Narayanan), Applied Science Publishers, London, pp. 77–102.

18. British Standards Institution (1982) *BS5400: Steel, Concrete, and Composite Bridges; Part 3: Code of Practice for the Design of Steel Bridges*, BSI, London.

19. SA (1998) *AS4100–1998 Steel Structures*, Standards Australia, Sydney, Australia.

5 In-plane bending of beams

5.1 Introduction

Beams are structural members which transfer the transverse loads they carry to the supports by bending and shear actions. Beams generally develop higher stresses than axially loaded members with similar loads, while the bending deflections are much higher. These bending deflections of a beam are often, therefore, a primary design consideration. On the other hand, most beams have small shear deflections, and these are usually neglected.

Beam cross-sections may take many different forms, as shown in Fig. 5.1, and these represent various methods of obtaining an efficient and economical member. Thus most steel beams are not of solid cross-section but have their material distributed more efficiently in thin walls. Thin-walled sections may be open, and while these tend to be weak in torsion, they are often cheaper to manufacture than the stiffer closed sections. Perhaps the most economic method of manufacturing steel beams is by hot-rolling but only a limited number of open cross-sections is available. When a suitable hot-rolled beam cannot be found, a substitute may be fabricated by connecting together a series of rolled plates, and this has become increasingly common. Fabricating techniques also allow the production of beams compounded from hot-rolled

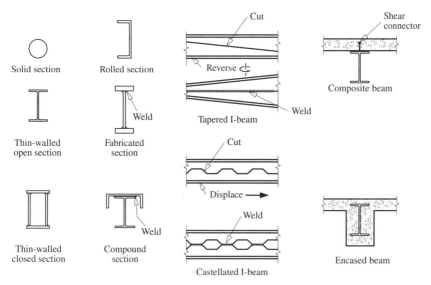

Fig. 5.1 Beam types

members and plates, and hybrid members in which the flange material is of a higher yield stress than the web. Tapered and castellated beams can also be fabricated from hot-rolled beams. In many cases, a steel beam is required to support a reinforced concrete slab, and in this case its strength may be increased by connecting the steel and concrete together so that they act compositely. The fire resistance of a steel beam may also be increased by encasing it in concrete. The final member cross-section chosen will depend on its suitability for the use intended, and on the overall economy.

The strength of a steel beam in the plane of loading depends on its section properties and on its yield stress f_y. When bending predominates in a determinate beam, the effective ultimate strength is reached when the most highly stressed cross-section becomes fully yielded so that it forms a plastic hinge. The moment M_p at which this occurs is somewhat higher than the first yield moment M_Y at which elastic behaviour nominally ceases, as shown in Fig. 5.2, and for hot-rolled I-beams this margin varies between 10% and 20% approximately. The attainment of the first plastic hinge forms the basis for the traditional method of design of beams in which the bending moment distribution is calculated from an elastic analysis.

However, in an indeterminate beam, a substantial redistribution of bending moment may occur after the first hinge forms, and failure does not occur until sufficient plastic hinges have formed to cause the beam to become a mechanism. The load which causes this mechanism to form provides the basis for the more rational method of plastic design.

When shear predominates, as in some heavily loaded deep beams of short

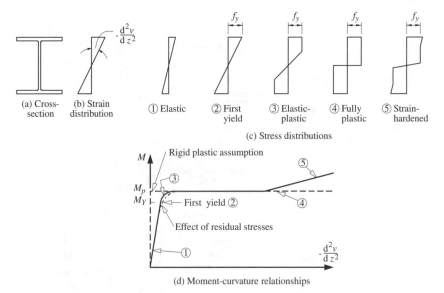

(a) Cross-section   (b) Strain distribution   ① Elastic   ② First yield   ③ Elastic-plastic   ④ Fully plastic   ⑤ Strain-hardened

(c) Stress distributions

(d) Moment-curvature relationships

Fig. 5.2 Moment–curvature relationships for steel beams

span, the ultimate strength is controlled by the shear force which causes complete plastification of the web. In the more common sections, this is close to the shear force which causes the nominal first yield in shear, and so the shear design is carried out for the shear forces determined from an elastic analysis. In this chapter, the in-plane behaviour and the design of beams are discussed. It is assumed that neither local buckling (which is treated in Chapter 4) nor lateral buckling (which is treated in Chapter 6) occurs. Beams with axial loads are discussed in Chapter 7, while the torsion of beams is treated in Chapter 10.

5.2 Elastic analysis of beams

The design of a steel beam is often preceded by an elastic analysis of the bending of the beam. One purpose of such an analysis is to determine the bending moment and shear force distributions throughout the beam, so that the maximum bending moments and shear forces can be found and compared with the moment and shear capacities of the beam. An elastic analysis is also required to determine the deflections of the beam so that these can be compared with the desirable limiting values.

The data required for an elastic analysis include both the distribution and the magnitudes of the applied loads and the geometry of the beam. In particular, the variation along the beam of the effective second moment of area I of the cross-section is needed to determine the deflections of the beam, and to determine the moments and shears when the beam is statically indeterminate. For this purpose, local variations in the cross-section such as those due to bolt holes may be ignored, but more general variations, including any general reductions arising from the use of the effective width concept for excessively thin compression flanges (see section 4.2.2.2), should be allowed for.

The bending moments and shear forces in statically determinate beams can be determined by making use of the principles of static equilibrium. These are fully discussed in standard textbooks on structural analysis [1, 2], as are various methods of analysing the deflections of such beams. On the other hand, the conditions of statics are not sufficient to determine the bending moments and shear forces in statically indeterminate beams, and the conditions of compatibility between the various elements of the beam or between the beam and its supports must also be used. This is done by analysing the deflections of the statically indeterminate beam. Many methods are available for this analysis, both manual and computer, and these are fully described in standard textbooks [3–8]. These methods can also be used to analyse the behaviour of structural frames in which the member axial forces are small.

While the deflections of beams can be determined accurately by using the methods referred to above, it is often sufficient to use approximate estimates for comparison with the desirable maximum values. Thus in simply supported or continuous beams, it is usually accurate enough to use the mid-span

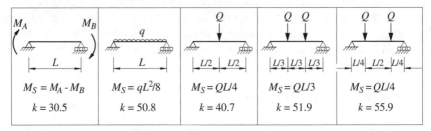

(a) Midspan deflection v_c and units

$$v_c = \frac{k M_S L^2}{I}$$

v_c - mm
M_S - kNm
L - m
I - cm^4

(b) Moments M_S and deflection coefficients k

Fig. 5.3 Mid-span deflections of steel beams

deflection (see Fig. 5.3). The mid-span deflection v_c (measured relative to the level of the left hand support) depends on the distribution of the applied load, the end moments M_A and M_B (taken as clockwise positive), and the sinking v_{AB} of the right hand support below the left hand support. This can be expressed as

$$v_c = \{kM_S + 30.5(M_A - M_B)\}\, \frac{L^2}{I} + \frac{v_{AB}}{2}. \tag{5.1}$$

In this equation, the deflections v_c and v_{AB} are expressed in mm, the moments M_S, M_A, M_B in kNm, the span L in m and the second moment of area I in cm^4. Expressions for the simple beam moments M_S and values of the coefficients k are given in Fig. 5.3 for a number of loading distributions. These can be combined together to find the central deflections caused by many other loading distributions.

5.3 Bending stresses in elastic beams

5.3.1 BENDING IN A PRINCIPAL PLANE

The distribution of the longitudinal bending stresses f in an elastic beam bent in a principal plane yz can be deduced from the experimentally confirmed assumption that plane sections remain substantially plane during bending, provided any shear lag effects which may occur in beams with very wide flanges (see section 5.4.5) are negligible. Thus the longitudinal strains ε vary linearly through the depth of the beam, as shown in Fig. 5.4, as do the longitudinal stresses f. It is shown in section 5.8.1 that the moment resultant M_x of these stresses is

$$M_x = -EI_x \frac{d^2 v}{dz^2}, \tag{5.2}$$

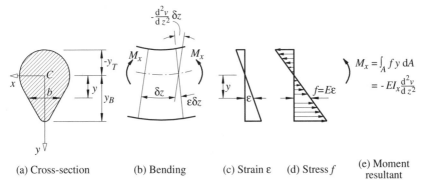

(a) Cross-section   (b) Bending   (c) Strain ε   (d) Stress f   (e) Moment resultant

Fig. 5.4 Elastic bending of beams

where the sign conventions for the moment M_x and the deflection v are as shown in Fig. 5.5, and the stress at any point in the section is

$$f = \frac{M_x y}{I_x} \tag{5.3}$$

in which tensile stresses are positive. In particular, the maximum stresses occur at the extreme fibres of the cross-section, and are given by

$$f_{max} = \frac{M_x y_T}{I_x} = -\frac{M_x}{Z_{xT}}$$

in compression, and

$$f_{max} = \frac{M_x y_B}{I_x} = -\frac{M_x}{Z_{xB}} \tag{5.4}$$

in tension, in which $Z_{xT} = -I_x/y_T$ and $Z_{xB} = I_x/y_B$ are the elastic section moduli for the top and bottom fibres, respectively, for bending about the x axis.

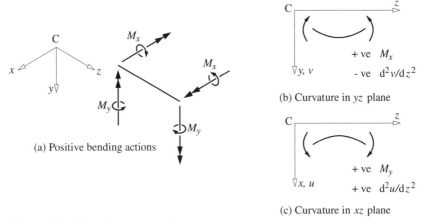

(a) Positive bending actions

(b) Curvature in yz plane
+ ve M_x
- ve d^2v/dz^2

(c) Curvature in xz plane
+ ve M_y
+ ve d^2u/dz^2

Fig. 5.5 Bending sign conventions

The corresponding equations for bending in the principal plane xz are

$$M_y = EI_y \frac{d^2u}{dz^2},\tag{5.5}$$

where the sign conventions for the moment M_y and the deflection u are also shown in Fig. 5.5,

$$f = -M_y x/I_y,\tag{5.6}$$

$$\left.\begin{array}{l} f_{max} = M_y/Z_{yL} \\ f_{max} = -M_y/Z_{yR} \end{array}\right\}\tag{5.7}$$

in which $Z_{yL} = -I_y/x_L$ and $Z_{yR} = I_y/x_R$ are the elastic section moduli for bending about the y axis.

Values of I_x, I_y, Z_x, Z_y for hot-rolled steel sections are given in [9], while values for other sections can be calculated as indicated in section 5.9 or in standard textbooks [1, 2]. Expressions for the properties of some thin-walled sections are given in Fig. 5.6 [10]. When a section has local holes, or excessive widths (see section 4.2.2.2), these properties may need to be reduced accordingly.

Worked examples of the calculation of cross-section properties are given in sections 5.12.1–5.12.4.

5.3.2 BIAXIAL BENDING

When a beam deflects only in a plane y_1z which is not a principal plane (see Fig. 5.7a), so that its curvature d^2u_1/dz^2 in the perpendicular x_1z plane is zero, the bending stresses

$$f = -Ey_1 d^2v_1/dz^2\tag{5.8}$$

have moment resultants

$$\left.\begin{array}{l} M_{x1} = -EI_{x1}d^2v_1/dz^2 \\ M_{y1} = -EI_{x1y1}d^2v_1/dz^2 \end{array}\right\},\tag{5.9}$$

where I_{x1y1} is the product second moment of area (see section 5.9). Thus the resultant bending moment

$$M = \sqrt{(M_{x1}^2 + M_{y1}^2)}\tag{5.10}$$

is inclined to the y_1z plane of bending, as shown in Fig. 5.7a.

The simplest method of analysing this or any other biaxial bending situation is to replace the moments M_{x1}, M_{y1} by their principal plane static equivalents M_x, M_y calculated from

$$\left.\begin{array}{l} M_x = M_{x1} \cos \alpha + M_{y1} \sin \alpha \\ M_y = -M_{x1} \sin \alpha + M_{y1} \cos \alpha \end{array}\right\}\tag{5.11}$$

in which α is the angle between the x_1, y_1 axes and the principal x, y axes, as shown in Fig. 5.7b. The bending stresses can then be determined from

$$f = \frac{M_x y}{I_x} - \frac{M_y x}{I_y} \tag{5.12}$$

in which I_x, I_y are the principal second moments of area. If the values of α, I_x, I_y are unknown, they can be determined from I_{x1}, I_{y1}, I_{x1y1} as shown in section 5.9.

Care needs to be taken to ensure that the correct signs are used for M_x and M_y in equation 5.12, as well as for x and y. For example, if the zed section shown in Fig. 5.8a is simply supported at both ends with a uniformly distributed load q acting in the plane of the web, then its principal plane

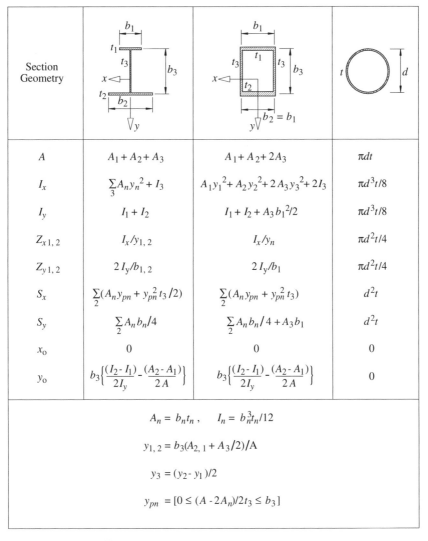

Section Geometry			
A	$A_1 + A_2 + A_3$	$A_1 + A_2 + 2A_3$	$\pi d t$
I_x	$\sum_3 A_n y_n^2 + I_3$	$A_1 y_1^2 + A_2 y_2^2 + 2 A_3 y_3^2 + 2 I_3$	$\pi d^3 t / 8$
I_y	$I_1 + I_2$	$I_1 + I_2 + A_3 b_1^2 / 2$	$\pi d^3 t / 8$
$Z_{x1,2}$	$I_x / y_{1,2}$	I_x / y_n	$\pi d^2 t / 4$
$Z_{y1,2}$	$2 I_y / b_{1,2}$	$2 I_y / b_1$	$\pi d^2 t / 4$
S_x	$\sum_2 (A_n y_{pn} + y_{pn}^2 t_3 / 2)$	$\sum_2 (A_n y_{pn} + y_{pn}^2 t_3)$	$d^2 t$
S_y	$\sum_2 A_n b_n / 4$	$\sum_2 A_n b_n / 4 + A_3 b_1$	$d^2 t$
x_o	0	0	0
y_o	$b_3 \left\{ \dfrac{(I_2 - I_1)}{2 I_y} - \dfrac{(A_2 - A_1)}{2 A} \right\}$	$b_3 \left\{ \dfrac{(I_2 - I_1)}{2 I_y} - \dfrac{(A_2 - A_1)}{2 A} \right\}$	0

$$A_n = b_n t_n , \quad I_n = b_n^3 t_n / 12$$

$$y_{1,2} = b_3 (A_{2,1} + A_3 / 2) / A$$

$$y_3 = (y_2 - y_1) / 2$$

$$y_{pn} = [0 \le (A - 2 A_n) / 2 t_3 \le b_3]$$

Fig. 5.6a Thin-walled section properties

Section Geometry	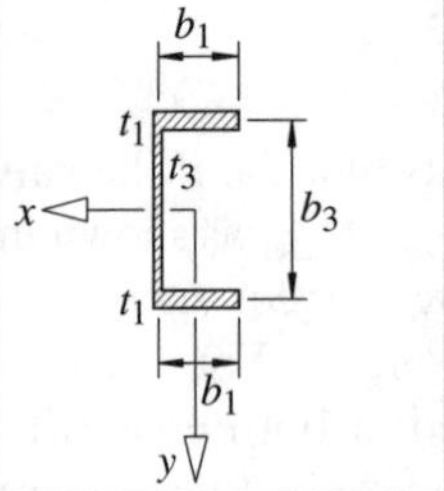	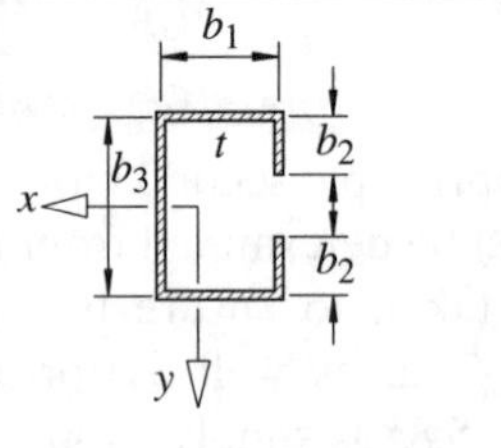	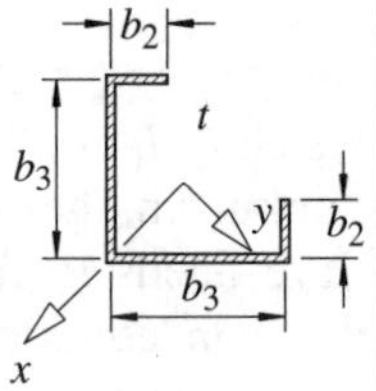
A	$2A_1 + A_3$	$2A_1 + 2A_2 + A_3$	$2\underset{2}{\Sigma} A_n$
I_x	$A_1 b_3^2/2 + I_3$	$\dfrac{A_1 b_3^2 + I_3 + I_2 + A_2(b_3-b_2)^2}{2}$	$A_2(b_3 - b_2/2)^2 + A_3 b_3^2/4 + \underset{2}{\Sigma} I_n$
I_y	$2A_1 x_1^2 + A_3 x_3^2 + 2I_1$	$2A_1 x_1^2 + 2A_2 x_2^2 + 2A_3 x_3^2 + 2I_1$	$A_2 b_3^2/4 + A_3 b_2^2/4 + \underset{2}{\Sigma} I_n$
Z_x	$2I_x/b_3$	$2I_x/b_3$	$\sqrt{2} I_x/b_3$
Z_y	$I_y/(b_1 - x_3)$ and I_y/x_3	I_y/x_2 and I_y/x_3	$2\sqrt{2} I_y/(b_2 + b_3)$
S_x	$A_1 b_3 + A_3 b_3/4$	$A_1 b_3 + A_2(b_3 - b_2) + A_3 b_3/4$	$(b_3^2 + 2b_2 b_3 - b_2^2)t/\sqrt{2}$
S_y	$(b_1 - x_p)^2 t_1 + x_p^2 t_1 + A_3 x_p$	$\{(b_1 - x_p)^2 + x_p^2 + 2b_2(b_1 - x_p) + b_3 x_p\}t$	$(b_2 + b_3)^2 t/2\sqrt{2}$
x_o	$x_3 + \dfrac{A_1 b_3^2 b_1}{4I_x}$	$x_3 + \left\{ b_3^2(b_1 + 2b_2) - 8b_2^3/3 \right\} \dfrac{A_1}{4I_x}$	$\dfrac{(b_2 + b_3)/2\sqrt{2} + (3b_3 - 2b_2)b_2^2 b_3 t}{3\sqrt{2} I_x}$
y_o	0	0	0
x_1	$b_1 A_3/2A$	$b_1/2 - x_3$	—
x_2	—	$b_1 - x_3$	—
x_3	$b_1 A_1/A$	$(b_1 + 2b_2) A_1/A$	—
x_p	$(A - 2A_3)/4t_1 \geq 0$	$(A - 2A_3)/4t \geq 0$	—
$A_n = b_n t_n, \quad I_n = b_n^3 t_n/12$			

Fig. 5.6b Thin-walled section properties

components q_x, q_y cause positive bending M_x and negative bending M_y, as indicated in Fig. 5.8b and c.

The deflections of the beam can also be determined from the principal plane moments by solving equations 5.2 and 5.5 in the usual way [1–8] for the principal plane deflections v and u, and by adding these vectorially.

Worked examples of the calculation of the elastic stresses and deflections in an angle section cantilever are given in section 5.12.5.

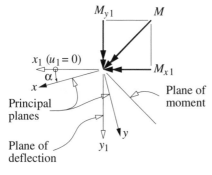

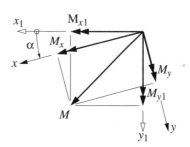

(a) Deflection in a non-principal plane

(b) Principal plane moments M_x, M_y

Fig. 5.7 Bending in a non-principal plane

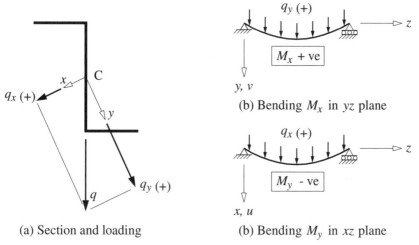

(a) Section and loading

(b) Bending M_x in yz plane

(b) Bending M_y in xz plane

Fig. 5.8 Biaxial bending of a zed beam

5.4 Shear stresses in elastic beams

5.4.1 SOLID CROSS-SECTIONS

A vertical shear force V_y acting parallel to the minor principal axis y of a section of a beam (see Fig. 5.9a) induces shear stresses τ_{zx}, τ_{zy} in the plane of the section. In solid section beams, these are usually assumed to act parallel to the shear force (i.e. $\tau_{zx} = 0$), and to be uniformly distributed across the width of the section, as shown in Fig. 5.9a. The distribution of the vertical shear stresses τ_{zy} can be determined by considering the horizontal equilibrium of an element of the beam as shown in Fig. 5.9b. As the bending normal stresses f vary with z, they create an imbalance of force in the z direction, which can only

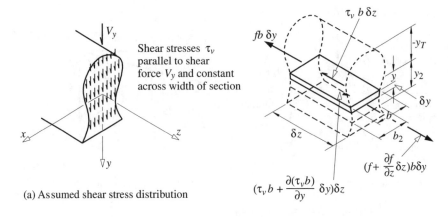

(a) Assumed shear stress distribution

(b) Horizontal equilibrium

Fig. 5.9 Shear stresses in solid sections

be compensated for by the horizontal shear stresses $\tau_{yz} = \tau_v$ which are equal to the vertical shear stresses τ_{zy}. It is shown in section 5.10.1 that the stress τ_v at a distance y_2 from the centroid where the section width is b_2 is given by

$$\tau_v = -\frac{V_y}{I_x b_2} \int_{y_T}^{y_2} by\, dy. \tag{5.13}$$

This can also be expressed as

$$\tau_v = -\frac{V_y A_2 \bar{y}_2}{I_x b_2} \tag{5.14}$$

in which A_2 is the area above b_2 and $\bar{y}_2$ the height of its centroid.

For the particular example of the rectangular section beam of width b and depth d shown in Fig. 5.10a, the width is constant, and so equation 5.13 becomes

$$\tau_v = \frac{V_y}{bd}\left(1.5 - \frac{6y^2}{d^2}\right). \tag{5.15}$$

This shear stress distribution is parabolic, as shown in Fig. 5.10b, and has a maximum value at the x axis of 1.5 times the average shear stress V_y/bd.

The shear stresses calculated from equation 5.13 or 5.14 are reasonably accurate for solid cross-sections, except near the unstressed edges where the shear stress is parallel to the edge rather than to the applied shear force.

5.4.2 THIN-WALLED OPEN CROSS-SECTIONS

The shear stress distributions in thin-walled open-section beams differ from those given by equation 5.13 or 5.14 for solid section beams in that the shear

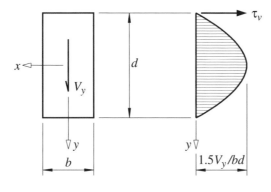

(a) Cross-section   (b) Shear stress distribution

Fig. 5.10 Shear stress distribution in a rectangular section

stresses are parallel to the wall of the section as shown in Fig. 5.11a instead of parallel to the applied shear force. Because of the thinness of the walls, it is quite accurate to assume that the shear stresses are uniformly distributed across the thickness t of the thin-walled section. Their distribution can be determined by considering the horizontal equilibrium of an element of the beam, as shown in Fig. 5.11b. It is shown in section 5.10.2 that the stress τ_v at a distance s from the end of the section caused by a vertical shear force V_y can be obtained from the shear flow

$$\tau_v t = -\frac{V_y}{I_x} \int_0^s yt \, ds. \tag{5.16}$$

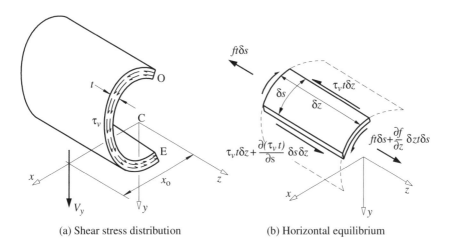

(a) Shear stress distribution   (b) Horizontal equilibrium

Fig. 5.11 Shear stresses in a thin-walled open section

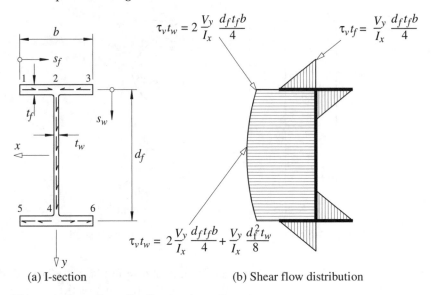

(a) I-section   (b) Shear flow distribution

Fig. 5.12 Shear flow distribution in an I-section

This can also be expressed as

$$\tau_v = \frac{-V_y A_s \bar{y}_s}{I_x t} \tag{5.17}$$

in which A_s is the area from the free end to the point s and $\bar{y}_s$ the height of the centroid of this area above the point s.

At a junction in the cross-section wall, such as that of the top flange and the web of the I-section shown in Figs 5.12 and 5.13, horizontal equilibrium of the zero area junction element requires

$$(\tau_v t_f)_{21} \delta z + (\tau_v t_f)_{23} \delta z = (\tau_v t_w)_{24} \delta z. \tag{5.18}$$

This can be thought of as an analogous flow continuity condition for the corresponding shear flows in each cross-section element at the junction, as shown in Fig. 5.13, so that

$$(\tau_v t_f)_{21} + (\tau_v t_f)_{23} = (\tau_v t_w)_{24}, \tag{5.19}$$

which is a particular example of the general junction condition

$$\Sigma(\tau_v t) = 0 \tag{5.20}$$

in which each shear flow is now taken as positive if it acts towards the junction.

A free end of a cross-section element is the special case of a one-element junction for which equation 5.20 reduces to

$$\tau_v t = 0. \tag{5.21}$$

This condition has already been used (at $s = 0$) in deriving equation 5.16.

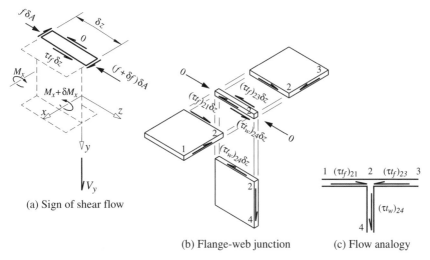

(a) Sign of shear flow

(b) Flange-web junction   (c) Flow analogy

Fig. 5.13 Horizontal equilibrium considerations for shear flow

An example of the analysis of the shear flow distribution in the I-section beam shown in Fig. 5.12a is given in section 5.12.6. The shear flow distribution is shown in Fig. 5.12b. It can be seen that the shear stress in the web is nearly constant and equal to its average value

$$\tau_{v(av)} = \frac{V_y}{d_f t_w}, \tag{5.22}$$

which provides the basis for the commonly used assumption that the applied shear is resisted only by the web. A similar result is obtained for the shear stress in the web of the channel section shown in Fig. 5.14 and analysed in section 5.12.7.

When a horizontal shear force V_x acts parallel to the x axis, additional shear stresses τ_h parallel to the walls of the section are induced. These can be obtained from the shear flow

$$\tau_h t = -\frac{V_x}{I_y} \int_0^s xt \, ds, \tag{5.23}$$

which is similar to equation 5.16 for the shear flow due to a vertical shear force. A worked example of the calculation of the shear flow distribution caused by a horizontal shear force is given in section 5.12.8.

5.4.3 SHEAR CENTRE

The shear stresses τ_v induced by the vertical shear force V_y exert a torque equal to $\int_0^E \tau_v t \rho \, ds$ about the centroid C of the thin-walled cross-section, as shown in

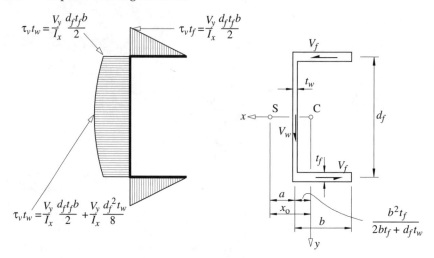

(a) Shear flow due to shear force V_y (b) Shear centre and flange and web shears

Fig. 5.14 Shear flow distribution in a channel section

Fig. 5.15, and are therefore statically equivalent to a vertical shear force V_y which acts at a distance x_0 from the centroid equal to

$$x_0 = \frac{1}{V_y} \int_0^E \tau_v t \rho \, ds. \qquad (5.24)$$

Similarly, the shear stresses τ_h induced by the horizontal shear force V_x are statically equivalent to a horizontal shear force V_x which acts at a distance y_0 from the centroid equal to

$$y_0 = -\frac{1}{V_x} \int_0^E \tau_h t \rho \, ds. \qquad (5.25)$$

The coordinates x_0, y_0 define the position of the shear centre S of the cross-section through which the resultant of the bending shear stresses must act. When any applied load does not act through the shear centre, as shown in Fig. 5.16, it induces another set of shear stresses in the section which are additional to those caused by the changes in the bending normal stresses described above (see equations 5.16 and 5.23). These additional shear stresses are statically equivalent to the torque exerted by the eccentric applied load about the shear centre. They can be calculated as shown in sections 10.2.1.4 and 10.3.1.2.

The position of the shear centre of the channel section shown in Fig. 5.14 is determined in section 5.12.8. Its x_0 coordinate is given by

$$x_0 = \frac{d_f^2 t_f b^2}{4I_x} + \frac{b^2 t_f}{2bt_f + d_f t_w}, \qquad (5.26)$$

while its y_0 coordinate is zero because the section is symmetrical about the x axis.

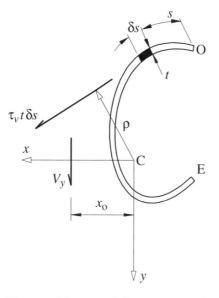

Fig. 5.15 Moment of shear stress τ_v about centroid

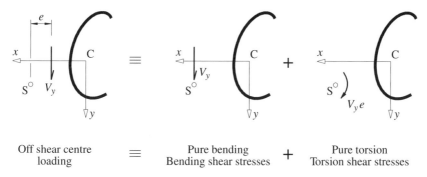

| Off shear centre loading | $\equiv$ | Pure bending Bending shear stresses | $+$ | Pure torsion Torsion shear stresses |

Fig. 5.16 Off shear centre loading

In Fig. 5.17, the shear centres of a number of thin-walled open sections are compared with their centroids. It can be seen that:

   (a) if the section has an axis of symmetry, then the shear centre and centroid lie on it;

   (b) if the section is of a channel type, then the shear centre lies outside the web and the centroid inside it;

   (c) if the section consists of a set of concurrent rectangular elements (tees, angles, and cruciforms), then the shear centre lies at the common point.

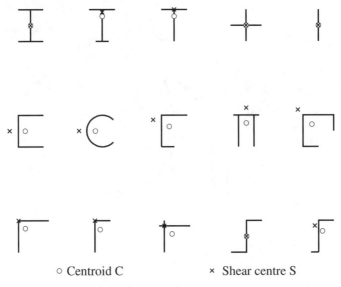

○ Centroid C   × Shear centre S

Fig. 5.17 Centroids and shear centres

A general matrix method for analysing the shear stress distribution and for determining the shear centres of thin-walled open-section beams has been prepared [11, 12].

Worked examples of the determination of the shear centre position are given in sections 5.12.8 and 5.12.10.

5.4.4 THIN-WALLED CLOSED CROSS-SECTIONS

The shear stress distribution in a thin-walled, closed-section beam is similar to that in an open-section beam, except that there is an additional constant shear flow $\tau_{vc}t$ around the section. This additional shear flow is required to prevent any discontinuity in the longitudinal warping displacements w which arise from the shear straining of the walls of the closed sections. To show this, consider the slit rectangular box whose shear flow distribution $\tau_{vo}t$ due to a vertical shear force V_y is as shown in Fig. 5.18a. Because the beam is not twisted, the longitudinal fibres remain parallel to the centroidal axis, so that the transverse fibres rotate through angles τ_{vo}/G equal to the shear strain in the wall, as shown in Fig. 5.19. These rotations lead to the longitudinal warping displacements w shown in Fig. 5.18b, the relative warping displacement at the slit being $\int_0^E (\tau_{vo}/G)\mathrm{d}s$.

However, when the box is not slit, as shown in Fig. 5.20, this relative warping displacement must be zero. Thus

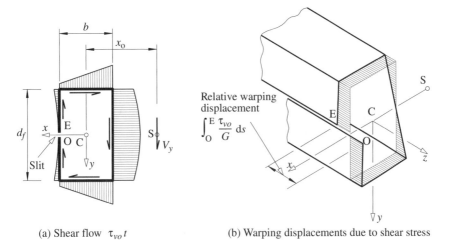

(a) Shear flow $\tau_{vo}t$ (b) Warping displacements due to shear stress

Fig. 5.18 Warping of a slit box

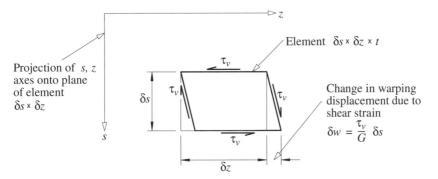

Fig. 5.19 Warping displacements due to shear

$$\oint \frac{\tau_v}{G} \, ds = 0 \tag{5.27}$$

in which total shear stress τ_v can be considered as the sum

$$\tau_v = \tau_{vc} + \tau_{vo} \tag{5.28}$$

of the slit box shear stress τ_{vo} (see Fig. 5.18a) and the shear stress τ_{vc} due to a constant shear flow $\tau_{vc}t$ circulating around the closed section (see Fig. 5.20a). Alternatively, equation 5.16 for the shear flow in an open section can be modified for the closed section to

$$\tau_v t = \tau_{vc}t - \frac{V_y}{I_x} \int_0^s yt \, ds, \tag{5.29}$$

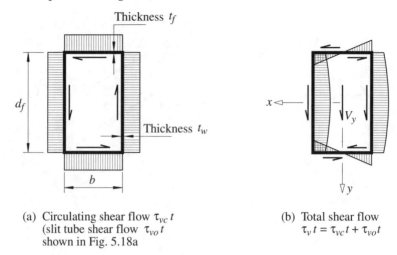

(a) Circulating shear flow $\tau_{vc}\, t$
    (slit tube shear flow $\tau_{vo}\, t$
    shown in Fig. 5.18a

(b) Total shear flow
    $\tau_v\, t = \tau_{vc}\, t + \tau_{vo}\, t$

Fig. 5.20 Shear stress distribution in a rectangular box

since it can no longer be said that τ_v is zero at $s = 0$, this not being a free end (the closed section has no free ends). Mathematically, $\tau_{vc}\, t$ is a constant of integration. Substituting equation 5.28 or 5.29 into equation 5.27 leads to

$$\tau_{vc}\, t = -\frac{\oint \tau_{vo}\ ds}{\oint (1/t)\ ds},$$
(5.30)

which allows the circulating shear flow $\tau_{vc}\, t$ to be determined.

The shear stress distribution in any single cell closed section can be obtained by using equations 5.29 and 5.30. The shear centre of the section can then be determined by using equations 5.24 and 5.25 as for open sections. For the particular case of the rectangular box shown in Fig. 5.20,

$$\tau_{vc}\, t = \frac{V_y}{I_x}\left(\frac{d_f^2 t_w}{8} + \frac{d_f t_f b}{4}\right),$$
(5.31)

and the resultant shear flow shown in Fig. 5.20b is symmetrical because of the symmetry of the cross-section, while the shear centre coincides with the centroid.

A worked example of the calculation of the shear stresses in a thin-walled closed section is given in section 5.12.11.

The shear stress distributions in multicell closed sections can be determined by extending this method, as indicated in section 5.10.3. A general matrix method of analysing the shear flows in thin-walled closed sections has been described [11, 12]. This can be used for both open and closed sections, including composite and asymmetric sections, and sections with open and closed parts. It can also be used to determine the centroid, principal axes, section constants, and the bending normal stress distribution.

5.4.5 SHEAR LAG

In the conventional theory of bending, shear strains are neglected so that it can be assumed that plane sections remain plane after loading. From this assumption follow the simple linear distributions of the bending strains and stresses discussed in section 5.3, and from these the shear stress distributions discussed in sections 5.4.1–5.4.4. The term shear lag [13] is related to some of the discrepancies between this approximate theory of the bending of beams and their real behaviour, and in particular, refers to the increases of the bending stresses near the flange-to-web junctions, and the corresponding decreases in the flange stresses away from these junctions.

The shear lag effects near the mid-point of the simply supported centrally loaded I-section beam shown in Fig. 5.21a are illustrated in Fig. 5.22. The shear stresses calculated by the conventional theory are as shown in Fig. 5.23a, and these induce the warping (longitudinal) displacements shown in Fig. 5.23b. The warping displacements of the web are almost linear, and these are responsible for the shear deflections [13] which are usually neglected when calculating the beam's deflections. The warping displacements of the flanges vary parabolically, and it is these which are responsible for most of the shear lag effect. The application of the conventional theory of bending at either side of a point in a beam where there is a sudden change in the shear force, such as at the mid-point of the beam of Fig. 5.21a, leads to two different distributions of warping displacement, as shown in Fig. 5.21b. These are clearly incompatible, and so changes are required in the bending stress distribution (and consequently in the shear stress distribution) to remove the incompatibility. These changes, which are illustrated in Fig. 5.22, constitute the shear lag effect.

Shear lag effects are usually very small except near points of highly concentrated load or at reaction points in short span beams with thin wide flanges.

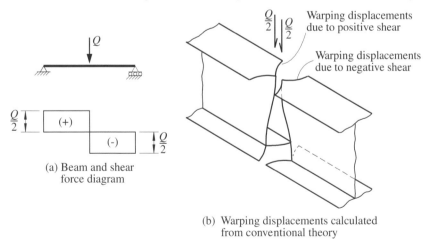

(a) Beam and shear force diagram

(b) Warping displacements calculated from conventional theory

Fig. 5.21 Incompatible warping displacements at a shear discontinuity

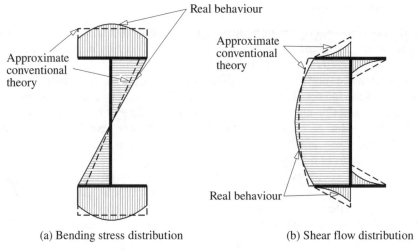

(a) Bending stress distribution   (b) Shear flow distribution

Fig. 5.22 Shear lag effects in an I-section beam

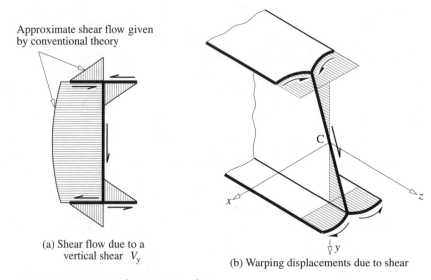

(a) Shear flow due to a
vertical shear V_y

(b) Warping displacements due to shear

Fig. 5.23 Warping of an I-section beam

In particular, shear lag effects may be significant in light-gauge, cold-formed sections [14] and in stiffened box girders [15–17]. Shear lag has no serious consequences in a ductile structure in which any premature local yielding leads to a favourable redistribution of stress. However, the increased stresses due to shear lag may be of consequence in a tension flange which is liable to brittle fracture or fatigue damage, or in a compression flange whose strength is controlled by its resistance to local buckling.

An approximate method of dealing with shear lag is to use an effective width concept, in which the actual width b of a flange is replaced by a reduced width b_e given by

$$\frac{b_e}{b} = \frac{\text{nominal bending stress}}{\text{maximum bending stress}}. \tag{5.32}$$

This is equivalent to replacing the actual flange bending stresses by constant stresses which are equal to the actual maximum stress and distributed over the effective flange area $b_e \times t$. Some values of effective widths are given in [14–16]. This approach is similar to that used to allow for the redistribution of stress which takes place in a thin compression flange after local buckling (see Chapter 4). However, the two effects of shear lag and local buckling are quite distinct, and should not be confused.

5.5 Plastic analysis of beams

5.5.1 GENERAL

As the load on a ductile steel beam is increased, the stresses in the beam also increase, until the yield stress is reached. With further increases in the load, yielding spreads through the most highly strained cross-section of the beam until it becomes fully plastic at a moment M_p. At this stage the section forms a plastic hinge which allows the beam segments on either side to rotate freely under the moment M_p. If the beam was originally statically determinate, this plastic hinge reduces it to a mechanism, and prevents it from supporting any additional load.

However, if the beam was statically indeterminate, the plastic hinge does not reduce it to a mechanism, and it can support additional load. This additional load causes a redistribution of bending moment, during which the moment at the plastic hinge remains fixed at M_p, while the moment at another highly strained cross-section increases until it forms a plastic hinge. This process is repeated until enough plastic hinges have formed to reduce the beam to a mechanism. The beam is then unable to support any further increase in load, and its ultimate strength is reached.

In the plastic analysis of beams, this mechanism condition is investigated to determine the ultimate strength. The principles and methods of plastic analysis are fully described in many textbooks [18–24], and so only a brief summary is given in the following subsections.

5.5.2 THE PLASTIC HINGE

The bending stresses in an elastic beam are distributed linearly across any section of the beam, as shown in Fig. 5.4, and the bending moment M is proportional to the curvature $-\mathrm{d}^2v/\mathrm{d}z^2$ (see equation 5.2). However, once the yield strain $\varepsilon_Y = f_y/E$ (see Fig. 5.24) of a steel beam is exceeded, the stress distribution is no longer linear, as indicated in Fig. 5.2c. Nevertheless, the strain distribution remains linear, and so the inelastic bending stress distributions are similar to the basic stress–strain relationship shown in Fig. 5.24, provided the influence of shear on yielding can be ignored (which is a reasonable assumption for many I-section beams). The moment resultant M of the bending stresses is no longer proportional to the curvature $-d^2v/dz^2$ but varies as shown in Fig. 5.2d. Thus the section becomes elastic–plastic when the yield moment $M_Y = f_y Z_x$ is exceeded, and the curvature increases rapidly as yielding progresses through the section. At high curvatures, the limiting situation is approached for which the section is completely yielded at the fully plastic moment

$$M_p = f_y S, \tag{5.33}$$

where S is the plastic section modulus. Methods of calculating the fully plastic moment M_p are discussed in section 5.11.1, and worked examples are given in sections 5.12.12 and 5.12.13. For solid rectangular sections, the shape factor S_x/Z_x is 1.5, but for rolled I-sections, S_x/Z_x varies between 1.1 and 1.2 approximately. Values of S_x for hot-rolled I-section members are given in [9].

In real beams, strain-hardening commences just before M_p is reached, and the real moment–curvature relationship rises above the fully plastic limit of M_p, as shown in Fig. 5.2d. On the other hand, high shear forces cause small reductions in M_p below $f_y S$, principally due to reductions in the plastic bending capacity of the web. This effect is discussed in section 4.5.2.

The approximate approach of the moment–curvature relationship shown in Fig. 5.2d to the fully plastic limit forms the basis of the simple rigid-plastic assumption for which the basic stress–strain relationship is replaced by the rectangular block shown by the dashed line in Fig. 5.24. Thus elastic strains are

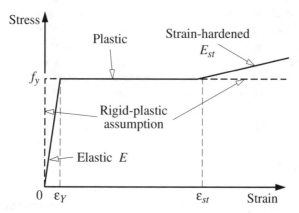

Fig. 5.24 Idealized stress–strain relationships for structural steel

completely ignored, as are the increased stresses due to strain-hardening. This assumption ignores the curvature of any elastic and elastic–plastic regions ($M < M_p$), and assumes that the curvature becomes infinite at any point where $M = M_p$.

The consequences of the rigid-plastic assumption on the theoretical behaviour of a simply supported beam with a central concentrated load are shown in Figs 5.25 and 5.26. In the real beam shown in Fig. 5.25a, there is a finite length of the beam which is elastic–plastic and in which the curvatures are large, while the remaining portions are elastic, and have small curvatures. However, according to the rigid-plastic assumption, the curvature becomes infinite at mid-span when this section becomes fully plastic ($M = M_p$), while the two halves of the beam have zero curvature and remain straight, as shown in Fig. 5.25b. The infinite curvature at mid-span causes a finite change θ in the beam slope, and so the deflected shape of the rigid-plastic beam closely approximates that of the real beam, despite the very different curvature distributions. The successive stages in the development of the bending moment diagram and the central deflection of the beam as the load is increased are summarized in Fig. 5.26b and c.

The infinite curvature at the point of full plasticity and the finite slope change θ predicted by the rigid-plastic assumption lead to the plastic hinge concept illustrated in Fig. 5.27a. The plastic hinge can assume any slope change θ once the full plastic moment M_p has been reached. This behaviour is contrasted with that shown in Fig. 5.27b for a frictionless hinge, which can assume any slope change θ at zero moment.

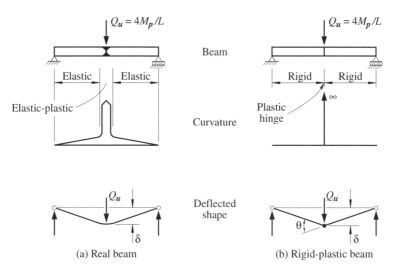

Fig. 5.25 Beam with full plastic moment M_p at mid-span

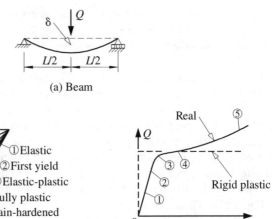

(a) Beam

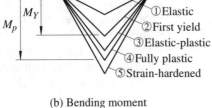

(b) Bending moment

(c) Deflection

Fig. 5.26 Behaviour of a simply supported beam

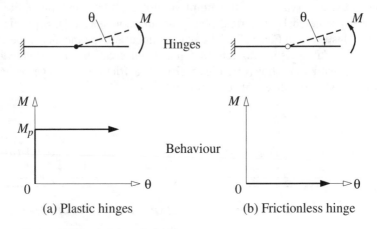

(a) Plastic hinges

(b) Frictionless hinge

Fig. 5.27 Plastic hinge behaviour

It should be noted that when the full plastic moment M_p of the simply supported beam shown in Fig. 5.26a is reached, the rigid-plastic assumption predicts that a two bar mechanism will be formed by the plastic hinge and the two frictionless support hinges, as shown in Fig. 5.25b, and that the beam will deform freely without any further increase in load, as shown in Fig. 5.26c. Thus the ultimate load of the beam is

$$Q_u = 4M_p/L, \tag{5.34}$$

which reduces the beam to a plastic collapse mechanism.

5.5.3 MOMENT REDISTRIBUTION IN INDETERMINATE BEAMS

The increase in the full plastic moment M_p of a beam over its nominal first yield moment M_Y is accounted for in elastic design (i.e. design based on an elastic analysis of the bending of the beam) by the use of M_p for the moment capacity of the cross-section. Thus elastic design might be considered to be 'first hinge' design. However, the feature of plastic design which distinguishes it from elastic design is that it takes into account the favourable redistribution of bending moment which takes place in an indeterminate structure after the first hinge forms. This redistribution may be considerable, and the final load at which the collapse mechanism forms may be significantly higher than that at which the first hinge is developed. Thus, a first hinge design based on an elastic analysis of the bending moment may significantly underestimate the ultimate strength.

The redistribution of bending moment is illustrated in Fig. 5.28 for a built-in beam with a concentrated load at a third point. This beam has two redundancies, and so three plastic hinges must form before it can be reduced to a collapse mechanism. According to the rigid-plastic assumption, all the bending moments remain proportional to the load until the first hinge forms at the left-hand support A at $Q = 6.75\ M_p/L$. As the load increases further, the moment at this hinge remains constant at M_p, while the moments at the load point and the right hand support increase until the second hinge forms at the load point B at $Q = 8.68\ M_p/L$. The moment at this hinge then remains constant at M_p while the moment at the right hand support C increases until the third and final hinge forms at this point at $Q = 9.00\ M_p/L$. At this load the beam becomes a

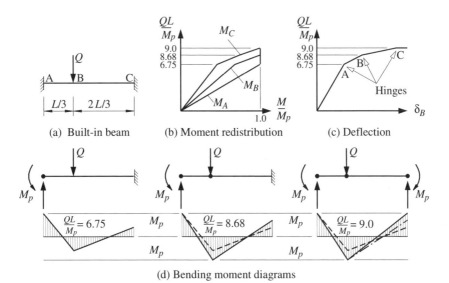

(a) Built-in beam   (b) Moment redistribution   (c) Deflection

(d) Bending moment diagrams

Fig. 5.28 Moment redistribution in a built-in beam

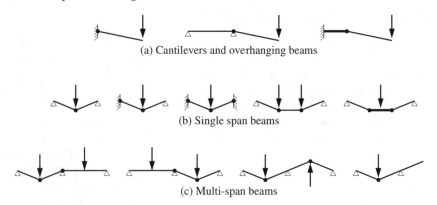

(a) Cantilevers and overhanging beams

(b) Single span beams

(c) Multi-span beams

Fig. 5.29 Beam plastic collapse mechanisms

mechanism, and so the ultimate load is $Q_u = 9.00\,M_p/L$ which is 33% higher than the first hinge load of $6.75\,M_p/L$. The redistribution of bending moment is shown in Fig. 5.28b and d, while the deflection of the load point (derived from an elastic–plastic assumption) is shown in Fig. 5.28c.

5.5.4 PLASTIC COLLAPSE MECHANISMS

A number of examples of plastic collapse mechanisms in cantilevers and single and multispan beams is shown in Fig. 5.29. Cantilevers and overhanging beams generally collapse as single bar mechanisms with a plastic hinge at the support. When there is a reduction in section capacity, the plastic hinge may form in the weaker section.

Single span beams generally collapse as two bar mechanisms, with a hinge (plastic or frictionless) at each support and a plastic hinge within the span. Sometimes general plasticity may occur along a uniform moment region.

Multispan beams generally collapse in one span only, as a local two bar mechanism, with a hinge (plastic or frictionless) at each support, and a plastic hinge within the span. Sometimes two adjacent spans may combine to form a three bar mechanism, with the common support acting as a frictionless pivot, and one plastic hinge forming within each span. A similar mechanism may form in an overhanging beam.

Potential locations for plastic hinges include supports, points of concentrated load, and points of cross-section change. The location of a plastic hinge in a beam with distributed load is often not well defined.

5.5.5 METHODS OF PLASTIC ANALYSIS

The purpose of the methods of plastic analysis is to determine the ultimate load at which a collapse mechanism first forms. Thus, it is only this final mechanism condition which must be found, and any intermediate load conditions can be ignored. A further important simplification arises from the fact that in its collapse condition, the beam is a mechanism, and can be analysed by statics, without any of the difficulties associated with the elastic analysis of a statically indeterminate beam.

The basic method of plastic analysis is to assume the locations of a series of plastic hinges and to investigate whether the three conditions of equilibrium, mechanism and plasticity are satisfied. The equilibrium condition is that the bending moment distribution defined by the assumed plastic hinges must be in static equilibrium with the applied loads and reactions. This condition applies to all beams, elastic or plastic. The mechanism condition is that there must be a sufficient number of plastic and frictionless hinges for the beam to form a mechanism. This condition is usually satisfied directly by the choices of hinges. The plasticity condition is that the full plastic moment of every cross-section must not be exceeded, so that

$$-M_p \leqslant M \leqslant M_p. \tag{5.35}$$

If the assumed plastic hinges satisfy these three conditions, then they are the correct ones to define the collapse mechanism of the beam, so that

$$\text{(Collapse mechanism) satisfies} \begin{pmatrix} \text{Equilibrium} \\ \text{Mechanism} \\ \text{Plasticity} \end{pmatrix} \tag{5.36}$$

The ultimate load can then be determined directly from the equilibrium conditions.

However, it is usually possible to assume more than one series of plastic hinges, and so while the assumed plastic hinges may satisfy the equilibrium and mechanism conditions, the plasticity condition (equation 5.35) may be violated. In this case, the load calculated from the equilibrium condition is greater than the true ultimate load ($Q_m > Q_u$). This forms the basis of the mechanism method of plastic analysis

$$\begin{pmatrix} \text{Mechanism method} \\ Q_m \geqslant Q_u \end{pmatrix} \text{ satisfies } \begin{pmatrix} \text{Equilibrium} \\ \text{Mechanism} \end{pmatrix}, \tag{5.37}$$

which provides an upper bound solution for the true ultimate load.

A lower bound solution ($Q_s \leqslant Q_u$) for the true ultimate load can be obtained by reducing the loads and bending moments obtained by the mechanism method proportionally (which ensures that the equilibrium condition remains satisfied) until the plasticity condition is satisfied everywhere. These reductions decrease the number of plastic hinges, and so the mechanism condition is not satisfied. This is the statical method of plastic analysis

$$\begin{pmatrix} \text{Statical method} \\ Q_s \leqslant Q_u \end{pmatrix} \text{ satisfies } \begin{pmatrix} \text{Equilibrium} \\ \text{Plasticity} \end{pmatrix}. \tag{5.38}$$

(a) Built-in beam   (b) Beam segments

(c) First mechanism and virtual displacements

(d) Correct mechanism and virtual displacements

Fig. 5.30 Plastic analysis of a built-in beam

If the upper and lower bound solutions obtained by the mechanism and statical methods coincide or are sufficiently close, then the beam may be designed immediately. If, however, these bounds are not precise enough, then the original series of hinges must be modified (and the bending moments determined in the statical analysis will provide some indication of how to do this), and the analysis repeated.

The use of the methods of plastic analysis is demonstrated in sections 5.11.2 and 5.11.3 for the examples of the built-in beam and the propped cantilever shown in Figs 5.30a and 5.31a. A worked example of the plastic collapse analysis of a non-uniform beam is given in section 5.12.14.

The limitations on the use of the method of plastic analysis in the BS5950 strength design of beams are discussed in section 5.6.2.

The application of the methods of plastic analysis to rigid-jointed frames is discussed in section 8.3.5.5. This application can only be made provided the effects of any axial forces in the members are small enough to preclude any instability effects and provided any significant decreases in the full plastic moments due to axial forces are accounted for (see section 7.2.2).

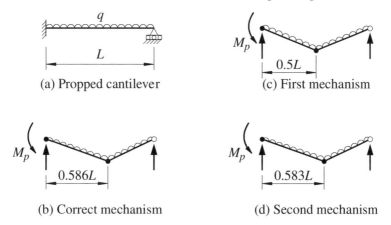

(a) Propped cantilever

(c) First mechanism

(b) Correct mechanism

(d) Second mechanism

Fig. 5.31 Plastic analysis of a propped cantilever

5.6 Strength design of beams

5.6.1 ELASTIC DESIGN OF BEAMS

5.6.1.1 General

For the elastic method of designing a beam for the strength limit state, the strength design loads are first obtained by multiplying the nominal loads (dead, imposed, or wind) by the appropriate load combination factors (see section 1.5.6). The distributions of the design bending moments and shear forces in the beam under the strength design load combinations are determined by an elastic bending analysis (when the structure is statically indeterminate), or by statics (when it is statically determinate). BS5950 permits a redistribution in each plastic or compact span of a continuous beam of up to 10% of the span's peak elastic moment, provided equilibrium is maintained.

The beam must then be designed to have sufficient capacity to resist these design moments and shears, as well as any concentrated forces arising from the factored loads and their reactions. The processes of checking a specified member or of designing an unknown member for the design actions are summarized in Fig. 5.32. In this figure, the design actions are distinguished by the addition of *, while the symbol p_y for the BS5950 design strength (usually taken as the specified minimum yield strength Y_s except for slender sections) replaces the symbol f_y used in earlier sections for the yield stress.

When a specified member is to be checked, the cross-section should first be classified as plastic, compact, semi-compact, or slender. This allows the effective section modulus to be determined and the design section moment capacity to be compared with the maximum design moment. Following this, the lateral bracing should be checked, then the web shear capacity, and then combined bending and shear should be checked at any cross-sections where both the

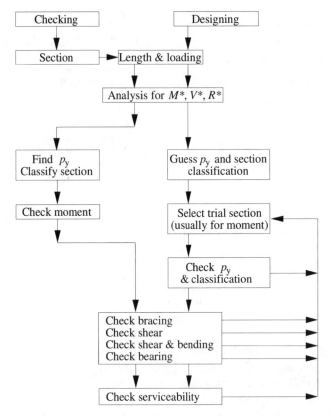

Fig. 5.32 Flow chart for the elastic design of beams

design moment and shear are high. Finally the web bearing capacity should be checked at reactions and concentrated load points.

When the member size is not known, a target value for the effective section modulus may be determined from the maximum design moment, and a trial section chosen whose effective section modulus exceeds this target. The remainder of the design process is then the same as that for checking a specified member. Usually the moment capacity governs the design, but when it does not, an iterative process may need to be followed, as indicated in Fig. 5.32.

The following subsections describe each of the BS5950 checking processes summarized in Fig. 5.32. A worked example of their application is given in section 5.12.15.

5.6.1.2 Section classification

The moment, shear and concentrated load bearing capacities of beams whose plate elements are slender may be significantly influenced by local buckling considerations (Chapter 4). Due to this, beam cross-sections are classified as plastic, compact, semi-compact, or slender depending on the ability of the elements to resist local buckling (section 4.7.2).

Plastic sections are unaffected by local buckling and are able to develop and maintain their fully plastic capacities until a collapse mechanism forms. Compact sections are able to form a first plastic hinge, but local buckling prevents subsequent moment redistribution. Semi-compact sections are able to reach the yield stress, but local buckling prevents full plastification of the cross-section. Slender sections have their capacities reduced below their first yield capacities by local buckling effects.

Sections are classified by first comparing the slenderness $\lambda_e = (b/t)\sqrt{(p_y/275)}$ of each compression element with the appropriate limits λ_L of Tables 11 or 12 of BS5950. These limits depend on the way in which the longitudinal edges of the element are supported (either one edge supported as for the flange outstand of an I-section, or two edges supported as for the internal flange element of a box section), the bending stress distribution (uniform compression as in a flange, or varying stresses as in a web), the type of section, and the residual stress distribution (associated for example, with hot-rolling or welding), as indicated in Figs 4.29 and 5.33.

The section classification is that of the lowest classification of its elements, with plastic being the highest possible and slender the lowest. No

Section description		Hot-rolled UB, UC	Welded PWG	Welded box	Cold-formed RHS
Section and element widths		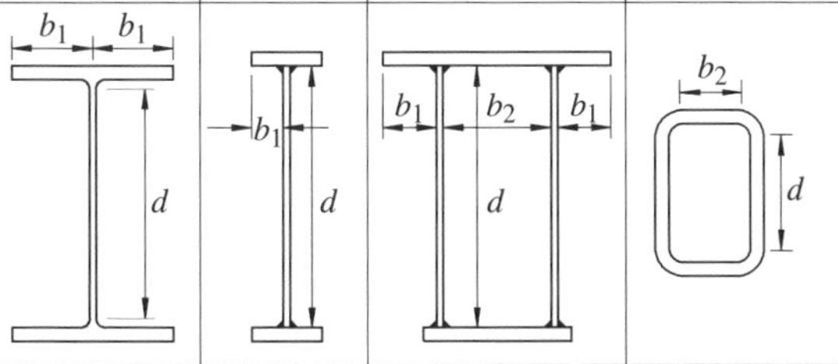			
Flange outstand b_1	Plastic	9	8	8	–
	Compact	10	9	9	–
	Semi-Compact	15	13	13	–
Flange b_2 supported along both edges	Plastic	–	–	28	72-d/$t\varepsilon$ < 26
	Compact	–	–	32	54-0.5 d/$t\varepsilon$< 28
	Semi-Compact	–	–	40	35
Web d	Plastic	80	80	80	56
	Compact	100	100	100	70
	Semi-Compact	120	120	120	105

Fig. 5.33 Local buckling limits $\lambda_L = (b/t)_L\sqrt{(p_y/275)}$ for some beam elements

UB's or UC's in S275 steel are slender, but welded I-section members may be slender.

Worked examples of section classification are given in sections 5.12.15 and 4.9.2–4.9.4.

5.6.1.3 Checking the moment capacity

For the moment capacity check of a beam that has adequate bracing against lateral buckling, the inequality

$$M^* \leqslant M_c \tag{5.39}$$

must be satisfied, in which M^* is the maximum design moment, and

$$M_c = p_y Z_e \tag{5.40}$$

is the section moment capacity, in which p_y is the design strength for the section and Z_e the effective section modulus.

For plastic and compact sections, the effective section modulus is given by the plastic section modulus S, so that

$$Z_e = S, \tag{5.41}$$

For semi-compact sections, the effective section modulus may simply be taken as the elastic section modulus Z, although an increased value may be calculated. For slender sections, the effective section modulus is reduced below the elastic section modulus Z, as discussed in section 4.7.2.

Worked examples of checking the moment capacity are given in sections 5.12.15 and 4.9.2–4.9.4, and a worked example of using the moment capacity check to design a suitable section is given in section 5.12.16.

5.6.1.4 Checking the lateral bracing

Beams with insufficient lateral bracing must also be designed to resist lateral buckling. The design of beams against lateral buckling is discussed in section 6.9. It is assumed in this chapter that there is sufficient bracing to prevent lateral buckling. While the adequacy of bracing for this purpose may be checked by determining the lateral buckling moment resistance as in section 6.9, BS5950 also gives simpler (and often conservative) rules for this purpose. For example, lateral buckling may be assumed to be prevented if the equivalent slenderness λ_{LT} (see section 6.4.2.1) of each segment between restraints is less than

$$\lambda_{L0} = 0.4\sqrt{(\pi^2 E/p_y)} = 34.3\sqrt{(275/p_y)} \tag{5.42}$$

for equal flanged I-section beams.

A worked example of checking the lateral bracing is given in section 5.12.15.

5.6.1.5 Checking the shear capacity

For the shear capacity check, the inequality

$$F_v^* \leqslant P_v \tag{5.43}$$

must be satisfied, in which F_v^* is the maximum design shear force, and P_v the shear capacity.

For a stocky web with $d/t \leqslant 62\sqrt{(275/p_y)}$ and for which the elastic shear stress distribution is approximately uniform (as in the case of an equal flanged I-section), the uniform shear capacity P_v is usually given by

$$P_v = 0.6p_y A_v \tag{5.44}$$

in which $0.6p_y$ is approximately equal to the shear yield stress $\tau_y = p_y/\sqrt{3}$ (section 1.3.1) and A_v is the area of the web. The design strength p_y should be taken as the minimum value for the section, while the web area A_v may be determined either from the overall depth D of a UB or UC, or from the web depth d of a welded section. All the webs of UB's and UC's in S275 steel satisfy $d/t \leqslant 62\sqrt{(275/p_y)}$. A worked example of checking the uniform shear capacity of a compact web is given in section 5.12.15.

For a stocky web with $d/t \leqslant 62\sqrt{(275/p_y)}$ and with a non-uniform elastic shear stress distribution (such as an unusual flanged I-section), the inequality

$$f_v^* \leqslant 0.7p_y \tag{5.45}$$

must be satisfied, in which f_v^* is the maximum design shear stress induced in the web by F_v^*, as determined by an elastic shear stress analysis (section 5.4).

Webs for which $d/t > 62\sqrt{(275/p_y)}$ have their shear capacities reduced by local buckling effects, as discussed in section 4.7.4. Worked examples of checking the shear capacity of such webs are given in sections 4.9.5 and 4.9.6.

5.6.1.6 Checking for bending and shear

High shear forces may reduce the capacity of a web to resist bending moment, as discussed in section 4.5. To allow for this, plastic and compact beams under combined bending and shear, in addition to satisfying equations 5.39 and 5.43, must also satisfy

$$M^* \leqslant p_y S\left\{1 - \left(\frac{2F_v^*}{P_v} - 1\right)^2 \frac{S_v}{S}\right\} \tag{5.46}$$

in which F_v^* and M^* are the design actions at the cross-section being checked and S_v is the plastic section modulus of the web. This condition need only be considered when $F_v^* > 0.6P_v$, as indicated in Fig. 4.32. It therefore need only be checked at cross-sections under high bending and shear, such as at the internal support of continuous beams. Even so, many beams have $F_v^* \leqslant 0.6P_v$ and so this check for combined bending and shear rarely governs.

5.6.1.7 Checking the bearing capacity

For the bearing check, the inequality

$$F_x^* \leq P_b \tag{5.47}$$

must be satisfied at all supports and points of concentrated load. In this equation, F_x^* is the design concentrated load or reaction, and P_b the bearing capacity of the web and its load-bearing stiffeners, if any. The bearing behaviour and design of webs and load-bearing stiffeners are discussed in sections 4.6 and 4.7.6.

Stocky webs with $d/t \leq 62\sqrt{(275/p_y)}$ often have sufficient bearing capacity and do not require the addition of load-bearing stiffeners. A worked example of checking the bearing capacity of a stocky unstiffened web is given in section 5.12.15, while a worked example of checking the bearing capacity of a slender web with load-bearing stiffeners is given in section 4.9.9.

5.6.2 PLASTIC DESIGN OF BEAMS

In the plastic method of designing for the strength limit state, the distributions of bending moment and shear force in a beam under factored loads are determined by a plastic bending analysis (section 5.5.5). The use of the plastic method is restricted to beams which satisfy certain limitations on material properties, cross-section form, and lateral bracing (additional limitations are imposed on members with axial forces, as discussed in section 7.2.4.2).

Generally, the material properties for plastic design are limited to ensure that the stress–strain curve has an adequate plastic plateau and exhibits strain-hardening, so that plastic hinges can be formed and maintained until the plastic collapse mechanism is fully developed.

The cross-section form for plastic design is limited by BS5950 to sections which are symmetrical about the axis perpendicular to the axis of bending. Only plastic sections may be used at plastic hinge locations, in order to ensure that local buckling effects (sections 4.2 and 4.3) do not reduce the ability of the I-section to maintain the full plastic moment until the plastic collapse mechanism is fully developed, while plastic or compact sections must be used elsewhere. The slenderness limits for plastic and compact beam flanges and webs are shown in Fig. 5.33.

The spacing of lateral braces is limited in plastic design to ensure that inelastic lateral buckling (see section 6.3) does not prevent the complete development of the plastic collapse mechanism. BS5950 requires the spacing L_m of braces at both ends of a plastic hinge segment to be limited. A conservative approximation for L_m is given by

$$L_m \leq 38r_y(36T/D)(275/p_y) \tag{5.48}$$

in which $r_y(= \sqrt{(I_y/A)}$ is the radius of gyration about the minor y axis, D is the overall depth of the section, and T is the flange thickness, although higher values may be calculated by allowing for the effects of moment gradient.

The resistance of a beam designed plastically is based principally on its plastic collapse load Q_u calculated by plastic analysis using the full plastic moment M_p of the beam. The plastic collapse load factor $\lambda_p = Q_u/Q^*$ must satisfy $\lambda_p \geq 1$.

The shear capacity of a beam analysed plastically is based on the fully plastic shear capacity of the web (see section 4.3.2.1). Thus the maximum shear F_v^* determined by plastic analysis under the design loads is limited to

$$F_v^* \leq 0.6 p_y A_v. \tag{5.49}$$

BS5950 also requires web stiffeners to be provided near all plastic hinge locations where an applied load or reaction exceeds 10% of the above limit.

The capacity of a beam web to transmit concentrated loads and reactions is discussed in section 4.6.2. The web should be designed as in section 4.7.6 using the design loads and reactions calculated by plastic analysis.

A worked example of the plastic design of a beam is given in section 5.12.17.

5.7 Serviceability design of beams

The design of a beam is often governed by the serviceability limit state, for which the behaviour of the beam should be so limited as to give a high probability that the beam will provide the serviceability necessary for it to carry out its intended function. The most common serviceability criteria are associated with the stiffness of the beam, which governs its deflections under load. These may need to be limited so as to avoid a number of undesirable situations: an unsightly appearance; cracking or distortion of elements fixed to the beam such as cladding, linings, and partitions; interference with other elements such as crane girders; or vibration under dynamic loads such as traffic, wind, or machinery loads.

A serviceability design should be carried out by making appropriate assumptions for the load types, combinations, and levels; for the structural response to load; and for the serviceability criteria. Because of the wide range of serviceability limit states, design rules are usually of limited extent. The given rules are often advisory rather than mandatory because of uncertainties as to the appropriate values to be used. These uncertainties result from the variable behaviour of real structures, and what is often seen to be the non-catastrophic (in terms of human life) nature of serviceability failures.

Serviceability design against unsightly appearance should be based on the total sustained load, and so should include the effects of dead and long-term imposed loads. Design against cracking or distortion of elements fixed to a beam should be based on the loads imposed after their installation, and will often include the unfactored imposed load or wind load, but exclude the dead load. Design against interference may need to include the effects of dead load as well as imposed and wind loads. Where imposed and wind loads act simultaneously, it may be appropriate to multiply the unfactored loads by combination factors such as 0.8. Design against vibration will require an assessment to be made of the dynamic nature of the loads.

Most stiffness serviceability limit states are assessed by using an elastic analysis to predict the static deflections of the beam. For vibration design, it may be sufficient in some cases to represent the dynamic loads by static equivalents and carry out an analysis of the static deflections. More generally, however, a dynamic elastic analysis will need to be made.

The serviceability deflection limits depend on the criterion being used. To avoid unsightly appearance, a value of $L/250$ ($L/125$ for cantilevers) might be used after making allowances for any precamber. Values as high as $L/360$ have traditionally been used for design against the cracking of plaster finishes, while a limit of $L/600$ has been used for crane gantry girders. A value of $H/300$ has been used for horizontal storey drift in buildings under wind load.

A worked example of serviceability design is given in section 5.12.18.

5.8 Appendix – bending stresses in elastic beams

5.8.1 BENDING IN A PRINCIPAL PLANE

When a beam bends in the yz principal plane, plane cross-sections rotate as shown in Fig. 5.4, so that sections δz apart become inclined to each other at $-(d^2v/dz^2)\delta z$, where v is the deflection in the y principal direction as shown in Fig. 5.5, and δz the length along the centroidal axis between the two cross-sections. The length between the two cross-sections at a distance y from the axis is greater than δz by $y(-d^2v/dz^2)\,\delta z$, so that the longitudinal strain is

$$\varepsilon = -y\frac{d^2v}{dz^2}.$$

The corresponding tensile stress $f = E\varepsilon$ is

$$f = -Ey\frac{d^2v}{dz^2},\tag{5.50}$$

which has the stress resultants

$$N = \int_A f\,dA = 0,$$

since $\int_A y\,dA = 0$ for centroidal axes (see section 5.9), and

$$M_x = \int_A fy\,dA,$$

whence

$$M_x = -EI_x\frac{d^2v}{dz^2}\tag{5.2}$$

since $\int_A y^2 \, dA = I_x$ (see section 5.9). If this is substituted into equation 5.50, then the bending tensile stress f is obtained as

$$f = \frac{M_x y}{I_x}. \tag{5.3}$$

5.8.2 ALTERNATIVE FORMULATION FOR BIAXIAL BENDING

When bending moments M_{x1}, M_{y1} acting about non-principal axes x_1, y_1 cause curvatures d^2v_1/dz^2, d^2u_1/dz^2 in the non-principal planes y_1z, x_1z, the stress f at a point x_1, y_1 is given by

$$f = -Ey_1 d^2v_1/dz^2 - Ex_1 d^2u_1/dz^2, \tag{5.51}$$

and the moment resultants of these stresses are

$$\left. \begin{array}{l} M_{x1} = -EI_{x1}d^2v_1/dz^2 - EI_{x1y1}d^2u_1/dz^2 \\ M_{y1} = EI_{x1y1}d^2v_1/dz^2 + EI_{y1}d^2u_1/dz^2 \end{array} \right\}. \tag{5.52}$$

These can be rearranged as

$$\left. \begin{array}{l} -E\dfrac{d^2v_1}{dz^2} = \dfrac{M_{x1}I_{y1} + M_{y1}I_{x1y1}}{I_{x1}I_{y1} - I_{x1y1}^2} \\[2mm] E\dfrac{d^2u_1}{dz^2} = \dfrac{M_{x1}I_{x1y1} + M_{y1}I_{x1}}{I_{x1}I_{y1} - I_{x1y1}^2} \end{array} \right\}, \tag{5.53}$$

which are much more complicated than the principal plane formulations of equations 5.9.

If equations 5.53 are substituted, then equation 5.51 can be expressed as

$$f = \left(\frac{M_{x1}I_{y1} + M_{y1}I_{x1y1}}{I_{x1}I_{y1} - I_{x1y1}^2} \right) y_1 - \left(\frac{M_{x1}I_{x1y1} + M_{y1}I_{x1}}{I_{x1}I_{y1} - I_{x1y1}^2} \right) x_1, \tag{5.54}$$

which is much more complicated than the principal plane formulation of equation 5.12.

5.9 Appendix – thin-walled section properties

The cross-section properties of many steel beams can be analysed with sufficient accuracy by using the thin-walled assumption demonstrated in Fig. 5.34, in which the cross-section is replaced by its mid-thickness line. While the thin-walled assumption may lead to small errors in some properties due to junction and thickness effects, its consistent use will avoid anomalies that arise when a more accurate theory is combined with the thin-walled theory, as might be the case in the determination of shear flows. The small errors are typically of the same order as those introduced by the common practice of ignoring the fillets or corner radii of hot-rolled sections.

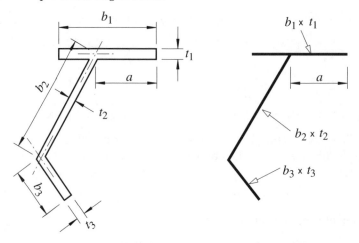

(a) Actual cross-section   (b) Thin-walled approximation

Fig. 5.34 Thin-walled cross-section assumption

The area A of a thin-walled section composed of a series of thin rectangles b_n wide and t_n thick ($b_n \gg t_n$) is given by

$$A = \int_A dA = \sum_n A_n \qquad (5.55)$$

in which $A_n = b_n t_n$. This approximation may make small errors of the order of $(t/b)^2$ at some junctions.

The centroid of a cross-section is defined by

$$\int_A x \, dA = \int_A y \, dA = 0$$

and for a thin-walled section these conditions become

$$\sum_n A_n x_n = \sum_n A_n y_n = 0 \qquad (5.56)$$

in which x_n, y_n are the centroidal distances of the centre of the rectangular element, as demonstrated in Fig. 5.35a. The position of the centroid can be found by adopting any convenient initial set of axes x_i, y_i as shown in Fig. 5.35a, so that

$$x_n = x_{in} - x_{ic},$$
$$y_n = y_{in} - y_{ic}.$$

When these are substituted into equation 5.56, the centroid position is found from

$$\left.\begin{array}{l} x_{ic} = \sum A_n x_{in}/A \\ y_{ic} = \sum A_n y_{in}/A \end{array}\right\}. \qquad (5.57)$$

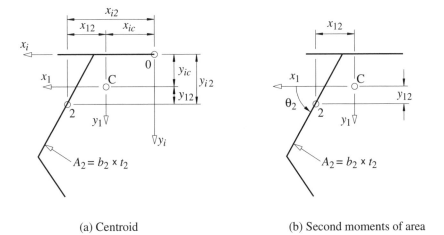

(a) Centroid   (b) Second moments of area

Fig. 5.35 Calculation of cross-section properties

The second moments of area about the centroidal axes are defined by

$$
\begin{aligned}
I_x &= \int_A y^2\, dA = \Sigma(A_n y_n^2 + I_n \sin^2 \theta_n) \\
I_y &= \int_A x^2\, dA = \Sigma(A_n x_n^2 + I_n \cos^2 \theta_n) \\
I_{xy} &= \int_A xy\, dA = \Sigma(A_n x_n y_n + I_n \sin \theta_n \cos \theta_n)
\end{aligned}
\qquad (5.58)
$$

in which θ_n is always the angle from the x axis to the rectangular element (positive from the x axis to the y axis), as shown in Fig. 5.35b, and

$$
I_n = b_n^3 t_n / 12. \qquad (5.59)
$$

These thin-walled approximations ignore small terms $b_n t_n^3/12$, while the use of the angle θ_n adjusts for the inclination of the element to the x, y axes.

The principal axis directions x, y are defined by the condition that

$$
I_{xy} = 0. \qquad (5.60)
$$

These directions can be found by considering a rotation of the centroidal axes from x_1, y_1 to x_2, y_2 through an angle α, as shown in Fig. 5.36a. The x_2, y_2 coordinates of any point are related to its x_1, y_1 coordinates as demonstrated in Fig. 5.36a by

$$
\begin{aligned}
x_2 &= x_1 \cos \alpha + y_1 \sin \alpha \\
y_2 &= -x_1 \sin \alpha + y_1 \cos \alpha
\end{aligned}
\qquad (5.61)
$$

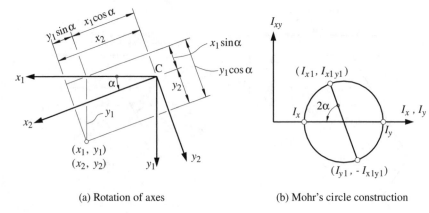

(a) Rotation of axes   (b) Mohr's circle construction

Fig. 5.36 Calculation of principal axis properties

The second moments of area about the x_2, y_2 axes are

$$
\left.\begin{aligned}
I_{x2} &= \int_A y_2^2 \, dA = I_{x1} \cos^2 \alpha - 2I_{x1y1} \sin \alpha \cos \alpha + I_{y1} \sin^2 \alpha \\
I_{y2} &= \int_A x_2^2 \, dA = I_{x1} \sin^2 \alpha + 2I_{x1y1} \sin \alpha \cos \alpha + I_{y1} \cos^2 \alpha \\
I_{x2y2} &= \int_A x_2 y_2 \, dA = \frac{1}{2}(I_{x1} - I_{y1}) \sin 2\alpha + I_{x1y1} \cos 2\alpha
\end{aligned}\right\} . \qquad (5.62)
$$

The principal axis condition of equation 5.60 requires $I_{x2y2} = 0$, so that

$$
\tan 2\alpha = -2I_{x1y1}/(I_{x1} - I_{y1}). \qquad (5.63)
$$

In this book, x and y are reserved exclusively for the principal centroidal axes.

The principal axis second moments of area can be obtained by using this value of α in equation 5.62, whence

$$
\left.\begin{aligned}
I_x &= \tfrac{1}{2}(I_{x1} + I_{y1}) + (I_{x1} - I_{y1})/2 \cos 2\alpha \\
I_y &= \tfrac{1}{2}(I_{x1} + I_{y1}) - (I_{x1} - I_{y1})/2 \cos 2\alpha
\end{aligned}\right\}. \qquad (5.64)
$$

These results are summarized by Mohr's circle constructed in Fig. 5.36b on the diameter whose ends are defined by (I_{x1}, I_{x1y1}) and $(I_{y1}, -I_{x1y1})$. The angles between this diameter and the horizontal axis are equal to 2α and $(2\alpha + 180°)$, while the intersections of the circle with the horizontal axis define the principal axis second moments of area I_x, I_y.

Worked examples of the calculations of cross-section properties are given in sections 5.12.1–5.12.4.

5.10 Appendix – shear stresses in elastic beams

5.10.1 SOLID CROSS-SECTIONS

A solid cross-section beam with a shear force V_y parallel to the y principal axis is shown in Fig. 5.9a. It is assumed that the resulting shear stresses also act parallel to the y axis and are constant across the width b of the section. The corresponding horizontal shear stresses $\tau_{yz} = \tau_v$ are in equilibrium with the horizontal bending stresses f. For the element $b \times \delta y \times \delta z$ shown in Fig. 5.9b, this equilibrium condition reduces to

$$b \frac{\partial f}{\partial z} \delta z \delta y + \frac{\partial (\tau_v b)}{\partial y} \delta y \delta z = 0,$$

whence

$$\tau_v = -\frac{1}{b_2} \int_{y_T}^{y_2} b \frac{\partial f}{\partial z} \, dy, \tag{5.65}$$

which satisfies the condition that the shear stress is zero at the unstressed boundary y_T. The bending normal stresses f are given by

$$f = \frac{M_x y}{I_x}$$

and the shear force V_y is

$$V_y = \frac{dM_x}{dz}$$

and so

$$\frac{\partial f}{\partial z} = \frac{V_y y}{I_x}. \tag{5.66}$$

Substituting this into equation 5.65 leads to

$$\tau_v = -\frac{V_y}{I_x b_2} \int_{y_T}^{y_2} by \, dy. \tag{5.13}$$

5.10.2 THIN WALLED OPEN CROSS-SECTIONS

A thin-walled open cross-section beam with a shear force V_y parallel to the y principal axis is shown in Fig. 5.11a. It is assumed that the resulting shear stresses act parallel to the walls of the section and are constant across the wall thickness. The corresponding horizontal shear stresses τ_v are in equilibrium with the horizontal bending stresses f. For the element $t \times \delta s \times \delta z$ shown in Fig. 5.11b, this equilibrium condition reduces to

$$\frac{\partial f}{\partial z} \delta z t \delta s + \frac{\partial (\tau_v t)}{\partial s} \delta s \delta z = 0,$$

and substitution of equation 5.66 and integration leads to

$$\tau_v t = -\frac{V_y}{I_x} \int_0^s yt \, ds, \tag{5.16}$$

which satisfies the condition that the shear stress is zero at the unstressed boundary $s = 0$.

The direction of the shear stress can be determined from the sign of the right hand side of equation 5.16. If this is positive, then the direction of the shear stress is the same as the positive direction of s, and vice-versa. The direction of the shear stress may also be determined from the direction of the corresponding horizontal shear force required to keep the out-of-balance bending force in equilibrium, as shown for example in Fig. 5.13a. Here a horizontal force $\tau_v t_f \delta z$ in the positive z direction is required to balance the resultant bending force $\delta f \delta A$ (due to the shear force $V_y = dM_x/dz$), and so the direction of the corresponding flange shear stress is from the tip of the flange towards its centre.

5.10.3 MULTI-CELL THIN-WALLED CLOSED SECTIONS

The shear stress distributions in multi-cell closed sections can be determined by extending the method discussed in section 5.4.4 for single cell closed sections. If the section consists of n junctions connected by m walls, then there are $(m - n + 1)$ independent cells. In each wall there is an unknown circulating shear flow $\tau_{vc}t$. There are $(m - n + 1)$ independent cell warping continuity conditions of the type (see equation 5.27)

$$\sum_{\text{cell}} \int_{\text{wall}} (\tau/G) \, ds = 0,$$

and $(n - 1)$ junction equilibrium equations of the type

$$\sum_{\text{junction}} (\tau t) = 0.$$

These equations can be solved simultaneously for the m circulating shear flows $\tau_{vc}t$.

5.11 Appendix – plastic analysis of beams

5.11.1 FULL PLASTIC MOMENT

The fully plastic stress distribution in a section bent about an x axis of symmetry is antisymmetric about that axis, as shown in Fig. 5.37a for a channel section example. The plastic neutral axis, which divides the cross-section into two equal areas so that there is no axial force resultant of the stress distribu-

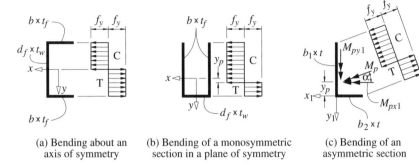

(a) Bending about an
axis of symmetry

(b) Bending of a monosymmetric
section in a plane of symmetry

(c) Bending of an
asymmetric section

Fig. 5.37 Full plastic moment

tion, coincides with the axis of symmetry. The moment resultant of the fully
plastic stress distribution is

$$M_{px} = f_y \sum_T A_n y_n - f_y \sum_C A_n y_n \qquad (5.67)$$

in which the summations $\sum_T$, $\sum_C$ are carried out for the tension and compression areas of the cross-section, respectively. Note that y_n is negative for the
compression areas of the channel section, so that the two summations in equation 5.67 are additive. The plastic section modulus $S_x = M_{px}/f_y$ is obtained from
equation 5.67 as

$$S_x = \sum_T A_n y_n - \sum_C A_n y_n. \qquad (5.68)$$

A worked example of the determination of the full plastic moment of a doubly
symmetric I-section is given in section 5.12.12.

The fully plastic stress distribution in a monosymmetric section bent in a
plane of symmetry is not antisymmetric, as can be seen in Fig. 5.37b for a
channel section example. The plastic neutral axis again divides the cross-
section into two equal areas so that there is no axial force resultant, but the
neutral axis no longer passes through the centroid of the cross-section. The
fully plastic moment M_{px} and the plastic section modulus S_x are given by equations 5.67 and 5.68. For convenience, any suitable origin may be used for computing y_n. A worked example of the determination of the full plastic moment
of a monosymmetric tee-section is given in section 5.12.13.

The fully plastic stress distribution in an asymmetric section is also not antisymmetric, as can be seen in Fig. 5.37c for an angle section example. The
plastic neutral axis again divides the cross-section into two equal areas. The
direction of the plastic neutral axis is perpendicular to the plane in which
the moment resultant acts. The full plastic moment is given by the vector addition of the moment resultants about any convenient x_1, y_1 axes, so that

$$M_p = \surd(M_{px1}^2 + M_{py1}^2), \qquad (5.69)$$

where M_{px1} is given by an equation similar to equation 5.67, and M_{py1} is given by

$$M_{py1} = -f_y \sum_T A_n x_{1n} + f_y \sum_C A_n x_{1n}. \tag{5.70}$$

The inclination α of the plastic neutral axis to the x_1 axis (and of the plane of the full plastic moment to the yz plane) can be obtained from

$$\tan \alpha = M_{py1}/M_{px1}. \tag{5.71}$$

5.11.2 BUILT-IN BEAM

5.11.2.1 General

The built-in beam shown in Fig. 5.30a has two redundancies, and so three plastic hinges must form before it can be reduced to a collapse mechanism. In this case the only possible plastic hinge locations are at the support points 1 and 4 and at the load points 2 and 3. Hence there are two possible mechanisms, one with a hinge at point 2 (Fig. 5.30c), and the other with a hinge at point 3 (Fig. 5.30d). Three different methods of analysing these mechanisms are presented in the following subsections.

5.11.2.2 Equilibrium equation solution

The equilibrium conditions for the beam express the relationships between the applied loads and the moments at these points. These relationships can be obtained by dividing the beam into the segments shown in Fig. 5.30b and expressing the end shears of each segment in terms of its end moments, so that

$$V_{12} = (M_2 - M_1)/(L/3),$$
$$V_{23} = (M_3 - M_2)/(L/3),$$
$$V_{34} = (M_4 - M_3)/(L/3).$$

For equilibrium, each applied load must be equal to the algebraic sum of the shears at the load point in question, and so

$$Q = V_{12} - V_{23} = \frac{1}{L}(-3M_1 + 6M_2 - 3M_3),$$

$$2Q = V_{23} - V_{34} = \frac{1}{L}(-3M_2 + 6M_3 - 3M_4),$$

which can be rearranged as

$$\left. \begin{array}{l} 4QL = -6M_1 + 9M_2 - 3M_4 \\ 5QL = -3M_1 + 9M_3 - 6M_4 \end{array} \right\}. \tag{5.72}$$

If the first mechanism chosen (incorrectly, as will be shown later) is that with plastic hinges at points 1, 2, and 4 (see Fig. 5.30c), so that

$$-M_1 = M_2 = -M_4 = M_p$$

then substitution of this into the first of equations 5.72 leads to $4QL = 18M_p$ and so

$$Q_u \leqslant 4.5M_p/L.$$

If this result is substituted into the second of equations 5.72, then

$$M_3 = 1.5M_p > M_p,$$

and the plasticity condition is violated. The statical method can now be used to obtain a lower bound by reducing Q until $M_3 = M_p$, whence

$$Q = \frac{4.5M_p}{L} \Big/ \frac{M_3}{M_p} = \frac{3M_p}{L}$$

and so

$$3M_p/L < Q_u < 4.5M_p/L.$$

The true collapse load can be obtained by assuming a collapse mechanism with plastic hinges at point 3 (the point of maximum moment for the previous mechanism) and at points 1 and 4. For this mechanism

$$-M_1 = M_3 = -M_4 = M_p,$$

and so, from the second of equations 5.72,

$$Q_u \leqslant 3.6M_p/L.$$

If this result is substituted into the first of equations 5.72, then

$$M_2 = 0.6M_p < M_p,$$

and so the plasticity condition is also satisfied. Thus

$$Q_u = 3.6M_p/L.$$

5.11.2.3 Graphical solution

A collapse mechanism can be analysed graphically by first plotting the known free moment diagram (for the applied loading on a statically determinate beam), and then the reactant bending moment diagram (for the redundant reactions) which corresponds to the collapse mechanism.

For the built-in beam (Fig. 5.38a), the free moment diagram for a simply supported beam is shown in Fig. 5.38c. It is obtained by first calculating the left hand reaction as

$$R_L = (Q \times 2L/3 + 2Q \times L/3)/L = 4Q/3$$

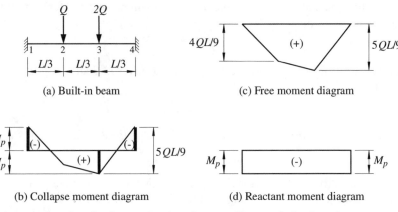

(a) Built-in beam

(c) Free moment diagram

(b) Collapse moment diagram

(d) Reactant moment diagram

Fig. 5.38 Graphical solution for the plastic collapse of a built-in beam

and then using this to calculate the moments

$$M_2 = (4Q/3) \times L/3 = 4QL/9,$$
$$M_3 = (4Q/3) \times 2L/3 - QL/3 = 5QL/9.$$

Both of the possible collapse mechanisms (Fig. 5.29c and d) require plastic hinges at the supports, and so the reactant moment diagram is one of uniform negative bending, as shown in Fig. 5.38d. When this is combined with the free moment diagram as shown in Fig. 5.38b, it becomes obvious that there must be a plastic hinge at point 3 and not at point 2, and that at collapse

$$2M_p = 5Q_uL/9,$$

so that

$$Q_u = 3.6M_p/L$$

as before.

5.11.2.4 Virtual work solution

The virtual work principle [4, 7, 8] can also be used to analyse each mechanism. For the first mechanism shown in Fig. 5.30c, the virtual external work done by the forces Q, $2Q$ during the incremental virtual displacements defined by the incremental virtual rotations $\delta\theta$ and $2\,\delta\theta$ shown is given by

$$\delta W = Q \times (\delta\theta \times 2L/3) + 2Q \times (\delta\theta \times L/3) = 4QL\delta\theta/3,$$

while the internal work absorbed at the plastic hinge locations during the incremental virtual rotations is given by

$$\delta U = M_p \times 2\,\delta\theta + M_p \times (2\,\delta\theta + \delta\theta) + M_p \times \delta\theta = 6M_p\delta\theta.$$

For equilibrium of the mechanism, the virtual work principle requires

$$\delta W = \delta U,$$

so that the upper bound estimate of the collapse load is

$$Q_u \leqslant 4.5 M_p/L$$

as in section 5.11.2.2.

Similarly, for the second mechanism shown in Fig. 5.30d,

$$\delta W = 5QL\delta\theta/3, \qquad \delta U = 6M_p\delta\theta,$$

so that

$$Q_u \leqslant 3.6 M_p/L$$

once more.

5.11.3 PROPPED CANTILEVER

5.11.3.1 General

The exact collapse mechanisms for beams with distributed loads are not always as easily obtained as those for beams with concentrated loads only, but suffi-ciently accurate solutions can usually be found without great difficulty. Three different methods of determining these mechanisms are presented in the following subsections for the propped cantilever shown in Fig. 5.31.

5.11.3.2 Equilibrium equation solution

If the plastic hinge is assumed to be at the mid-span of the propped cantilever as shown in Fig. 5.31c (the other plastic hinge must obviously be located at the built-in support), then equilibrium considerations of the left and right segments of the beam allow the left and right reactions to be determined from

$$R_L L/2 = 2M_p + (qL/2)L/4,$$
$$R_R L/2 = M_p + (qL/2)L/4,$$

while for overall equilibrium

$$qL = R_L + R_R.$$

Thus

$$qL = (4M_p/L + qL/4) + (2M_p/L + qL/4),$$

so that

$$q_u \leqslant 12M_p/L^2.$$

The left-hand reaction is therefore given by

$$R_L = 4M_p/L + (12M_p/L^2)L/4 = 7M_p/L.$$

The bending moment variation along the propped cantilever is therefore given by

$$M = -M_p + (7M_p/L)z - (12M_p/L^2)z^2/2,$$

which has a maximum value of $25M_p/24 > M_p$ at $z = 7L/12$. Thus the lower bound is given by

$$q_u \geq (12M_p/L^2)/(25/24) \approx 11.52M_p/L^2,$$

and so

$$\frac{11.52M_p}{L^2} < q_u < \frac{12M_p}{L^2}.$$

If desired, a more accurate solution can be obtained by assuming that the hinge is located at a distance $7L/12(=0.583L)$ from the built-in support (i.e. at the point of maximum moment for the previous mechanism), whence

$$\frac{11.6567M_p}{L^2} < q_u < \frac{11.6575M_p}{L^2}.$$

These are very close to the exact solution of

$$q_u = (6 + 4\sqrt{2})M_p/L^2 \approx 11.6569M_p/L^2$$

for which the hinge is located at a distance $(2 - \sqrt{2})L \approx 0.5858L$ from the built-in support.

5.11.3.3 Graphical solution

For the propped cantilever of Fig. 5.31, the parabolic free moment diagram for a simply supported beam is shown in Fig. 5.39c. All the possible collapse mechanisms have a plastic hinge at the left hand support and a frictionless hinge at the right hand support, so that the reactant moment diagram is triangular, as shown in Fig. 5.39d. A first attempt at combining this with the free moment diagram so as to produce a trial collapse mechanism is shown in Fig. 5.39b. For this mechanism, the interior plastic hinge occurs at mid-span, so that

$$M_p + M_p/2 = qL^2/8,$$

which leads to

$$q = 12M_p/L^2$$

as before.

Figure 5.39b indicates that in this case, the maximum moment occurs to the right of mid-span, and is greater than M_p, so that the above solution is an upper bound. A more accurate solution can be obtained by trial and error, by increasing the slope of the reactant moment diagram until the moment of the left

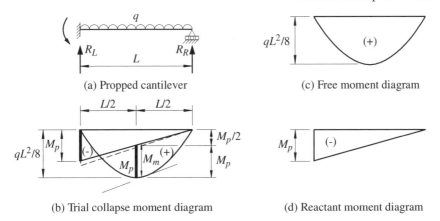

(a) Propped cantilever

(c) Free moment diagram

(b) Trial collapse moment diagram

(d) Reactant moment diagram

Fig. 5.39 Graphical analysis of the plastic collapse of a propped cantilever

hand support is equal to the interior maximum moment, as indicated by the dashed line in Fig. 5.39b.

5.11.3.4 Virtual work solution

For the first mechanism for the propped cantilever shown in Fig. 5.31c, for virtual rotations $\delta\theta$,

$$\delta W = qL^2\delta\theta/4, \qquad \delta U = 3M_p\delta\theta, \qquad q_u \leqslant 12M_p/L^2$$

as before.

5.12 Worked examples

5.12.1 EXAMPLE 1 – PROPERTIES OF A PLATED UB SECTION

Problem.   The 610×229 UB 125 section shown in Fig. 5.40a is strengthened by welding a 300 mm $\times 20$ mm plate to each flange. Determine the section properties I_x and Z_x.

Solution.   Because of symmetry of the section, the centroid of the plated UB is at the web centre. Adapting equation 5.58,

$$I_x = 98\,500 + 2 \times 300 \times 20 \times [(611.9 + 20)/2]^2/10^4 = 218\,290 \text{ cm}^4$$
$$Z_x = 218\,290/[(611.9 + 2 \times 20)/(2 \times 10)] = 6697 \text{ cm}^3$$

5.12.2 EXAMPLE 2 – PROPERTIES OF A TEE-SECTION

Problem.   Determine the section properties I_x and $Z_{x\,\text{min}}$ of the welded tee-section shown in Fig. 5.40b.

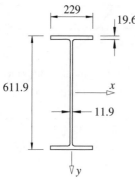

(a) 610 x 229 UB 125

$I_x = 98500 \text{ cm}^4$

$S_x = 3673 \text{ cm}^3$

(b) Tee-section

Fig. 5.40 Worked examples 1 and 2

Solution. Using the thin-walled assumption and equations 5.57 and 5.58,

$$y_i = \frac{(80 + 5/2) \times 10 \times (80 + 5/2)/2}{(80 \times 5) + (80 + 5/2) \times 10} = 27.8 \text{ mm,}$$

$$I_x = (80 \times 5) \times 27.8^2 + (82.5^3 \times 10/12) + (82.5 \times 10) \times (82.5/2 - 27.8)^2 \text{ mm}^4$$
$$= 92.63 \text{ cm}^4$$

$$Z_{x\,\text{min}} = 92.63/((82.5 - 27.8)/10) = 16.93 \text{ cm}^3$$

Alternatively, the equations of Fig. 5.6 may be used.

5.12.3 EXAMPLE 3 – PROPERTIES OF AN ANGLE SECTION

Problem. Determine the properties A, I_{x1}, I_{y1}, I_{x1y1}, I_x, I_y and α of the thin-walled angle section shown in Fig. 5.41c.

Solution. Using the thin-walled assumption, the angle section is replaced by two rectangles 142.5 ($= 150 - 15/2$) $\times$ 15 and 82.5 ($= 90 - 15/2$) $\times$ 15, as shown in Fig. 5.41d. If the initial origin is taken at the intersection of the legs and the initial x_i, y_i axes parallel to the legs, then using equations 5.55 and 5.57:

$$A = (142.5 \times 15) + (82.5 \times 15) = 3375 \text{ mm}^2$$
$$x_{ic} = \{(142.5 \times 15) \times 0 + (82.5 \times 15) \times (-82.5/2)\}/3375 = -15.1 \text{ mm}$$
$$y_{ic} = \{(142.5 \times 15) \times (-142.5/2) + (82.5 \times 15) \times 0\}/3375 = -45.1 \text{ mm}$$

Transferring the origin to the centroid and using equations 5.58:

$$I_{x1} = (142.5 \times 15) \times (-142.5/2 + 45.1)^2 + (142.5^3 \times 15/12) \times \sin^2 90°$$
$$+ (82.5 \times 15) \times (45.1)^2 + (82.5^3 \times 15/12) \times \sin^2 180°$$
$$= 7.596 \times 10^6 \text{ mm}^4 = 759.6 \text{ cm}^4$$

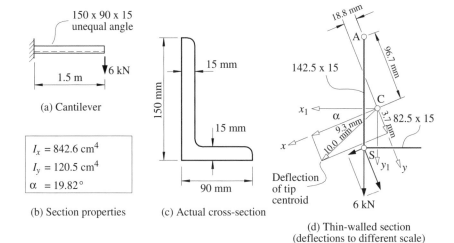

Fig. 5.41 Worked examples 3 and 5

$$I_{y1} = (142.5 \times 15) \times (15.1)^2 + (142.5^3 \times 15/12) \times \cos^2 90°$$
$$+(82.5 \times 15) \times (-82.5/2 + 15.1)^2 + (82.5^3 \times 15/12) \times \cos^2 180°$$
$$= 2.035 \times 10^6 \text{ mm}^4 = 203.5 \text{ cm}^4$$
$$I_{x1y1} = (142.5 \times 15) \times (-142.5/2 + 45.1) \times 15.1 + (142.5^3 \times 15/12)$$
$$\times \sin 90° \cos 90° + (82.5 \times 15) \times (45.1) \times (-82.5/2 + 15.1)$$
$$+ (82.5^3 \times 15/12) \times \sin 180° \cos 180°$$
$$= -2.303 \times 10^6 \text{ mm}^4 = -230.3 \text{ cm}^4$$

Using equation 5.63,

$$\tan 2\alpha = -2 \times (-230.3)/(759.6 - 203.5) = 0.8283$$

whence $2\alpha = 39.63°$ and $\alpha = 19.82°$.
  Using Fig. 5.30b, the Mohr's circle centre is at

$$(I_{x1} + I_{y1})/2 = (759.6 + 203.5)/2 = 481.6 \text{ cm}^4$$

and the Mohr's circle diameter is

$$\sqrt{\{(I_{x1} - I_{y1})^2 + (2I_{x1y1})^2\}} = \sqrt{\{(759.6 - 203.5)^2 + (2 \times 230.3)^2\}}$$
$$= 722.1 \text{ cm}^4$$

and so

$$I_{x1} = 481.6 + 722.1/2 = 842.6 \text{ cm}^4$$
$$I_{y1} = 481.6 - 722.1/2 = 120.5 \text{ cm}^4$$

5.12.4 EXAMPLE 4 – PROPERTIES OF A ZED SECTION

Problem.   Determine the properties I_{x1}, I_{y1}, I_{x1y1}, I_x, I_y, and α of the thin-walled Z-section shown in Fig. 5.42a.

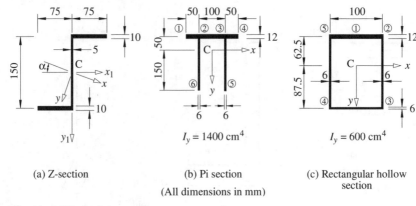

(a) Z-section   (b) Pi section   (c) Rectangular hollow
                                                      section

(All dimensions in mm)

Fig. 5.42 Worked examples 4, 9, 10, and 11

Solution.

$I_{x1} = 2 \times (75 \times 10 \times 75^2) + (150^3 \times 5/12) = 9844 \times 10^3 \text{ mm}^4 = 984.4 \text{ cm}^4$,
$I_{y1} = 2 \times (75 \times 10 \times 37.5^2) + 2 \times (75^3 \times 10/12) = 2813 \times 10^3 \text{ mm}^4 = 281.3 \text{ cm}^4$,
$I_{x1y1} = 75 \times 10 \times 37.5 \times (-75) + 75 \times 10 \times (-37.5) \times 75$
$= -4219 \times 10^3 \text{ mm}^4 = -421.9 \text{ cm}^4$.

The centre of Mohr's circle (see 5.37b) is at

$$\frac{I_{x1} + I_{y1}}{2} = \left[\frac{984.4 + 281.3}{2} \right] = 632.8 \text{ cm}^4.$$

The diameter of the circle is

$$10^3 \times \sqrt{[(984.4 - 281.3)^2 + (2 \times 4.219)^2]} = 1098.3 \text{ cm}^4,$$

and so:

$$I_x = 632.8 + 1098.3/2 = 1182 \text{ cm}^4,$$
$$I_y = 632.8 - 1098.3/2 = 83.7 \text{ cm}^4,$$
$$\tan 2\alpha = \frac{2 \times 421.9}{984.4 - 281.3} = 1.200, \qquad \alpha = 25.1°.$$

5.12.5 EXAMPLE 5 – BIAXIAL BENDING OF AN ANGLE SECTION CANTILEVER

Problem. Determine the maximum elastic stress and deflection of the angle section cantilever shown in Fig. 5.41 whose section properties were determined in section 5.12.3.

Solution for maximum stress. The shear centre (see section 5.4.3) load of 6 kN acting in the plane of the long leg of the angle has principal plane components of:

$$6 \cos 19.82° = 5.645 \text{ kN parallel to the } y \text{ axis},$$
$$6 \sin 19.82° = 2.034 \text{ kN parallel to the } x \text{ axis},$$

as indicated in Fig. 5.41d. These load components cause principal axis moment components at the support of:

$$M_x = -5.645 \times 1.5 = -8.467 \text{ kNm about the } x \text{ axis},$$
$$M_y = +2.034 \times 1.5 = +3.052 \text{ kNm about the } y \text{ axis}.$$

The maximum elastic bending stress occurs at the point A ($x_{1A} = 15.1$ mm, $y_{1A} = -97.4$ mm) shown in Fig. 5.41d. The coordinates of this point can be obtained by using equation 5.61 as:

$$x = 15.1 \cos 19.82° - 97.4 \sin 19.82° = -18.8 \text{ mm},$$
$$y = -15.1 \sin 19.82° - 97.4 \cos 19.82° = -96.7 \text{ mm}.$$

The maximum stress at A can be obtained by using equation 5.12 as

$$f_{max} = \frac{(-8.467 \times 10^6) \times (-96.7)}{842.6 \times 10^4} - \frac{(3.052 \times 10^6) \times (-18.8)}{120.5 \times 10^4} = 146.9 \text{ N/mm}^2.$$

The stresses at the other leg ends should be checked to confirm that the maximum stress is at A.

Solution for maximum deflection. The maximum deflection occurs at the tip of the cantilever. Its components u and v can be calculated using $QL^3/3EI$. Thus

$$v = (5.645 \times 10^3) \times (1500)^3/(3 \times 205\,000 \times 842.6 \times 10^4) = 3.7 \text{ mm},$$
$$u = (2.034 \times 10^3) \times (1500)^3/(3 \times 205\,000 \times 120.5 \times 10^4) = 9.3 \text{ mm}.$$

The resultant deflection can be obtained by vector addition as

$$\delta = \sqrt{(3.7^2 + 9.3^2)} = 10.0 \text{ mm}.$$

Using non-principal plane properties. Problems of this type are sometimes incorrectly analysed on the basis that the plane of loading can be assumed to be a principal plane. When this approach is used, the maximum stress calculated for the cantilever is

$$f_{max} = \frac{(6 \times 10^3) \times 1500 \times (97.4)}{759.6 \times 10^4} = 115 \text{ N/mm}^2,$$

which seriously underestimates the correct value by more than 20%. The correct non-principal plane formulations which are needed to solve the example in this way are given in section 5.8.2. The use of these formulations should be avoided since their excessive complication makes them very error-prone. The much simpler principal plane formulations of equations 5.9–5.12 should be used instead.

It should be noted that in this book x and y are reserved exclusively for the principal axis directions, instead of also being used for directions parallel to the legs.

5.12.6 EXAMPLE 6 – SHEAR STRESSES IN AN I-SECTION

Problem.   Determine the shear flow distribution and the maximum shear stress in the I-section shown in Fig. 5.12a.

Solution.   Applying equation 5.16 to the left hand half of the top flange,

$$\tau_v t_f = -(V_y/I_x)\int_0^{s_f}(-d_f/2)t_f ds_f = (V_y/I_x)d_f t_f s_f/2,$$

which is linear, as shown in Fig. 5.12b.

At the flange web junction,

$$(\tau_v t_f)_{12} = (V_y/I_x)d_f t_f b/4,$$

which is positive, and so the shear flow acts into the flange–web junction, as shown in Fig. 5.12a. Similarly, for the right hand half of the top flange

$$(\tau_v t_f)_{23} = (V_y/I_x)d_f t_f b/4,$$

and again the shear flow acts into the flange–web junction.

For horizontal equilibrium at the flange–web junction, the shear flow from the junction into the web is obtained from equation 5.20 as

$$(\tau_v t_w)_{24} = (V_y/I_x)d_f t_f b/4 + (V_y/I_x)d_f t_f b/4 = (V_y/I_x)d_f t_f b/2.$$

Adapting equation 5.16 for the web,

$$
\begin{aligned}
\tau_v t_w &= (V_y/I_x)(d_f t_f b/2) - (V_y/I_x)\int_0^{s_w}(-d_f/2 + s_w)t_w ds_w \\
&= (V_y/I_x)(d_f t_f b/2) + (V_y/I_x)(d_f s_w/2 - s_w^2/2)t_w,
\end{aligned}
$$

which is parabolic, as shown in Fig. 5.12b. At the web centre, $(\tau_v t_w)_{d_f/2} = (V_y/I_x)(d_f t_f b/2 + d_f^2 t_w/8)$ which is the maximum shear flow.

At the bottom of the web,

$$
\begin{aligned}
(\tau_v t_w)_{d_f} &= (V_y/I_x)(d_f t_f b/2 + d_f^2 t_w/2 - d_f^2 t_w/2) \\
&= (V_y/I_x)(d_f t_f b/2) = (\tau_v t_w)_{24},
\end{aligned}
$$

which provides a symmetry check.

The maximum shear stress is

$$(\tau_v)_{d_f/2} = (V_y/I_x)(d_f t_f b/2 + d_f^2 t_w/8)/t_w.$$

A vertical equilibrium check is provided by finding the resultant of the web shear flow as

$$
\begin{aligned}
V_w &= \int_0^{d_f}(\tau_v t_w)ds_w \\
&= (V_y/I_x)[d_f t_f b s_w/2 + d_f t_w s_w^2/4 - s_w^3 t_w/6]_0^{d_f} \\
&= (V_y/I_x)(d_f^2 t_f b/2 + d_f^3 t_w/12) \\
&= V_y,
\end{aligned}
$$

when $I_x = d_f^2 t_f b/2 + d_f^3 t_w/12$ is substituted.

5.12.7 EXAMPLE 7 – SHEAR STRESS IN A CHANNEL SECTION

Problem.   Determine the shear flow distribution in the channel section shown in Fig. 5.14.

Solution.   Applying equation 5.15 to the top flange,

$$\tau_v t_f = -(V_y/I_x)\int_0^{s_f}(-d_f/2)t_f\,ds_f = (V_y/I_x)d_f t_f s_f/2,$$
$$(\tau_v t_f)_b = (V_y/I_x)d_f t_f b/2,$$
$$\tau_v t_w = (V_y/I_x)(d_f t_f b/2 - \int_0^{s_w}(-d_f/2 + s_w)t_w\,ds_w)$$
$$= (V_y/I_x)(d_f t_f b/2 + d_f s_w t_w/2 - s_w^2 t_w/2),$$
$$(\tau_v t_w)_{d_f} = (V_y/I_x)(d_f t_f b/2 + d_f^2 t_w/2 - d_f^2 t_w/2)$$
$$= (V_y/I_x)(d_f t_f b/2) = (\tau_v t_f)_b,$$

which provides a symmetry check.

A vertical equilibrium check is provided by finding the resultant of the web flows as

$$V_w = \int_0^{d_f}(\tau_v t_w)ds_w$$
$$= (V_y/I_x)[d_f t_f b s_w/2 + d_f s_w^2 t_w/4 - s_w^3 t_w/6]_0^{d_f}$$
$$= (V_y/I_x)(d_f^2 t_f b/2 + d_f^3 t_w/12)$$
$$= V_y,$$

when $I_x = d_f^2 t_f b/2 + d_f^3 t_w/12$ is substituted.

5.12.8 EXAMPLE 8 – SHEAR CENTRE OF A CHANNEL SECTION

Problem.   Determine the position of the shear centre of the channel section shown in Fig. 5.14.

Solution.   The shear flow distribution in the channel section was analysed in section 5.12.7 and is shown in Fig. 5.14a. The resultant flange shear forces shown in Fig. 5.14b are obtained from

$$V_f = \int_0^b (\tau_v t_f)ds_f$$
$$= (V_y/I_x)[d_f t_f s_f^2/4]_0^b$$
$$= (V_y/I_x)(d_f t_f b^2/4),$$

while the resultant web shear force is $V_w = V_y$ (see section 5.12.7).

These resultant shear forces are statically equivalent to a shear force V_y acting through the point S which is a distance

$$a = \frac{V_f d_f}{V_w} = \frac{d_f^2 t_f b^2}{4I_x}$$

from the web, and so

$$x_0 = \frac{d_f^2 t_f b^2}{4I_x} + \frac{b^2 t_f}{2bt_f + d_f t_w}.$$

5.12.9 EXAMPLE 9 – SHEAR STRESSES IN A PI-SECTION

Problem. The pi shaped section shown in Fig. 5.42b has a shear force of 100 kN acting parallel to the x axis. Determine the shear stress distribution.

Solution. Using equation 5.23, the shear flow in the segment 12 is

$$(\tau_h \times 12)_{12} = \frac{-100 \times 10^3}{1400 \times 10^4} \int_0^{s_1} (-100 + s_1) \times 12 \times ds_1,$$

whence

$$(\tau_h)_{12} = 0.007143 \times (100s_1 - s_1^2/2) \text{ N/mm}^2.$$

Similarly, the shear flow in the segment 62 is

$$(\tau_h \times 6)_{62} = \frac{-100 \times 10^3}{1400 \times 10^4} \int_0^{s_6} (-50) \times 6 \times ds_6,$$

whence

$$(\tau_h)_{62} = 0.007143 \times (50s_6) \text{ N/mm}^2.$$

For horizontal equilibrium of the longitudinal shear forces at junction 2, the shear flows in the segments 12, 62, and 23 must balance (equation 5.20), and so

$$[(\tau_h \times 12)_{23}]^{s_3=0} = [(\tau_h \times 12)_{12}]^{s_1=50} + [(\tau_h \times 6)_{62}]^{s_6=200}$$

$$= 0.007143 \left[\left(100 \times 50 - \frac{50^2}{2}\right) \times 12 + 50 \times 200 \times 6 \right]$$

$$= 0.007143 \times 8750.$$

The shear flow in the segment 23 is

$$(\tau_h \times 12)_{23} = 0.007143 \times 12 \times 8750$$

$$- \frac{100 \times 10^3}{1400 \times 10^4} \times \int_0^{s_2} (-50 + s_2) \times 12 \times ds_2,$$

whence

$$(\tau_h)_{23} = 0.007143(8750 + 50s_2 - s_2^2/2) \text{ N/mm}^2.$$

To verify this, the resultant of the shear stresses in segments 12, 23 and 34 is

$$\int \tau_h t \, ds = 2 \times 0.007143 \times 12 \int_0^{50} (100s_1 - s_1^2/2)ds_1$$

$$+ 0.007143 \times 12 \times \int_0^{100} (8750 + 50s_2 - s_2^2/2)ds_2$$

$$= 0.007143 \times 12\{2 \times (100 \times 50^2/2 - 50^3/6)$$
$$+ (8750 \times 100 + 50 \times 100^2/2 - 100^3/6)\} \text{ N}$$
$$= 100 \text{ kN},$$

which is equal to the shear force.

5.12.10 EXAMPLE 10 – SHEAR CENTRE OF A PI-SECTION

Problem. Determine the position of the shear centre of the pi-section shown in Fig. 5.42b.

Solution. The resultant of the shear stresses in the segment 62 is

$$\int \tau_h t \, ds = 0.007143 \times 6 \int_0^{200} (50s_6) \times ds_6$$
$$= 0.007143 \times 6 \times 50 \times 200^2/2 \text{ N}$$
$$= 42.86 \text{ kN}.$$

The resultant of the shear stresses in segment 53 is equal and opposite to this.

The torque about the centroid exerted by the shear stresses is equal to the sum of the torques of the force in flange 1234 and the forces in webs 62 and 53. Thus

$$\int_A \tau_h t\rho \, ds = (100 \times 50) + (2 \times 42.86 \times 50) \text{ kNmm}$$
$$= 9286 \text{ kNmm}.$$

For the applied shear to be statically equivalent to the shear stresses, it must exert an equal torque about the centroid, and so

$$-100 \times y_0 = 9286,$$

whence

$$y_0 = -92.9 \text{ mm}.$$

Because of symmetry, the shear centre lies on the y axis, and so $x_0 = 0$.

5.12.11 EXAMPLE 11 – SHEAR STRESSES IN A BOX SECTION

Problem. The rectangular box section shown in Fig. 5.42c has a shear force of 100 kN acting parallel to the x axis. Determine the maximum shear stress.

Solution. If initially a slit is assumed at point 1 on the axis of symmetry so that $(\tau_{ho}t)_1 = 0$, then

$$(\tau_{ho}t)_{12} = \frac{-100 \times 10^3}{6 \times 10^6} \int_0^{s_1} s_1 \times 12 \times ds_1 = -0.1s_1^2,$$

$$(\tau_{ho}t)_2 = -0.1 \times 50^2 = -0.1 \times 2500 \text{ N/mm},$$

$$(\tau_{ho}t)_{23} = -0.1 \times 2500 - \frac{100 \times 10^3}{6 \times 10^6} \int_0^{s_2} 50 \times 6 \times ds_2 = -0.1(2500 + 50s_2),$$

$$(\tau_{ho}t)_3 = -0.1(2500 + 50 \times 150) = -0.1 \times 10\,000 \text{ N/mm},$$

$$(\tau_{ho}t)_{34} = -0.1 \times 10\,000 - \frac{100 \times 10^3}{6 \times 10^6} \int_0^{s_3} (50 - s_3) \times 6 \times ds_3$$

$$= -0.1[10\,000 + (50s_3 - s_3^2/2)]$$

$$(\tau_{ho}t)_4 = -0.1[10\,000 + (50 \times 100 - 100^2/2)] = -0.1 \times 10\,000 \text{ N/mm}$$

$$= (\tau_{ho}t)_3,$$

which is a symmetry check.

The remainder of the shear stress distribution can therefore be obtained by symmetry.

The circulating shear flow $\tau_{hc}t$ can be determined by adapting equation 5.30. For this

$$\oint \tau_{ho}\, ds = 2 \times (0.1 \times 50^3)/3)/12$$

$$+ 2 \times (-0.1) \times (2500 \times 150 + 50 \times 150^2/2)/6$$

$$+ (-0.1) \times (10\,000 \times 100 + 50 \times 100^2/2 - 100^3/6)/6$$

$$= (-0.1) \times (6\,000\,000)/12 \text{ N/mm},$$

$$\oint \frac{1}{t}\, ds = \frac{100}{12} + \frac{150}{6} + \frac{100}{6} + \frac{150}{6}$$

$$= 900/12.$$

Thus

$$\tau_{hc}t = (-0.1) \times \frac{(6\,000\,000)}{12} \times \frac{12}{900}$$

$$= 666.7 \text{ N/mm}.$$

The maximum shear stress occurs at the centre of the segment 34, and adapting equation 5.28,

$$(\tau_h)_{max} = \{-0.1 \times (10\,000 + 50 \times 50 - 50^2/2) + 666.7\}/6 \text{ N/mm}^2$$

$$= -76.4 \text{ N/mm}^2.$$

5.12.12 EXAMPLE 12 – FULLY PLASTIC MOMENT OF A PLATED UB

Problem. A 610×229 UB 125 beam (Fig. 5.40a) of S275 steel is strengthened by welding a 300 mm $\times$ 20 mm plate of S275 steel to each flange. Compare the full plastic moments of the unplated and the plated sections.

Unplated section.   With $T = 19.6$ mm, Table 9 of BS5950 gives $p_y = 265$ N/mm^2.

$$M_{pu} = p_y S_x = 265 \times 3673 \times 10^3 \text{ Nmm} = 973 \text{ kNm.}$$

Plated section.   With $t_p = 20$ mm, Table 9 of BS5950 gives $p_y = 265$ N/mm^2. Because of the symmetry of the cross-section, the plastic neutral axis is at the centroid. Adapting equation 5.67,

$$M_{pp} = \{973 + 2 \times 265 \times (300 \times 20) \times (611.9/2 + 20/2)/10^6\} \text{ kNm}$$
$$= 1978 \text{ kNm} \approx 2 \times M_{pu}.$$

5.12.13 EXAMPLE 13 – FULLY PLASTIC MOMENT OF A TEE-SECTION

Problem.   The welded tee-section shown in Fig. 5.40b is fabricated from two S275 steel plates. Compare the first yield and fully plastic moments.

First Yield Moment.   The minimum elastic section modulus was determined in section 5.12.2 as $Z_{x \text{ min}} = 16.93$ cm^3. With $t_{max} = 10$ mm, Table 9 of BS5950 gives $p_y = 275$ N/mm^2. $M_{Yx \text{ min}} = p_y Z_{x \text{ min}} = 275 \times 16.93 \times 10^3$ Nmm $= 4.655$ kNm.

Full Plastic Moment.   For the plastic neutral axis to divide the cross-section into two equal areas, using the thin-walled assumption leads to

$$(80 \times 5) + (y_p \times 10) = (82.5 - y_p) \times 10$$

so that

$$y_p = (82.5 \times 10 - 80 \times 5)/(2 \times 10) = 21.3 \text{ mm}$$

Using equation 5.67,

$$M_{px} = 275 \times [(80 \times 5) \times 21.3 + (21.3 \times 10)$$
$$\times 21.3/2 + (82.5 - 21.3)^2 \times 10/2] \text{ Nmm}$$
$$= 8.117 \text{ kNm}$$

Alternatively, the section properties $Z_{x \text{ min}}$ and S_x can be calculated using Fig. 5.6, and used to calculate $M_{Yx \text{ min}}$ and M_{px}.

5.12.14 EXAMPLE 14 – PLASTIC COLLAPSE OF A NON-UNIFORM BEAM

Problem.   The two span continuous beam shown in Fig. 5.43a is a 610×229 UB 125 of S275 steel with two flange plates 300 mm $\times$ 20 mm of S275 steel extending over a central length of 12 m. Determine the value of the applied loads Q at plastic collapse.

Solution.   The full plastic moments of the beam (unplated and plated) were determined in section 5.12.12 as $M_{pu} = 973$ kNm and $M_{pp} = 1978$ kNm.

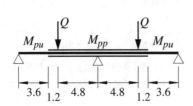

(a) Continuous beam
(lengths in m)

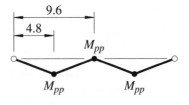

(c) First mechanism

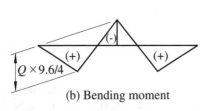

(b) Bending moment

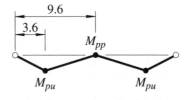

(d) Second mechanism

Fig. 5.43 Worked example 14

The bending moment diagram is shown in Fig. 5.43b. Two plastic collapse mechanisms are possible, both with a plastic hinge $(-M_{pp})$ at the central support. For the first mechanism (Fig. 5.43c), plastic hinges (M_{pp}) occur in the plated beam at the load points but for the second mechanism (Fig. 5.43d), plastic hinges (M_{pu}) occur in the unplated beam at the changes of cross-section.

For the first mechanism shown in Fig. 5.43c, inspection of the bending moment diagram shows that

$$M_{pp} + M_{pp}/2 = Q_1 \times 9.6/4$$

so that

$$Q_1 = (3 \times 1978/2) \times 4/9.6 = 1236 \text{ kN}.$$

For the second mechanism shown in Fig. 5.43d, inspection of the bending moment diagram shows that

$$M_{pu} + 3.6M_{pp}/9.6 = (3.6/4.8) \times Q_2 \times 9.6/4,$$

so that

$$Q_2 = [973 + (3.6/9.6) \times 1978] \times (4.8/3.6) \times 4/9.6 = 953 \text{ kN},$$

which is less than $Q_1 = 1236$ kN. Thus plastic collapse occurs at $Q = 953$ kN by the second mechanism.

5.12.15 EXAMPLE 15 – CHECKING A SIMPLY SUPPORTED BEAM

Problem. The simply supported 610×229 UB 125 of S275 steel shown in Fig. 5.44a has a span of 6.0 m and is laterally braced at 1.5 m intervals. Check the

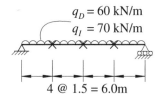

(a) Example 15

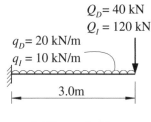

(c) Example 16

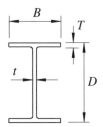

D = 611.9 mm T = 19.6 mm
B = 229 mm t = 11.9 mm
r = 12.7 mm
S_x = 3673 cm^3
r_y = 49.7 mm
I_x = 98500 cm^4
(b) 610 x 229 UB 125

Fig. 5.44 Worked examples 15 and 16

adequacy of the beam for a nominal uniformly distributed dead load of 60 kN/m together with a nominal uniformly distributed imposed load of 70 kN/m.

Classifying the section.

$$T = 19.6 \text{ mm}, p_y = 265 \text{ N/mm}^2 \hfill \text{T9}$$
$$\varepsilon = \surd(275/265) = 1.019 \hfill \text{T11}$$
$$b/(T\varepsilon) = (229/2)/(19.6 \times 1.019)$$
$$= 5.73 < 9 \text{ and the flange is plastic.} \hfill \text{T11}$$
$$d/(t\varepsilon) = (611.9 - 2 \times 19.6 - 2 \times 12.7)/(11.9 \times 1.019)$$
$$= 45.1 < 80 \text{ and the web is plastic.} \hfill \text{T11}$$

(Note the general use of the minimum p_y obtained for the flange.)

Checking for moment.

$$q^* = (1.4 \times 60) + (1.6 \times 70) = 196 \text{ kN/m} \hfill \text{T2}$$
$$M^* = 196 \times 6^2/8 = 882 \text{ kNm}$$
$$M_c = 265 \times 3673 \times 10^4 \text{ Nmm} \hfill 4.2.5.2$$
$$= 973.3 \text{ kNm} > 882 \text{ kNm} = M^*$$

which is satisfactory.

Checking for lateral bracing.

$$u = 0.9 \hfill 4.3.6.8$$
$$\lambda = L/r_y = 1500/49.7 = 30.2 \hfill 4.3.6.7$$
$$x = D/T = 611.9/19.6 = 31.2 \hfill 4.3.6.8$$
$$\lambda/x = 30.2/31.2 = 0.967$$
$$\eta = 0.5 \hfill 4.3.6.7$$

$$v = 0.99 \qquad \text{T19}$$
$$\beta_w = 1.0 \qquad \text{4.3.6.9}$$
$$\lambda_{LT} = uv\lambda\sqrt{\beta_w} = 0.9 \times 0.99 \times 30.2 \times \sqrt{(1.0)} = 26.9 \qquad \text{4.3.6.7}$$
$$\lambda_{L0} = 35.0 \qquad \text{T16}$$

or using equation 5.42,

$$\lambda_{L0} = 34.3\sqrt{(275/265)}$$
$$= 34.9 > 26.9 = \lambda_{LT} \text{ and the bracing is satisfactory.}$$

Checking for shear.

$$F_v^* = 196 \times 6/2 = 588 \text{ kN}$$
$$P_v = 0.6 \times 265 \times (611.9 \times 11.9) \text{ N} \qquad \text{4.2.3}$$
$$= 1158 \text{ kN} > 588 \text{ kN} = F_v^*$$

which is satisfactory.

Checking for bending and shear. The maximum M^* occurs at mid-span where $F_v^* = 0$, and the maximum F_v^* occurs at the support where $M^* = 0$, and so there is no need to check for combined bending and shear. (Note that in any case, $0.6P_v = 0.6 \times 1158 = 694.7 \text{ kN} > 588 \text{ kN} = F_v^*$ and so the combined bending and shear condition does not operate.)

Checking for bearing.

$$F_x^* = 196 \times 6/2 = 588 \text{ kN}$$

The stiff bearing length b_1 required can be determined from the bearing yield condition

$$F_x^* \leqslant P_{bw} = (b_1 + nk)tp_{yw} \qquad \text{4.5.2.1}$$
$$t = 11.9 \text{ mm}, p_{yw} = 275 \text{ N/mm}^2 \qquad \text{T9}$$
$$k = T + r = 19.6 + 12.7 = 32.3 \text{ mm} \qquad \text{4.5.2.1}$$

If the end of the stiff bearing is at the end of the member, then $b_e = 0$ and $n = 2$. Thus

$$b_1 \geqslant \frac{588 \times 10^3}{11.9 \times 275} - 2 \times 32.3 = 115.1 \text{ mm} \qquad \text{4.5.2.1}$$

The web buckling bearing width b_1 required can be determined from the bearing buckling condition

$$F_x^* \leqslant P_x = 25\varepsilon t P_{bw}/\sqrt{\{(b_1 + nk)d\}} \qquad \text{4.5.3.1}$$

with

$$P_{bw} = (b_1 + nk)tp_{yw} \qquad \text{4.5.2.1}$$

Thus

$$\sqrt{(b_1 + nk)} \geqslant F_x^*\sqrt{d}/(25\varepsilon t^2 p_{yw})$$
$$\geqslant 588 \times 10^3 \times \sqrt{611.9}/\{25 \times \sqrt{(275/275)} \times 11.9^2 \times 275\}$$
$$= 14.94$$

so that

$$b_1 \geqslant 14.94^2 - 2 \times 32.3 = 158.6 \text{ mm} (> \text{the yield length of } 104.1 \text{ mm}).$$

5.12.16 EXAMPLE 16 – DESIGNING A CANTILEVER

Problem.   A 3.0 m long cantilever has the nominal imposed and dead loads (which include allowances for self-weight) shown in Fig. 5.44c. Design a suitable UB section in S275 steel if the cantilever has sufficient bracing to prevent lateral buckling.

Selecting a trial section. $M_x^* = 1.4 \times (40 \times 3 + 20 \times 3^2/2) + 1.6 \times (120 \times 3 + 10 \times 3^2/2) = 942$ kNm. Assume that $p_y = 265$ N/mm^2 and that the section is compact.

$$S_x \geqslant M_x^*/p_y = 942 \times 10^6/(265 \times 10^3) = 3555 \text{ cm}^3. \qquad 4.2.5.2$$

Choose a 610 × 229 UB 125 with $S_x = 3673$ cm$^3 > 3555$ cm^3.

Checking the trial section.   The trial section must now be checked by classifying the section and checking for moment, shear, moment and shear, and bearing. This process is similar to that used in section 5.12.15 for checking a simply supported beam. If the section chosen fails any of the checks, then a new section must be chosen.

5.12.17 EXAMPLE 17 – CHECKING A PLASTICALLY ANALYSED BEAM

Problem.   Check the adequacy of the non-uniform beam shown in Fig. 5.43 and analysed plastically in section 5.12.14. The beam and its flange plates are of S275 steel, and the concentrated loads have nominal dead load components of 250 kN (which include allowances for self-weight), and imposed load components of 400 kN.

Classifying the sections.
The 610 × 229 UB 125 was found to be plastic in section 5.12.15.
    For the 300 × 20 plate,

$$t_p = 20 \text{ mm}, p_y = 265 \text{ N/mm}^2 \qquad \qquad \text{T9}$$

and

$$\varepsilon = \sqrt{(275/265)} = 1.019 \qquad \qquad \text{T11}$$

For the 300 × 20 plate as two outstands

$$\begin{aligned} b_p/(t_p\varepsilon) &= (300/2)/(19.6 \times 1.019) & \text{Fig. 6(a)} \\ &= 7.51 < 9 \text{ and the outstands are plastic.} & \text{T11} \end{aligned}$$

For the internal element of the 300 × 20 plate,

$b_p/(t_p\varepsilon) = 229/(20 \times 1.019)$ Fig. 6(b)
       $= 11.24 < 28$ and the internal element is plastic.   T11

Thus both the unplated and the plated sections are plastic.

Checking for plastic collapse.   The design loads are $Q^* = (1.4 \times 250) + (1.6 \times 400) = 920$ kN. The plastic collapse load based on the nominal full plastic moments of the beam was calculated in section 5.12.14 as $Q = 953$ kN, and so the resistance is adequate.

Checking for lateral bracing.   For the unplated segment, using equation 5.48,

$L_m < (38 \times 49.7) \times (36 \times 19.6/611.9) \times (275/265) = 2260$ mm   5.3.3(a)

Thus a brace should be provided at the change of section (a plastic hinge location), and a second brace between this and the support.

For the plated segment,

$$r_y = \sqrt{\left\{ \frac{(3932 \times 10^4) + (2 \times 300^3 \times 20/12)}{(159 \times 10^2) + (2 + 300 \times 20)} \right\}} = 68.1 \text{ mm}$$

and using equation 5.48,

$L_m < (38 \times 68.1) \times \{36 \times (19.6 + 20)/(611.9 + 2 \times 20)\} \times (275/265)$
       < 5873 mm.   5.3.3(a)

Thus a single brace should be provided in the plated segment. A satisfactory location is at the load point.

However, if the approximate method allowing for moment gradient of Clause 5.3.3(b) is used, it will be found that it is only necessary to provide a single brace in each span, at the change of section.

Checking for shear.   Under the collapse loads $Q = 953$ kN, the end reactions are

$$R_E = \{(953 \times 4.8) - 1978\}/9.6 = 270.5 \text{ kN}$$

and the central reaction is

$$R_C = 2 \times 953 - 2 \times 270.5 = 1365 \text{ kN}$$

so that the maximum shear at collapse is

$$F = 1365/2 = 682.5 \text{ kN}.$$

The design maximum shear is

$F_v^* = 682.5 \times (920/953) = 658.9$ kN
$P_v = 0.6 \times 265 \times (611.9 \times 11.9)$ N   4.2.3
       $= 1158$ kN > 658.9 kN $= F_v^*$,

which is satisfactory.

Checking for bending and shear.   At the central support,

$$F_v^* = 658.9 \text{ kN}$$

and

$$0.6P_v = 0.6 \times 1158 = 694.8 \text{ kN} > 658.9 \text{ kN} = F_v^*$$

and so there is no reduction in the shear capacity.   4.2.5.3

Checking for bearing.   Web stiffeners are required for beams analysed plastically at all plastic hinge points where the design shear exceeds 10% of the web capacity, or $0.1 \times 1158 = 115.8$ kN. Thus stiffeners are required at the hinge locations at the points of cross-section change and at the interior support. Load bearing stiffeners should also be provided at the load points but are not required at the outer supports (see section 5.12.15). An example of the design of load bearing stiffeners is given in section 4.9.9.

5.12.18 EXAMPLE 18 – SERVICEABILITY OF A SIMPLY SUPPORTED BEAM

Problem.   Check the imposed load deflection of the 610 × 229 UB 125 of Fig. 5.44a for a serviceability limit of $L/360$.

Solution.   The central deflection v_c of a simply supported beam with uniformly distributed load q can be calculated using

$$v_c = \frac{5qL^4}{384EI} = \frac{5 \times 70 \times 6000^4}{384 \times 205\,000 \times 98\,500 \times 10^4} = 5.8 \text{ mm.}$$

(The same result can be obtained using Fig. 5.3.)
    $L/360 = 6000/360 = 16.7$ mm > 5.8 mm $= v_c$ and so the beam is satisfactory.

5.13 Unworked examples

5.13.1 EXAMPLE 19 – SECTION PROPERTIES

Determine the principal axis properties I_x, I_y of the half box section shown in Fig. 5.45a.

5.13.2 EXAMPLE 20 – ELASTIC STRESSES AND DEFLECTIONS

The box section beam shown in Fig. 5.45b is to be used as a simply supported beam over a span of 20.0 m. The erection procedure proposed is to fabricate the beam as two half boxes (the right half as in Fig. 5.45a), to erect them separately, and to pull them together before making the longitudinal connections between the flanges. The total construction load is estimated to be 10 kN/m × 20 m.

(a) Determine the maximum elastic bending stress and deflection of one
    half box.

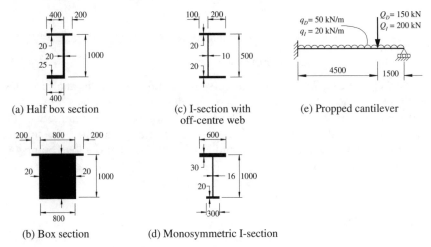

(a) Half box section

(c) I-section with off-centre web

(e) Propped cantilever

(b) Box section

(d) Monosymmetric I-section

Fig. 5.45 Unworked examples

(b) Determine the horizontal distributed force required to pull the two half boxes together.

(c) Determine the maximum elastic bending stress and deflection after the half boxes are pulled together.

5.13.3 EXAMPLE 21 – SHEAR STRESSES

Determine the shear stress distribution caused by a vertical shear force of 500 kN in the section shown in Fig. 5.45c.

5.13.4 EXAMPLE 22 – SHEAR CENTRE

Determine the position of the shear centre of the section shown in Fig. 5.45c.

5.13.5 EXAMPLE 23 – SHEAR CENTRE OF A BOX SECTION

Determine the shear flow distribution caused by a horizontal shear force of 2000 kN in the box section shown in Fig. 5.45b, and determine the position of its shear centre.

5.13.6 EXAMPLE 24 – PLASTIC MOMENT OF A MONOSYMMETRIC I-SECTION

Determine the fully plastic moment and the shape factor of the monosymmetric I-beam shown in Fig. 5.45d if the yield stress is 265 N/mm^2.

5.13.7 EXAMPLE 25 – ELASTIC DESIGN

Determine a suitable hot-rolled I-section member of S275 steel for the propped cantilever shown in Fig. 5.45e by using the elastic method of design.

5.13.8 EXAMPLE 26 – PLASTIC COLLAPSE ANALYSIS

Determine the plastic collapse mechanism of the propped cantilever shown in Fig. 5.45e.

5.13.9 EXAMPLE 27 – PLASTIC DESIGN

Use the plastic design method to determine a suitable hot-rolled I-section of S275 steel for the propped cantilever shown in Fig. 5.45e.

5.14 References

1. Popov, E.P. (1968) *Introduction to Mechanics of Solids*, Prentice-Hall Englewood Cliffs, New Jersey.
2. Hall, A.S. (1984) *An Introduction to the Mechanics of Solids*, 2nd edition, John Wiley, Sydney.
3. Pippard, A.J.S. and Baker, J.F. (1968) *The Analysis of Engineering Structures*, 4th edition, Edward Arnold, London.
4. Norris, C.H., Wilbur, J.B. and Utku, S. (1976) *Elementary Structural Analysis*, 3rd edition, McGraw-Hill, New York.
5. Harrison, H.B. (1973) *Computer Methods in Structural Analysis*, Prentice-Hall, Englewood Cliffs, New Jersey.
6. Harrison, H.B. (1990) *Structural Analysis and Design, Parts 1 and 2*, 2nd edition, Pergamon Press, Oxford.
7. Ghali, A. and Neville, A.M. (1989) *Structural Analysis – A Unified Classical and Matrix Approach*, Chapman and Hall, London.
8. Coates, R.C., Coutie, M.G. and Kong, F.K. (1988) *Structural Analysis*, 3rd edition, Van Nostrand Reinhold (UK), Wokingham.
9. Steel Construction Institute (1997) *Steelwork Design Guide to BS5950: Part 1: 1990, Volume 1 – Section Properties and Member Capacities*, 5th edition, SCI, Ascot.
10. Bridge, R.Q. and Trahair, N.S. (1981) Bending, shear, and torsion of thin-walled beams, *Steel Construction*, **15**, No. 1, 2–18.
11. Hancock, G.J. and Harrison, H.B. (1972) A general method of analysis of stresses in thin-walled sections with open and closed parts, *Civil Engineering Transactions*, Institution of Engineers, Australia, **CE14**, No. 2, pp. 181–8.
12. Papangelis, J.P. and Hancock, G.J. (1997) *THIN-WALL – Cross-section Analysis and Finite Strip Buckling Analysis of Thin-Walled Structures*, Centre for Advanced Structural Engineering, University of Sydney.
13. Timoshenko, S.P. and Goodier, J.N. (1970) *Theory of Elasticity*, 3rd edition, McGraw-Hill, New York.
14. Winter, G. (1940) Stress distribution in and equivalent width of flanges of wide,

thin-walled steel beams, *Technical Note 784*, National Advisory Committee for Aeronautics.

15. Abdel-Sayed, G. (1969) Effective width of steel deckplates in bridges, *Journal of the Structural Division, ASCE*, **95**, No. ST7, pp. 1459–74.

16. Malcolm, D.J. and Redwood, R.G. (1970) Shear lag in stiffened box girders, *Journal of the Structural Division, ASCE*, **96**, No. ST7, pp. 1403–19.

17. Kristek, V. (1983) Shear lag in box girders, *Plated Structures. Stability and Strength*, (ed. R. Narayanan), Applied Science Publishers, London, pp. 165–94.

18. Baker, J.F. and Heyman, J. (1969) *Plastic Design of Frames – 1. Fundamentals*, Cambridge University Press, Cambridge.

19. Heyman, J. (1971) *Plastic Design of Frames – 2. Applications*, Cambridge University Press, Cambridge.

20. Horne, M.R. (1978) *Plastic Theory of Structures*, 2nd edition, Pergamon Press, Oxford.

21. Neal, B.G. (1977) *The Plastic Methods of Structural Analysis*, 3rd edition, Chapman and Hall, London.

22. Beedle, L.S. (1958) *Plastic Design of Steel Frames*, John Wiley, New York.

23. Horne, M.R. and Morris, L.J. (1981) *Plastic Design of Low-Rise Frames*, Granada, London.

24. Davies, J.M. and Brown, B.A. (1996) *Plastic Design to BS5950*, Blackwell Science Ltd, Oxford.

6 Lateral buckling of beams

6.1 Introduction

In the discussion given in Chapter 5 of the in-plane behaviour of beams, it was assumed that when a beam is loaded in its stiffer principal plane, it deflects only in that plane. If the beam does not have sufficient lateral stiffness or lateral support to ensure that this is so, then it may buckle out of the plane of loading, as shown in Fig. 6.1. The load at which this buckling occurs may be

Fig. 6.1 Lateral buckling of a cantilever

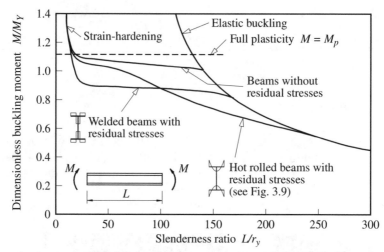

Fig. 6.2 Lateral buckling strengths of simply supported I-beams

substantially less than the beam's in-plane load carrying capacity, as indicated in Fig. 6.2.

For an idealized perfectly straight elastic beam, there are no out-of-plane deformations until the applied moment M reaches the elastic buckling moment M_{ob}, when the beam buckles by deflecting laterally and twisting, as shown in Fig. 6.1. These two deformations are interdependent: when the beam deflects laterally, the applied moment has a component which exerts a torque about the deflected longitudinal axis which causes the beam to twist. This behaviour, which is important for long unrestrained I-beams whose resistances to lateral bending and torsion are low, is called elastic flexural–torsional buckling.

The failure of a perfectly straight slender beam is initiated when the additional stresses induced by elastic buckling cause first yield. However, a perfectly straight beam of intermediate slenderness may yield before the elastic buckling moment is reached, because of the combined effects of the in-plane bending stresses and any residual stresses, and may subsequently buckle inelastically, as indicated in Fig. 6.2. For very stocky beams, the inelastic buckling moment may be higher than the in-plane plastic collapse moment M_p, in which case the moment capacity of the beam is not affected by lateral buckling.

In this chapter, the behaviour and design of beams which fail by lateral buckling and yielding are discussed. It is assumed that local buckling of the compression flange or of the web (which is dealt with in Chapter 4) does not occur. The behaviour and design of beams bent about both principal axes, and of beams with axial loads, are discussed in Chapter 7.

6.2 Elastic beams

6.2.1 BUCKLING OF STRAIGHT BEAMS

6.2.1.1 Simply supported beams with equal end moments

A perfectly straight elastic beam which is loaded by equal and opposite end moments is shown in Fig. 6.3. The beam is simply supported at its ends so that lateral deflection and twist rotation are prevented, while the flange ends are free to rotate in horizontal planes so that the beam ends are free to warp (see section 10.8.3). The beam will buckle at a moment M_{ob} when a deflected and twisted equilibrium position, such as that shown in Fig. 6.3, is possible. It is shown in section 6.10.1.1 that this position is given by

$$u = \frac{M_{ob}}{\pi^2 EI_y/L^2}\phi = \delta \sin\frac{\pi z}{L}, \tag{6.1}$$

where δ is the undetermined magnitude of the central deflection, and that the elastic buckling moment is given by

$$M_{ob} = M_{yz}, \tag{6.2}$$

where

$$M_{yz} = \sqrt{\left\{\left(\frac{\pi^2 EI_y}{L^2}\right)\left(GJ + \frac{\pi^2 EI_w}{L^2}\right)\right\}}, \tag{6.3}$$

where EI_y is the minor axis flexural rigidity, GJ the torsional rigidity, and EI_w the warping rigidity of the beam. Equation 6.3 shows that the resistance to

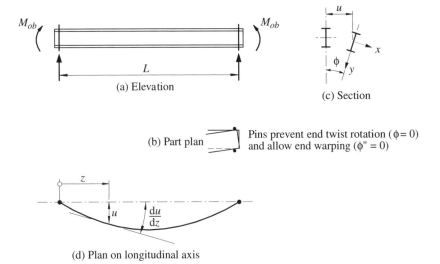

(a) Elevation

(c) Section

(b) Part plan

Pins prevent end twist rotation ($\phi = 0$) and allow end warping ($\phi'' = 0$)

(d) Plan on longitudinal axis

Fig. 6.3 Buckling of a simply supported I-beam

buckling depends on the geometric mean of the flexural resistance $(\pi^2 EI_y/L^2)$ and the torsional resistance $(GJ + \pi^2 EI_w/L^2)$.

Equation 6.3 ignores the effects of the major axis curvature $d^2v/dz^2 = -M_{ob}/EI_x$, and produces conservative estimates of the elastic buckling moment equal to $\sqrt{[(1 - EI_y/EI_x)\{1 - (GJ + \pi^2 EI_w/L^2)/2EI_x\}]}$ times the true value. This correction factor, which is just less than unity for most beam sections but may be significantly less than unity for column sections, is usually neglected in design. Nevertheless, its value approaches zero as I_y approaches I_x so that the true elastic buckling moment approaches infinity. Thus an I-beam in uniform bending about its weak axis does not buckle, which is intuitively obvious. Research [1] has indicated that in some other cases the correction factor may be close to unity, and that it is prudent to ignore the effect of major axis curvature.

6.2.1.2 Beams with unequal end moments

A simply supported beam with unequal major axis end moments M and $\beta_m M$ is shown in Fig. 6.4a. It is shown in section 6.10.1.2 that the value of the end moment M_{ob} at elastic flexural–torsional buckling can be expressed in the form of

$$M_{ob} = \alpha_m M_{yz} \tag{6.4}$$

in which the moment modification factor α_m which accounts for the effect of the non-uniform distribution of the major axis bending moment can be closely approximated by

$$\alpha_m = 1.75 + 1.05\beta_m + 0.3\beta_m^2 \leqslant 2.56, \tag{6.5}$$

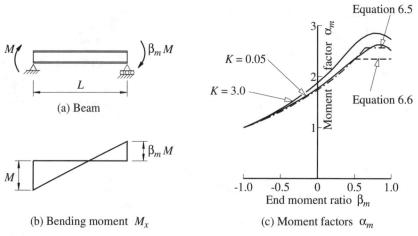

(a) Beam

(b) Bending moment M_x

(c) Moment factors α_m

Fig. 6.4 Buckling of beams with unequal end moments

or by

$$1/\alpha_m = 0.57 - 0.33\beta_m + 0.10\beta_m^2 \geqslant 0.43. \tag{6.6}$$

These approximations form the basis of a very simple method of predicting the buckling of the segments of a beam which is loaded only by concentrated loads applied through transverse members preventing local lateral deflection and twist rotation. In this case, each segment between load points may be treated as a beam with unequal end moments, and its elastic buckling moment may be estimated by using equation 6.4 and either equation 6.5 or 6.6 and by taking L as the segment length. Each buckling moment so calculated corresponds to a particular buckling load parameter for the complete load set, and the lowest of these parameters gives a conservative approximation of the actual buckling load parameter. This simple method ignores any buckling interactions between the segments. A more accurate method which accounts for these interactions is discussed in section 6.6.2.2.

6.2.1.3 Beams with central concentrated loads

A simply supported beam with a central concentrated load Q acting at a distance $-y_Q$ above the centroidal axis of the beam is shown in Fig. 6.5a. When the beam buckles by deflecting laterally and twisting, the line of action of the load moves with the central cross-section, but remains vertical, as shown in Fig. 6.5c. The case when the load acts above the centroid is more dangerous than that of centroidal loading because of the additional torque $-Qy_Q\phi_{L/2}$ which increases the twisting of the beam and decreases its resistance to buckling.

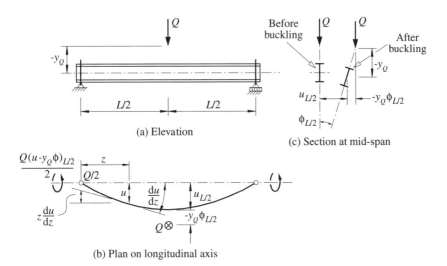

(a) Elevation

(c) Section at mid-span

(b) Plan on longitudinal axis

Fig. 6.5 I-beam with a central concentrated load

It is shown in section 6.10.1.3 that the dimensionless buckling load $QL^2/\sqrt{(EI_yGJ)}$ varies as shown in Fig. 6.6 with the beam parameter $K = \sqrt{(\pi^2 EI_w/GJL^2)}$ and the dimensionless height ε of the point of application of the load given by

$$\varepsilon = \frac{y_Q}{L}\sqrt{\left(\frac{EI_y}{GJ}\right)}. \tag{6.7}$$

For centroidal loading ($\varepsilon = 0$), the elastic buckling moment $M_{ob} = QL/4$ increases with the beam parameter K in much the same way as does the buckling moment of beams with equal and opposite end moments (see equation 6.3). Thus the elastic buckling moment can be written in the form

$$M_{ob} = QL/4 = \alpha_m M_{yz} \tag{6.8}$$

in which the moment modification factor α_m which accounts for the effect of the non-uniform distribution of major axis bending moment is approximately equal to 1.35.

The elastic buckling load also varies with the load height parameter ε, and although the resistance to buckling is high when the load acts below the centroidal axis, it decreases significantly as the point of application rises, as shown in Fig. 6.6. For equal flanged I-beams, the parameter ε can be transformed into

$$\frac{2y_Q}{d_f} = \frac{\pi}{K}\varepsilon, \tag{6.9}$$

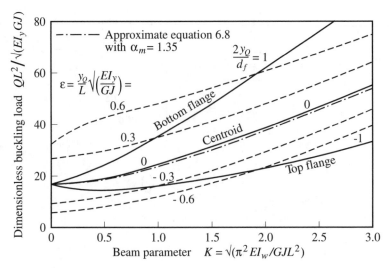

Fig. 6.6 Elastic buckling of beams with central concentrated loads

where d_f is the distance between flange centroids. The variation of the buckling load with $2y_Q/d_f$ is shown by the solid lines in Fig. 6.6, and it can be seen that the differences between top $(2y_Q/d_f = -1)$ and bottom $(2y_Q/d_f = 1)$ flange loading increase with the beam parameter K. This effect is therefore more important for deep beam-type sections of short span than for shallow column-type sections of long span. Approximate expressions for the variations of the moment modification factor α_m with the beam parameter K which account for the dimensionless load height $2y_Q/d_f$ for equal flanged I-beams are given in [2]. Alternatively, the maximum moment at elastic buckling $M_{ob} = QL/4$ may be approximated by using

$$\frac{M_{ob}}{M_{yz}} = \alpha_m \left\{ \sqrt{\left[1 + \left(\frac{0.4\alpha_m y_Q N_{oy}}{M_{yz}} \right)^2 \right]} + \frac{0.4\alpha_m y_Q N_{oy}}{M_{yz}} \right\}, \tag{6.10}$$

and $\alpha_m \approx 1.35$, in which

$$N_{oy} = \pi^2 EI_y/L^2. \tag{6.11}$$

6.2.1.4 Other loading conditions

The effect of the distribution of the applied load along the length of a simply supported beam on its elastic buckling strength has been investigated numerically by many methods, including those discussed in [3–5]. A particularly powerful computer method is the finite element method [6–10], while the finite integral method [11, 12], which allows accurate numerical solutions of the coupled minor axis bending and torsion equations to be obtained, has been used extensively. Many particular cases have been studied [13–16], and tabulations of elastic buckling loads are available [2, 3, 5, 13, 17], as is a user-friendly computer program [18] for analysing elastic flexural–torsional buckling.

Some approximate solutions for the maximum moments M_{ob} at elastic buckling of simply supported beams which are loaded along their centroidal axes are given in Fig. 6.7 by the moment modification factors α_m in the equation

$$M_{ob} = \alpha_m M_{yz}. \tag{6.12}$$

It can be seen that the more dangerous loadings are those which produce more nearly constant distributions of major axis bending moment, and that the worst case is that of equal and opposite end moments for which $\alpha_m = 1.0$.

For other beam loadings than those shown in Fig. 6.7, the moment modification factor α_m may be approximated by using

$$\alpha_m = \frac{1.75 M_m}{\sqrt{(M_2^2 + M_3^2 + M_4^2)}} \leqslant 2.5 \tag{6.13}$$

in which M_m is the maximum moment, M_2, M_4, the moments at the quarter points, and M_3 the moment at the mid-point of the beam.

Beam	Moment distribution	α_m	Range
$M\left(\text{⟿}\right)\beta_m M$	$\beta_m M$	$1.75 + 1.05\beta_m + 0.3\beta_m^2$	$-1 \le \beta_m \le 0.6$
		2.5	$0.6 < \beta_m \le 1$
$Q\downarrow \quad \downarrow Q$ $\vdash 2a \dashv$	$-\frac{QL}{2}(1-\frac{2a}{L})$	$1.0 + 0.35(1-2a/L)^2$	$0 \le \frac{2a}{L} \le 1$
$Q\downarrow$ $\vdash a \dashv$	$-\frac{QL}{4}\{1-(2a/L)^2\}$	$1.35 + 0.4(2a/L)^2$	$0 \le \frac{2a}{L} \le 1$
$Q\downarrow \quad \searrow \frac{3\beta_m QL}{16}$ $\vdash L/2 \vdash L/2 \dashv$	$3\beta_m QL/16$; $-\frac{QL}{4}(1-3\beta_m/8)$	$1.35 + 0.15\beta_m$	$0 \le \beta_m < 0.9$
		$-1.2 + 3.0\beta_m$	$0.9 \le \beta_m \le 1$
$\frac{\beta_m QL}{8}\left(\begin{array}{c}Q\downarrow\\ \vdash L/2 \vdash L/2 \dashv\end{array}\right)\frac{\beta_m QL}{8}$	$\beta_m QL/8$; $-\frac{QL}{4}(1-\beta_m/2)$	$1.35 + 0.36\beta_m$	$0 \le \beta_m \le 1$
q $\beta_m qL^2/8$	$\beta_m qL^2/8$; $-\frac{qL^2}{8}(1-\beta_m/4)^2$	$1.13 + 0.10\beta_m$	$0 \le \beta_m \le 0.7$
		$-1.25 + 3.5\beta_m$	$0.7 \le \beta_m \le 1$
$\frac{\beta_m qL^2}{12}\left(\begin{array}{c}q\\ \end{array}\right)\frac{\beta_m qL^2}{12}$	$\beta_m qL^2/12$; $-\frac{qL^2}{8}(1-2\beta_m/3)$	$1.13 + 0.12\beta_m$	$0 \le \beta_m \le 0.75$
		$-2.38 + 4.8\beta_m$	$0.75 \le \beta_m \le 1$

Fig. 6.7 Moment modification factors for simply supported beams

6.2.1.5 Cantilevers

The support conditions of cantilevers differ from those of simply supported beams in that a cantilever is usually completely fixed at one end and completely free at the other. The elastic buckling solution for a cantilever in uniform bending caused by an end moment M which rotates ϕ_L with the end of the cantilever [16] can be obtained from the solution given by equations 6.2 and 6.3 for simply supported beams by replacing the beam length L by twice the cantilever length $2L$, whence

$$M_{ob} = \sqrt{\left\{\left(\frac{\pi^2 EI_y}{4L^2}\right)\left(GJ + \frac{\pi^2 EI_w}{4L^2}\right)\right\}}. \tag{6.14}$$

This procedure is similar to the effective length method used to obtain the buckling load of a cantilever column (see Fig. 3.14).

Cantilevers with other loading conditions are not so easily analysed, but numerical solutions are available [16, 19–21]. The particular case of a cantilever with an end concentrated load Q is discussed in section 6.10.1.4, and plots of the dimensionless elastic buckling moments $QL^2/\sqrt{(EI_yGJ)}$ for bottom flange, centroidal, and top flange loading are given in Fig. 6.8, together with plots of the dimensionless elastic buckling moments $qL^3/2\sqrt{(EI_yGJ)}$ of cantilevers with uniformly distributed loads q.

6.2.2 BENDING AND TWISTING OF CROOKED BEAMS

Real beams are not perfectly straight, but have small initial curvatures and twists which cause them to bend and twist at the beginning of loading. If a

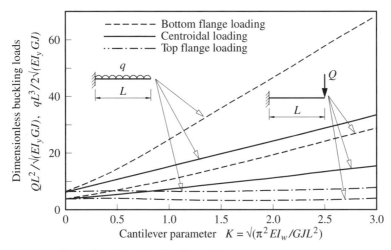

Fig. 6.8 Elastic buckling loads of cantilevers

simply supported beam with equal and opposite end moments M has an initial curvature and twist which are given by

$$\frac{u_0}{\delta_0} = \frac{\phi_0}{\theta_0} = \sin \frac{\pi z}{L} \tag{6.15}$$

in which the central initial lack of straightness δ_0 and twist rotation θ_0 are related by

$$\frac{\delta_0}{\theta_0} = \frac{M_{yz}}{\pi^2 EI_y/L^2}, \tag{6.16}$$

then the deformations of the beam are given by

$$\frac{u}{\delta} = \frac{\phi}{\theta} = \sin \frac{\pi z}{L} \tag{6.17}$$

in which

$$\frac{\delta}{\delta_0} = \frac{\theta}{\theta_0} = \frac{M/M_{yz}}{1 - M/M_{yz}}, \tag{6.18}$$

as shown in section 6.10.2. The variations of the dimensionless central deflection δ/δ_0 and twist θ/θ_0 are shown in Fig. 6.9, and it can be seen that deformation begins at the commencement of loading, and increases rapidly as the elastic buckling moment M_{yz} is approached.

The simple load–deformation relationships of equations 6.17 and 6.18 are of the same forms as those of equations 3.6 and 3.7 for compression members with sinusoidal initial curvature. It follows that the Southwell plot technique

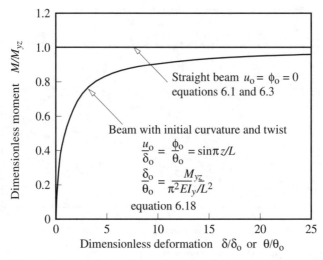

Fig. 6.9 Lateral deflection and twist of a beam with equal end moments

for extrapolating the elastic buckling loads of compression members from experimental measurements (see section 3.2.2) may also be used for beams.

As the deformations increase with the applied moments M, so do the stresses. It is shown in section 6.10.2 that the limiting moment M_L at which a beam without residual stresses first yields is given by

$$\frac{M_L}{M_Y} = \left(\frac{1.25 + M_{yz}/M_Y}{2}\right) - \sqrt{\left[\left(\frac{1.25 + M_{yz}/M_Y}{2}\right)^2 - \frac{M_{yz}}{M_Y}\right]} \quad (6.19)$$

in which $M_Y = f_y Z_x$ is the nominal first yield moment, when the central lack of straightness δ_0 is given by

$$\frac{\delta_0 N_{oy}}{M_{yz}} = \theta_0 = \frac{Z_y/Z_x}{1 + (d_f/2)(N_{oy}/M_{yz})} \frac{1}{4} \frac{M_Y}{M_{yz}} \quad (6.20)$$

in which N_{oy} is given by equation 6.11. Equation 6.19 is similar to equations 3.9 and 3.11 for the limiting axial force in an elastic compression member. The variation of the dimensionless limiting moment M_L/M_Y is shown in Fig. 6.10, in which the ratio $\sqrt{(M_Y/M_{yz})}$ plotted along the horizontal axis is equivalent to the modified slenderness ratio used in Fig. 3.4 for an elastic compression member. Figure 6.10 shows that the limiting moments of short beams approach the yield moment M_Y, while for long beams the limiting moments approach the elastic buckling moment M_{yz}.

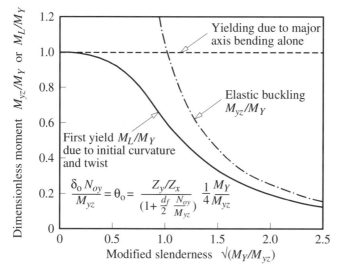

Fig. 6.10 Buckling and yielding of beams

6.3 Inelastic beams

The solution for the buckling moment M_{yz} of a perfectly straight simply supported I-beam with equal end moments given by equations 6.2 and 6.3 is only valid while the beam remains elastic. In a short span beam, yielding occurs before the ultimate moment is reached, and significant portions of the beams are inelastic when buckling commences. The effective rigidities of these inelastic portions are reduced by yielding, and consequently, the buckling moment is also reduced.

For beams with equal and opposite end moments ($\beta_m = -1$), the distribution of yield across the section does not vary along the beam, and when there are no residual stresses, the inelastic buckling moment can be calculated from a modified form of equation 6.3 as

$$M_I = \sqrt{\left\{\left[\frac{\pi^2(EI_y)_e}{L^2}\right]\left[(GJ)_e + \frac{\pi^2(EI_w)_e}{L^2}\right]\right\}} \tag{6.21}$$

in which the subscripted quantities ( )$_e$ are the reduced inelastic rigidities which are effective at buckling. Estimates of these rigidities can be obtained by using the tangent moduli of elasticity (see section 3.3.1) which are appropriate to the varying stress levels throughout the section. Thus the values of E and G are used in the elastic areas, while the strain-hardening moduli E_{st} and G_{st} are used in the yielded and strain-hardened areas (see section 3.3.4). When the effective rigidities calculated in this way are used in equation 6.21, a lower bound estimate of the buckling moment is determined (section 3.3.3). The variation of the dimensionless buckling moment M/M_Y with the ratio L/r_y of a typical stress-relieved rolled steel section is shown in Fig. 6.2. In the inelastic range, the buckling moment increases almost linearly with decreasing slenderness from the first yield moment $M_Y = f_y Z_x$ to the full plastic moment $M_p = f_y S_x$, which is reached after the flanges are fully yielded, and buckling is controlled by the strain-hardening moduli E_{st}, G_{st}.

The inelastic buckling moment of a beam with residual stresses can be obtained in a similar manner, except that the pattern of yielding is not symmetrical about the section major axis, so that a modified form of equation 6.69 for a monosymmetric I-beam must be used instead of equation 6.21. The inelastic buckling moment varies markedly with both the magnitude and the distribution of the residual stresses. The moment at which inelastic buckling initiates depends mainly on the magnitude of the residual compressive stresses at the compression flange tips, where yielding causes significant reductions in the effective rigidities $(EI_y)_e$ and $(EI_w)_e$. The flange tip residual stresses are comparatively high in hot-rolled beams, especially those with high ratios of flange to web area, and so inelastic buckling is initiated comparatively early in these beams, as shown in Fig. 6.2. The residual stresses in hot-rolled beams decrease away from the flange tips (see Fig. 3.9 for example), and so the extent of yielding increases and the effective rigidities steadily decrease as the applied moment increases. Because of this, the inelastic buckling moment decreases in an approximately linear fashion as the slenderness increases, as shown in Fig. 6.2.

In beams fabricated by welding flange plates to web plates, the compressive residual stresses at the flange tips, which increase with the welding heat input, are usually somewhat smaller than those in hot-rolled beams, and so the initiation of inelastic buckling is delayed, as shown in Fig. 6.2. However, the variations of the residual stresses across the flanges are nearly uniform in welded beams, and so, once flange yielding is initiated, it spreads quickly through the flange with little increase in moment. This causes large reductions in the inelastic buckling moments of stocky beams, as indicated in Fig. 6.2.

When a beam has a more general loading than that of equal and opposite end moments, the in-plane bending moment varies along the beam, and so when yielding occurs its distribution also varies. Because of this the beam acts as if non-uniform, and the torsion equilibrium equation becomes more complicated. Nevertheless, numerical solutions have been obtained for some hot-rolled beams with a number of different loading arrangements [22, 23], and some of these (for unequal end moments M and $\beta_m M$) are shown in Fig. 6.11, together with approximate solutions given by

$$\frac{M_I}{M_p} = 0.7 + \frac{0.3(1 - 0.7M_p/M_{ob})}{(0.61 - 0.3\beta_m + 0.07\beta_m^2)} \tag{6.22}$$

in which M_{ob} is given by equations 6.4 and 6.5.

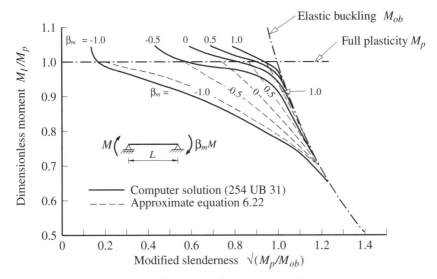

Fig. 6.11 Inelastic buckling of beams with unequal end moments

In this equation, the effects of the bending moment distribution are included in both the elastic buckling resistance $M_{ob} = \alpha_m M_{yz}$ through the use of the end moment ratio β_m in the moment modification factor α_m, and also through the direct use of β_m in equation 6.22. This latter use causes the inelastic buckling moments M_I to approach the elastic buckling moment M_{ob} as the end moment ratio increases towards $\beta_m = 1$.

The most severe case is that of equal and opposite end moments ($\beta_m = -1$), for which yielding is constant along the beam so that the resistance to lateral buckling is reduced everywhere. Less severe cases are those of beams with unequal end moments M and $\beta_m M$ with $\beta_m > 0$, where yielding is confined to short regions near the supports, for which the reductions in the section properties are comparatively unimportant. The least severe case is that of equal end moments that bend the beam in double curvature ($\beta_m = 1$), for which the moment gradient is steepest and the regions of yielding are most limited.

The range of modified slenderness $\sqrt{(M_p/M_{ob})}$ for which a beam can reach the full plastic moment M_p depends very much on the loading arrangement. An approximate expression for the limit of this range for beams with end moments M and $\beta_m M$ can be obtained from equation 6.22 as

$$\sqrt{\left(\frac{M_p}{M_{ob}}\right)_p} = \sqrt{\left(\frac{0.39 + 0.30\beta_m - 0.07\beta_m^2}{0.70}\right)}. \tag{6.23}$$

In the case of a simply supported beam with an unbraced central concentrated load, yielding is confined to a small central portion of the beam, so that any reductions in the section properties are limited to this region. Inelastic buckling can be approximated by using equation 6.22 with $\beta_m = -0.7$ and $\alpha_m = 1.35$.

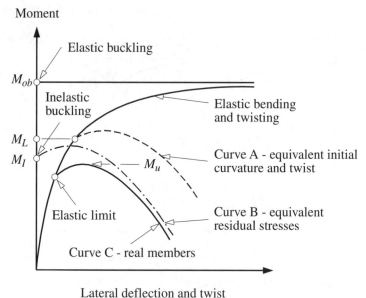

Fig. 6.12 Behaviour of real beams

6.4 Real beams

6.4.1 BEHAVIOUR OF REAL BEAMS

Real beams differ from the ideal beams analysed in section 6.2.1 in much the same way as do real compression members (see section 3.4.1). Thus any small imperfections such as initial curvature, twist, eccentricity of load, or horizontal load components cause the beam to behave as if it had an equivalent initial curvature and twist (see section 6.2.2), as shown by curve A in Fig. 6.12. On the other hand, imperfections such as residual stresses or variations in material properties cause the beam to behave as shown by curve B in Fig. 6.12. The behaviour of real beams having both types of imperfection is indicated by curve C in Fig. 6.12, which shows a transition from the elastic behaviour of a beam with curvature and twist to the inelastic post-buckling behaviour of a beam with residual stresses.

6.4.2 DESIGN RULES

6.4.2.1 Simply supported beams in uniform bending

It is possible to develop a refined analysis of the behaviour of real beams which includes the effects of all types of imperfection. However, the use of such an analysis is unwarranted because the magnitudes of the imperfections are uncertain. Instead, design rules are often based on a simple analysis for one type of equivalent imperfection which allows approximately for all imperfections.

For BS5950, the buckling resistance moment $M_b = p_b Z_{ex}$ of a hot-rolled equal flanged beam in uniform bending is calculated from the uniform bending elastic buckling moment

$$M_E = \frac{M_{yz}}{\sqrt{(1 - EI_y/EI_x)}} = p_E Z_{ex} \qquad (6.24)$$

and the section moment capacity $M_{cx} = p_y Z_{ex}$ by using the relationship

$$\frac{M_b}{M_{cx}} = \left(\frac{1 + (1 + \eta_{LT})M_E/M_{cx}}{2} \right) - \sqrt{\left\{ \left(\frac{1 + (1 + \eta_{LT})M_E/M_{cx}}{2} \right)^2 - \frac{M_E}{M_{cx}} \right\}} \qquad (6.25)$$

in which

$$\eta_{LT} = 0.007 \sqrt{\left\{ \frac{\pi^2 E}{p_y} \right\}} \left(\sqrt{\left(\frac{M_{cx}}{M_E} \right)} - 0.4 \right) \geqslant 0 \qquad (6.26)$$

Equation 6.25 is similar to equation 6.19 for the first yield in a beam with initial curvature and twist, except that the section capacity M_{cx} is introduced in place of the yield moment M_Y, and the factor 1.25 in equation 6.25 which arises from the initial curvature and twist definition of equation 6.20 is replaced by $(1 + \eta_{LT}M_E/M_{cx})$ in which the initial curvature and twist is redefined by equation 6.26.

Equations 6.25 and 6.26 (for $E = 205\,000$ N/mm^2 and $p_y = 275$ N/mm^2) are compared with experimental results in Fig. 6.13.

For very slender beams with high values of modified slenderness $\sqrt{(M_{cx}/M_E)}$, the uniform bending buckling resistance moment M_b shown in Fig. 6.13

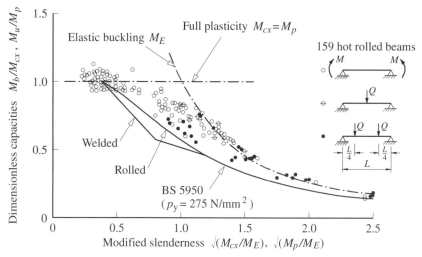

Fig. 6.13 Moment capacities of beams in near-uniform bending

approaches the elastic buckling moment M_E, while for stocky beams the resistance moment M_b reaches the nominal section capacity M_{cx}, and so is governed by yielding or local buckling, as discussed in section 4.7.2. For beams of intermediate slenderness, equation 6.25 provides a transition between these limits, which is a little below the lower bound of the experimental results shown in Fig. 6.13.

BS5950 presents the results of using equations 6.24–6.26 in Table 16, which gives values of the bending strength

$$p_b = M_b/Z_{ex} = \text{fn}\ (p_y, \lambda_{LT}) \tag{6.27}$$

in which p_y is the design strength and the equivalent slenderness λ_{LT} is given by

$$\lambda_{LT} = \sqrt{(\pi^2 E/p_y)}\sqrt{(M_{cx}/M_E)} = uv\lambda\sqrt{\beta_w} \tag{6.28}$$

in which for the uniform bending of equal flanged hot-rolled I-beams

$$u = \{4S_x^2(1 - EI_y/EI_x)/(A^2 h_s^2)\}^{1/4} \tag{6.29a}$$
$$v = \{1 + 0.05(\lambda/x)^2\}^{-1/4} \tag{6.29b}$$
$$\lambda = L_E/r_y \tag{6.29c}$$
$$x = 0.566 h_s(A/J)^{1/2} \tag{6.29d}$$
$$\beta_w = Z_{ex}/S_x \tag{6.29e}$$

in which h_s is the distance between flange shear centres, L_E is the effective length, $r_y = \sqrt{(I_y/A)}$ is the minor axis radius of gyration, and Z_{ex} is the effective section modulus.

Alternatively, the design calculations may be simplified for hot-rolled sections with equal flanges by using Table 20, which gives

$$p_b = \text{fn}\ (p_y, D/T, \sqrt{(\beta_w)}L_E/r_y) \tag{6.30}$$

in which D is the overall depth of the section and T is the flange thickness.

BS5950 uses a similar method for determining the buckling resistance moment M_b of a welded equal flanged beam in uniform bending, except that a different formulation is used for the Perry factor η_{LT} in order to allow for the decrease in the inelastic buckling resistance caused by the welding residual stresses (see section 6.3). The values of M_b/M_{cx} for welded beams are compared in Fig. 6.13 with the corresponding values for hot-rolled beams. Values of the bending strength p_b are given in Table 17, but there is no equivalent of the simple method for hot-rolled beams.

6.4.2.2 Unequal end moments

BS5950 allows for the effects of unequal end moments M and $-\beta M$ on the buckling moment resistance of a hot-rolled equal flanged beam by introducing an equivalent uniform moment factor m_{LT} determined from

$$m_{LT} = \frac{0.2M_{max} + 0.15M_2 + 0.5M_3 + 0.15M_4}{M_{max}} \geqslant 0.44 \tag{6.31}$$

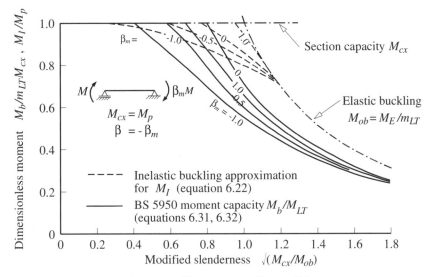

Fig. 6.14 Moment resistances of beams according to BS5950

in which M_{max} is the absolute value of the maximum moments in the length L and M_2, M_3, and M_4 are the absolute values of the moments at the quarter point, mid point, and three quarter point of the length L. This equation gives value of m_{LT} which are close to the values of $1/\alpha_m$ given by equation 6.6 (note the sign change resulting from the different end moment ratio definition of $\beta = -\beta_m$). Values of m_{LT} are given in Table 18.

The beam is satisfactory when

$$M_x^* \leqslant M_b/m_{LT} \leqslant M_{cx} \tag{6.32}$$

in which M_x^* is the maximum design moment in the beam and M_b is determined using equation 6.27. Values of $M_b/(m_{LT}M_{cx})$ are shown in Fig. 6.14 together with the approximate inelastic buckling resistances given by equation 6.22. It can be seen that the moment capacities are reduced below the inelastic buckling resistances, thus providing allowances for the effects of geometrical imperfections on the strengths of real beams.

6.4.2.3 Transverse loads which are not destabilizing

Destabilizing gravity loads are loads which act at the top flange at an unrestrained cross-section, and which are free to deflect laterally with the beam. Gravity loads acting at the centroid or bottom flange are not destabilizing.

For transverse loads which are not destabilizing, BS5950 allows for the effect of a non-uniform bending moment distribution on the buckling moment resistance of a hot-rolled equal flanged beam by using the equivalent uniform moment factor m_{LT} determined from equation 6.31 in equation 6.32. These

values of m_{LT} are close to the values of $1/\alpha_m$ obtained from Fig. 6.7. Values of m_{LT} are given in Table 18.

6.4.2.4 Destabilizing transverse loads

Destabilizing gravity loads which act at the top flange at an unrestrained cross-section decrease a beam's resistance to elastic buckling, as shown in Fig. 6.6. BS5950 allows for destabilizing loads by increasing the effective length L_E (Table 13) and by requiring $m_{LT} = 1$.

6.4.2.5 Cantilevers

Cantilevers which are free to deflect laterally and twist at the unsupported end are treated by BS5950 as equivalent beams of the same length with $m_{LT} = 1$, but with even higher increases in the effective length L_E when the loads are destabilizing (Table 14).

6.5 Effective lengths of beams

6.5.1 SIMPLE SUPPORTS AND RIGID RESTRAINTS

In the previous sections it was assumed that the beam was supported laterally only at its ends. When a beam with equal and opposite end moments has an additional rigid support at its centre which prevents lateral deflection and twist rotation, its buckled shape is given by

$$\frac{\phi}{(\phi)_{L/4}} = \frac{u}{(u)_{L/4}} = \sin \frac{\pi z}{L/2}, \tag{6.33}$$

and its elastic buckling moment is given by

$$M_{ob} = \sqrt{\left\{ \left(\frac{\pi^2 E I_y}{(L/2)^2} \right) \left(GJ + \frac{\pi^2 E I_w}{(L/2)^2} \right) \right\}}. \tag{6.34}$$

When a beam has several rigid supports which prevent local lateral deflection and twist rotation, the beam is divided into a series of segments. The elastic buckling of each segment may be approximated by using its length as the effective length L_E. One segment will be the most critical, and the elastic buckling of this segment will provide a conservative estimate of the buckling capacity of the whole beam.

The end supports of a beam may also differ from simple supports. For example, both ends of the beam may be rigidly built-in against lateral rotation about the minor axis and against end warping. If the beam has equal and opposite end moments, then its buckled shape is given by

$$\frac{\phi}{(\phi)_{L/2}} = \frac{u}{(u)_{L/2}} = \frac{1}{2} \left(1 - \cos \frac{\pi z}{L/2} \right), \tag{6.35}$$

and its elastic buckling moment is given by equation 6.34.

In general, the elastic buckling moment M_{ob} of a restrained beam with equal and opposite end moments can be expressed as

$$M_{ob} = \sqrt{\left\{\left(\frac{\pi^2 EI_y}{L_E^2}\right)\left(GJ + \frac{\pi^2 EI_w}{L_E^2}\right)\right\}} \tag{6.36}$$

in which L_E is the effective length.

In BS5950, variations in the end restraint conditions are accounted for by using the effective length

$$L_E = k_e L, \tag{6.37}$$

instead of the member length L. Thus the elastic buckling moment M_E for uniform bending is obtained by substituting M_{ob} from equation 6.36 for M_{yz} in equation 6.24. This value is then used in equation 6.25 to determine the uniform bending buckling moment resistance M_b. Some simple approximations for the effective length ratios of beams and cantilevers with end restraints are given in Tables 13 and 14 of BS5950.

The use of the effective length concept can be extended to beams with loading conditions other than equal and opposite end moments. In general, the maximum moment M_{ob} at elastic buckling depends on the beam section, its loading, and its restraints, so that

$$\frac{M_{ob}L}{\sqrt{(EI_y GJ)}} = fn\left(\frac{\pi^2 EI_w}{GJL^2}, \text{loading}, \frac{2y_Q}{d_f}, \text{restraints}\right). \tag{6.38}$$

A partial separation of the effect of the loading from that of the restraint conditions may be achieved for beams with *centroidal* loading by approximating the maximum moment M_{obo} at elastic buckling by using

$$M_{obo} = \alpha_m \sqrt{\left\{\left(\frac{\pi^2 EI_y}{(L_E^2)}\right)\left(GJ + \frac{\pi^2 EI_w}{(L_E^2)}\right)\right\}} \tag{6.39}$$

for which it is assumed that the factor α_m depends only on the in-plane bending moment distribution, and that the effective length L_E depends only on the restraint conditions. This approximation allows the α_m factors calculated for simply supported beams (see section 6.2.1) and the effective lengths L_E determined for restrained beams with equal and opposite end moments ($\alpha_m = 1$) to be used more generally. Thus the value of M_{obo}/α_m may be used for M_{yz} in equation 6.24 to determine the buckling resistance moment M_b.

Unfortunately, the form of equation 6.39 is not suitable for beams with loads acting away from the centroid. A method of using accurate buckling solutions for such beams is presented in section 6.6.

6.5.2 INTERMEDIATE RESTRAINTS

In the previous subsection it was shown that the elastic buckling moment of a simply supported I-beam is substantially increased when a restraint is provided

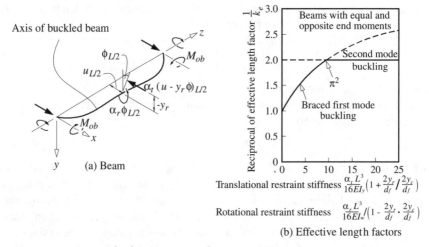

(b) Effective length factors

Fig. 6.15 Beam with elastic intermediate restraints

which prevents the centre of the beam from deflecting laterally and twisting. This restraint need not be completely rigid, but may be elastic, provided its translational and rotational stiffnesses exceed certain minimum values.

The case of the beam with equal and opposite end moments M shown in Fig. 6.15a is analysed in [24]. This beam has a central translational restraint of stiffness α_t (where α_t is the ratio of the lateral force exerted by the restraint to the lateral deflection of the beam measured at the height y_r of the restraint), and a central torsional restraint of stiffness α_r (where α_r is the ratio of the torque exerted by the restraint to the twist of the beam). It is shown in [24] that the elastic buckling moment M_{ob} can be expressed in the standard form of equation 6.36 when the stiffnesses α_t, α_r are related to each other and the effective length ratio k_e through

$$\frac{\alpha_t L^3}{16EI_y}\left(1 + \frac{2y_r/d_f}{2y_c/d_f}\right) = \frac{\left(\dfrac{\pi}{2k_e}\right)^3 \cot \dfrac{\pi}{2k_e}}{\dfrac{\pi}{2k_e}\cot\dfrac{\pi}{2k_e} - 1}, \tag{6.40}$$

$$\frac{\alpha_r L^3}{16EI_w}\bigg/\left(1 - \frac{2y_r}{d_f}\frac{2y_c}{d_f}\right) = \frac{\left(\dfrac{\pi}{2k_e}\right)^3 \cot \dfrac{\pi}{2k_e}}{\dfrac{\pi}{2k_e}\cot\dfrac{\pi}{2k_e} - 1} \tag{6.41}$$

in which

$$y_c = \frac{M_{ob}}{\pi^2 EI_y/(k_e L)^2} = \frac{d_f}{2}\sqrt{(1 + k_e^2/K^2)} \tag{6.42}$$

$$L_E = k_e L \tag{6.43}$$

in which L_E is the effective length and $K = \sqrt{(\pi^2 EI_w/GJL^2)}$.

These relationships are shown graphically in Fig. 6.15b, and are similar to that shown in Fig. 3.15c for compression members with intermediate restraints. It can be seen that the effective length ratio k_e varies from 1 when the restraints are of zero stiffness to 0.5 when

$$\frac{\alpha_t L^3}{16 E I_y}\left(1 + \frac{2y_r/d_f}{2y_c/d_f}\right) = \frac{\alpha_r L^3}{16 E I_w}\bigg/\left(1 - \frac{2y_r}{d_f}\frac{2y_c}{d_f}\right) = \pi^2. \tag{6.44}$$

If the restraint stiffnesses exceed these values, then the beam buckles in the second mode with zero central deflection and twist at a moment which corresponds to $k_e = 0.5$. When the height $-y_r$ of the translational restraint above the centroid is equal to $d_f^2 4 y_c$ the required rotational stiffness α_r given by equation 6.44 is zero. Since y_c is never less than $d_f/2$ (see equation 6.42), it follows that a top flange translational restraint of stiffness

$$\alpha_L = \frac{16\pi^2 E I_y / L^3}{1 + d_f/2y_c} \tag{6.45}$$

is always sufficient to brace the beam into the second mode. This minimum stiffness can be expressed as

$$\alpha_L = \frac{8 M_{ob}}{L d_f}\frac{1}{1 + \sqrt{(1 + 4K^2)}}, \tag{6.46}$$

and the greatest value of this is

$$\alpha_L = \frac{4 M_{ob}}{L d_f}. \tag{6.47}$$

The flange force Q_f at elastic buckling can be approximated by

$$Q_f = M_{ob}/d_f, \tag{6.48}$$

and so the minimum top flange translational stiffness can be approximated by

$$\alpha_L = 4 Q_f/L. \tag{6.49}$$

This is of the same form as equation 3.31 for the minimum stiffness of intermediate restraints for compression members.

There are no restraint stiffness requirements in BS5950. This is supported by the finding [25] that compression member restraints which are capable of transmitting 2.5% of the force in the compression member are invariably stiff enough to ensure second mode buckling. BS5950 generally requires any group of restraining elements to be capable of transmitting 2.5% of the compression flange force in the beam being restrained, and any individual element to be capable of transmitting 1% of the flange force.

The influence of intermediate restraints on beams with central concentrated and uniformly distributed loads has also been studied, and many values of the minimum restraint stiffnesses required to cause the beams to buckle as if rigidly braced have been determined [16, 24, 26, 27]. The effects of diaphragm

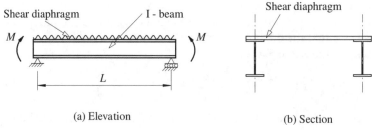

(a) Elevation

(b) Section

Fig. 6.16 Diaphragm braced I-beams

bracing on the lateral buckling of simply supported beams with equal and opposite end moments (see Fig. 6.16) have also been investigated [16, 28, 29], and a simple method of determining whether a diaphragm is capable of providing full bracing has been developed [28].

6.5.3 ELASTIC END RESTRAINTS

When a beam forms part of a rigid-jointed structure, the adjacent members elastically restrain the ends of the beam (i.e. they induce restraining moments which are proportional to the end rotations). These restraining actions significantly modify the elastic buckling load of the beam. Four different types of restraining moment may act at each end of a beam, as shown in Fig. 6.17. They are:

(a) the major axis end moment M which provides restraint about the major axis,
(b) the bottom flange end moment M_B, and
(c) the top flange end moment M_T, which provide restraints about the minor axis and against end warping,
(d) the axial torque T_0 which provides restraint against end twisting.

The major axis end restraining moments M vary directly with the applied loads, and can be determined by a conventional in-plane bending analysis. The degree of restraint experienced at one end of the beam depends on the major axis stiffness α_x of the adjacent member (which is defined as the ratio of the end moment to the end rotation). This may be expressed by the ratio R_1 of the actual restraining moment to the maximum moment required to prevent major axis end rotation. Thus R_1 varies from 0 when there is no restraining moment to 1 when there is no end rotation. For beams which are symmetrically loaded and restrained, the restraint parameter R_1 is related to the stiffness α_x of each adjacent member by

$$\alpha_x = \frac{EI_x}{L} \frac{2R_1}{1 - R_1}. \tag{6.50}$$

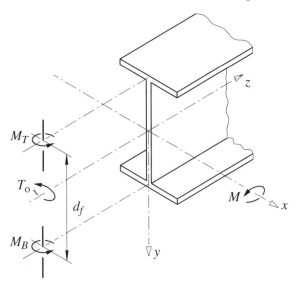

Fig. 6.17 End restraining moments

Although the major axis end moments are independent of the buckling deformations, they do affect the buckling load of the beam because of their effect on
the in-plane bending moment distribution (see section 6.2.1.4). Many particular
cases have been studied, and tabulations of buckling loads are available [5, 12,
13, 30–32].

On the other hand, the flange end moments M_B and M_T remain zero until
the buckling load of the beam is reached, and then increase in proportion to
the flange end rotations. Again, the degree of end restraint can be expressed by
the ratio of the actual restraining moment to the maximum value required to
prevent end rotation. Thus, the minor axis end restraint parameter R_2 (which
describes the relative magnitude of the restraining moment $M_B + M_T$) varies
between 0 and 1, and the end warping restraint parameter R_4 (which describes
the relative magnitude of the differential flange end moments $(M_T - M_B)/2$)
varies from 0 when the ends are free to warp to 1 when end warping is prevented. The particular case of symmetrically restrained beams with equal and
opposite end moments (Fig. 6.18) is analysed in section 6.11.1. It is assumed for
this that the minor axis and end warping restraints take the form of equal rotational restraints which act at each flange end and whose stiffnesses are such
that

$$\frac{\text{Flange end moment}}{\text{Flange end rotation}} = \frac{-EI_y}{L}\frac{R}{1-R} \qquad (6.51)$$

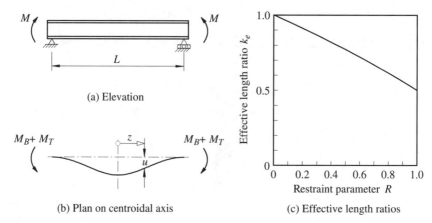

(a) Elevation

(b) Plan on centroidal axis

(c) Effective length ratios

Fig. 6.18 Elastic buckling of end restrained beams

in which case

$$R_2 = R_4 = R. \tag{6.52}$$

It is shown in section 6.11.1 that the moment M_{ob} at which the restrained beam buckles elastically is given by equations 6.30 and 6.37 when the effective length ratio k_e is the solution of

$$\frac{R}{1 - R} = \frac{-\pi}{2k_e} \cot \frac{\pi}{2k_e}. \tag{6.53}$$

It can be seen from the solutions of this equation shown in Fig. 6.18c that the effective length ratio k_e decreases from 1 to 0.5 as the restraint parameter R increases from 0 to 1. These solutions are exactly the same as those obtained from the compression member effective length chart of Fig. 3.20a when

$$k_1 = k_2 = \frac{1 - R}{1 - 0.5R}, \tag{6.54}$$

which suggests that the effective length ratios k_e for beams with unequal end restraints may be approximated by using the values given by Fig. 3.20a.

The elastic buckling of symmetrically restrained beams with unequal end moments has also been analysed [33], while solutions have been obtained for many other minor axis and end warping restraint conditions [16, 26, 29–31]. These provide a basis for the approximate effective length ratios used in BS5950.

The end torques T_0 which resist end twist rotations also remain zero until elastic buckling occurs, and then increase with the end twist rotations. It has been assumed that the ends of all the beams discussed so far are rigidly restrained against end twist rotations. When the end restraints are elastic instead of rigid, some end twist rotation occurs during buckling and the elastic buckling load is reduced. Analytical studies [21] of beams in uniform bending with elastic torsional end restraints have shown that the reduced buckling moment M_{yzr} can be approximated by

$$\frac{M_{yzr}}{M_{yz}} = \sqrt{\left\{ \frac{1}{(4.9 + 4.5K^2)R_3 + 1} \right\}} \qquad (6.55)$$

in which $1/R_3$ is the dimensionless stiffness of the torsional end restraints given by

$$\frac{1}{R_3} = \frac{\alpha_{rz}L}{GJ} \qquad (6.56)$$

in which α_{rz} is the ratio of the restraining torque T_0 to the end twist rotation $(\phi)_0$. Reductions for other loading conditions can be determined from the elastic buckling solutions in [16, 21, 31] or by using a computer program [18] to carry out an elastic buckling analysis.

Another situation for which end twist rotation is not prevented is illustrated in Fig. 6.19, where the bottom flange of a beam is simply supported at its end and prevented from twisting but the top flange is unrestrained. In this case, beam buckling may be accentuated by distortion of the cross-section which results in the web bending shown in Fig. 6.19. Studies of the distortional buckling [34–36] of beams such as that shown in Fig. 6.19 have suggested that the reduction in the buckling capacity can be allowed for approximately by using an effective length ratio

$$k_t = 1 + (d_f/6L)(t_f/t_w)^3(1 + b_f/d_f)/2 \qquad (6.57)$$

in which t_f and t_w are the flange and web thicknesses and b_f the flange width.

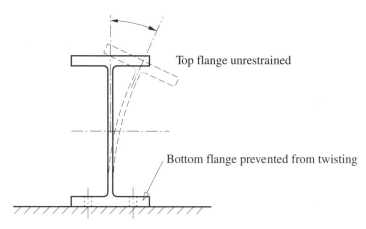

Top flange unrestrained

Bottom flange prevented from twisting

Fig. 6.19 End distortion of a beam with an unrestrained top flange

In BS5950 equation 6.57 is simplified to

$$k_t = 1 + 2D/L \qquad (6.58)$$

for beams which are supported in this way at both ends, in which D is the depth of the beam.

The preceding discussion has dealt with the effects of each type of end restraint, but most beams in rigid-jointed structures have all types of elastic restraint acting simultaneously. Many such cases of combined restraints have been analysed, and tabular, graphical, or approximate solutions are given in [12, 16, 30, 32, 37].

6.6 Design by buckling analysis

6.6.1 DESIGN METHOD

The effective length concept developed in section 6.5 is essentially a convenient method of expressing the variations of the maximum moment M_{ob} at elastic buckling with the beam geometry, loading and restraints. Its use in the design of rigid-jointed structures is an example of a more general approach to the analysis and design of structures whose strengths are governed by the interaction between yielding and buckling. Another example of this approach was developed in section 3.6, in which the relationship between the ultimate strength of a simply supported uniform member in uniform compression and its squash and elastic buckling loads was used to determine the in-plane ultimate strengths of other compression members.

The application of this general approach to beams and flexural frames which fail by yielding and flexural–torsional buckling is based on the dependence of the ultimate moment capacity M_u of a simply supported beam with equal and opposite end moments on its full plastic moment M_p and its elastic buckling moment M_{yz}. Thus equation 6.25 which relates the resistance moment M_b of BS5950 to the section capacity M_{cx} and the uniform bending elastic buckling moment M_E can be applied more generally, provided these moments are calculated for the section of the beam which is closest to full plasticity or local buckling. In this method, the elastic buckling moment M_{ob} is used in place of the uniform bending value M_{yz} in equations 6.24–6.26 to determine the nominal moment capacity M_b. Care must be taken not to allow a second time (through the use of an m_{LT} value less than 1.0) for the effects of the moment distribution which have already been accounted for in the determination of M_{ob}. For this reason, $m_{LT} = 1.0$ should be used in this method.

However, this procedure takes no account of the effects of the moment distribution on inelastic buckling, and instead relies on the most conservative conversion from M_{cx} and M_{ob} to M_b. A more rational approach is permitted in the Australian standard AS4100 [38], which allows the elastic buckling value of α_m

($\equiv 1/m_{LT}$) to be used provided the value $M_{oa} = M_{ob}/\alpha_m$ is used in place of M_{yz} in equation 6.24, with the subsequent use of $\alpha_m M_b$ ($\equiv M_b/m_{LT}$) as in equation 6.32. This approach relies on the separation of the effects of moment distribution and restraint discussed in section 6.5.1, and demonstrated in equation 6.39.

The first step in the method of design by buckling analysis is to determine the load at which elastic flexural–torsional buckling takes place, so that the elastic buckling moment M_{ob} can be calculated. A number of general computer programs which can be used for this have been developed [6, 8, 10, 18, 39–41]. However, these are not yet widely used, and so information which has already been obtained on the elastic buckling of continuous beams, rigid frames, and monosymmetric and non-uniform beams is presented in the following sections and subsections, where approximate methods of analysis are discussed.

6.6.2 CONTINUOUS BEAMS

Perhaps the simplest type of rigid-jointed structure is the continuous beam which can be regarded as a series of segments which are rigidly connected together at points where lateral deflection and twist are prevented (i.e. at supports and intermediate brace points). In general, the interaction between the segments during buckling depends on the loading pattern for the whole beam, and so also do the magnitudes of the buckling loads. This behaviour is similar to the in-plane buckling behaviour of rigid frames discussed in section 8.3.5.2.

6.6.2.1 Beams with only one segment loaded

When only one segment of a continuous beam is loaded, its elastic buckling load may be evaluated approximately when tabulations for segments with elastic end restraints are available [12, 30, 32]. To do this it is first necessary to determine the end restraint parameters R_1, R_2, R_4 ($R_3 = 0$ because it is assumed that twisting is prevented at the supports and brace points) from the stiffnesses of the adjacent unloaded segments. For example, when only the centre span of the symmetrical three span continuous beam shown in Fig. 6.20a is loaded ($Q_1 = 0$), the end spans provide elastic restraints which depend on their stiffnesses. By analysing the major axis bending, minor axis bending, and differential flange bending of the end spans, it can be shown [31] that

$$R_1 = R_2 = R_4 = \frac{1}{1 + 2L_1/3L_2} \tag{6.59}$$

approximately. With these values, the elastic buckling load for the centre span can be determined from the tabulations in [30] or [32]. Some elastic buckling loads (for beams of narrow rectangular section for which $K = 0$) determined in this way are shown in a non-dimensional form in Fig. 6.21.

A similar procedure can be followed when only the outer spans of

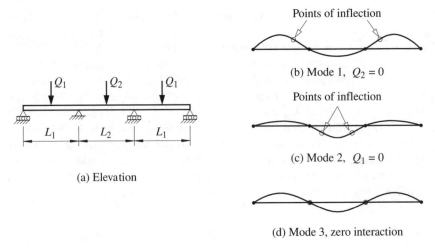

(a) Elevation

Points of inflection

(b) Mode 1, $Q_2 = 0$

Points of inflection

(c) Mode 2, $Q_1 = 0$

(d) Mode 3, zero interaction

Fig. 6.20 Buckling modes for a symmetrical three span continuous beam

symmetrical three span continuous beams shown in Fig. 6.20 are loaded ($Q_2 = 0$). In this case the restraint parameters [12] are given by

$$R_1 = R_2 = R_4 = \frac{1}{1 + 3L_2/2L_1} \tag{6.60}$$

approximately. Some dimensionless buckling loads (for beams of narrow rectangular section) determined by using these restraint parameters in the tabulations of [12] are shown in Fig. 6.21. Similar diagrams have been produced [42] for two span beams of narrow rectangular section.

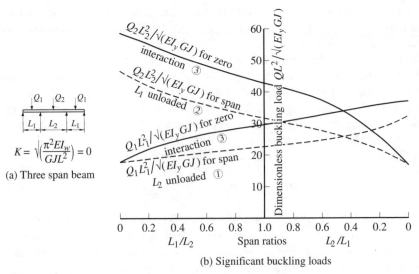

(a) Three span beam

$$K = \sqrt{\left(\frac{\pi^2 EI_w}{GJL^2}\right)} = 0$$

(b) Significant buckling loads

Fig. 6.21 Significant buckling loads of symmetrical three span beams

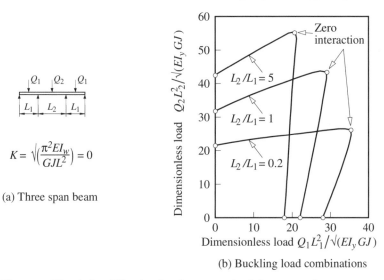

(a) Three span beam

(b) Buckling load combinations

Fig. 6.22 Elastic buckling loads of symmetrical three span beams

6.6.2.2 Beams with general loading

When more than one segment of a continuous beam is loaded, the buckling loads can be determined by analysing the interaction between the segments. This has been done for a number of continuous beams of narrow rectangular section [42], and the results for some symmetrical three span beams are shown in Fig. 6.22. These indicate that as the loads Q_1 on the end spans increase from zero, so does the buckling load Q_2 of the centre span until a maximum value is reached, and that a similar effect occurs as the centre span load Q_2 increases from zero.

The results shown in Fig. 6.22 suggest that the elastic buckling load interaction diagram can be closely and safely approximated by drawing straight lines as shown in Fig. 6.23 between the following three significant load combinations:

(1) When only the end spans are loaded ($Q_2 = 0$), they are restrained during buckling by the centre span, and the buckled shape has inflection points in the end spans, as shown in Fig. 6.20b. In this case the buckling loads Q_1 can be determined by using $R_1 = R_2 = R_4 = 1/(1 + 3L_2/2L_1)$ in the tabulations of [12].

(2) When only the centre span is loaded ($Q_1 = 0$), it is restrained by the end spans, and the buckled shape has inflection points in the centre span, as shown in Fig. 6.20c. In this case the buckling load Q_2 can be determined by using $R_1 = R_2 = R_4 = 1/(1 + 2L_1/3L_2)$ in the tabulations of [30, 32].

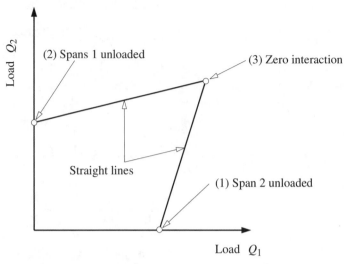

Fig. 6.23 Straight line approximation

(3) Between these two extremes exists the zero interaction load combination for which the buckled shape has inflection points at the internal supports, as shown in Fig. 6.20d, and each span buckles as if unrestrained in the buckling plane. In this case the buckling loads can be determined by using

$$R_2 = R_4 = 0, \tag{6.61}$$

$$R_{11} = \frac{1 + Q_2 L_2^2 / Q_1 L_1^2}{1 + 3L_2/2L_1}, \tag{6.62}$$

$$R_{12} = \frac{1 + Q_1 L_1^2 / Q_2 L_2^2}{1 + 2L_1/3L_2} \tag{6.63}$$

in the tabulations of [12, 30, 32]. Some zero interaction load combinations (for three span beams of narrow rectangular section) determined in this way are shown in Fig. 6.21, while other zero interaction combinations are given in [42].

Unfortunately, this comparatively simple approximate method has proved too complex for use in routine design, possibly because the available tabulations of elastic buckling loads are not only insufficient to cover all the required loading and restraint conditions, but also too detailed to enable them to be easily used. Instead, an approximate method of analysis [43] is often used, in which the effects of lateral continuity between adjacent segments are ignored and each segment is regarded as being simply supported laterally. Thus the elastic buckling of each segment is analysed for its in-plane moment distribution (the moment modification factors of equation 6.13 or Fig. 6.7 may be used) and for an effective length L_E equal to the segment length L. The so-determined elastic buckling moment of each segment is then used to evaluate a corresponding beam load set, and the lowest of these is taken as the elastic buckling load set. This method produces a lower bound estimate which is sometimes remarkably close to the true buckling load set.

However, this is not always the case, and so a much more accurate but still reasonably simple method has been developed [33]. In this method, the accuracy of the lower bound estimate (obtained as described above) is improved by allowing for the interactions between the critical segment and the adjacent segments at buckling. This is done by using a simple approximation for the destabilizing effects of the in-plane bending moments on the stiffnesses of the adjacent segments, and by approximating the restraining effects of these segments on the critical segment by using the effective length chart of Fig. 3.20a for braced compression members to estimate the effective length of the critical beam segment. A step by step summary [33] is as follows:

(1) Determine the properties EI_y, GJ, EI_w, L of each segment.
(2) Analyse the in-plane bending moment distribution through the beam, and determine the moment modification factors α_m for each segment from equation 6.13 or Fig. 6.7.
(3) Assume all effective length ratios k_e are equal to unity.
(4) Calculate the maximum moment M_{ob} in each segment at elastic buckling from

$$M_{ob} = \alpha_m \sqrt{\left\{ \left(\frac{\pi^2 EI_y}{L_E^2} \right) \left(GJ + \frac{\pi^2 EI_w}{L_E^2} \right) \right\}} \tag{6.64}$$

with $L_E = L$, and the corresponding beam buckling loads Q_s.
(5) Determine a lower bound estimate of the beam buckling load as the lowest value Q_{ms} of the loads Q_s, and identify the segment associated with this as the critical segment 12. (This is the approximate method [43] described in the preceding paragraph.)
(6) If a more accurate estimate of the beam buckling load is required, then use the values Q_{ms} and Q_{rs1}, Q_{rs2} calculated in step 5 together with Fig. 6.24 (which is similar to Fig. 3.18 for braced compression members) to approximate the stiffness α_{r1}, α_{r2} of the segments adjacent to the critical segment 12.
(7) Calculate the stiffness of the critical segment 12 from

$$2EI_{ym}/L_m.$$

(8) Calculate the stiffness ratios k_1, k_2 from

$$k_{1,2} = \frac{2EI_{ym}/L_m}{0.5\alpha_{r1,r2} + 2EI_{ym}/L_m} \tag{6.65}$$

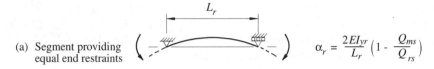

(a) Segment providing
equal end restraints
$$\alpha_r = \frac{2EI_{yr}}{L_r}\left(1 - \frac{Q_{ms}}{Q_{rs}}\right)$$

(b) Segment hinged
at far end
$$\alpha_r = \frac{3EI_{yr}}{L_r}\left(1 - \frac{Q_{ms}}{Q_{rs}}\right)$$

(c) Segment fixed
at far end
$$\alpha_r = \frac{4EI_{yr}}{L_r}\left(1 - \frac{Q_{ms}}{Q_{rs}}\right)$$

Fig. 6.24 Stiffness approximations for restraining segments

(9) Determine the effective length ratio k_e for the critical segment 12 from Fig. 3.20a, and the effective length $L_E = k_e L$.
(10) Calculate the elastic buckling moment M_{ob} of the critical segment 12 using L_E in equation 6.64, and from this the corresponding improved approximation of the elastic buckling load Q_{ob} of the beam.

It should be noted that while the calculations for the lower bound (the first five steps) are made for all segments, those for the improved estimate are only made for the critical segment, and so comparatively little extra effort is involved. The development and application of this method is further described in [33].

6.6.3 RIGID FRAMES

Under some conditions, the elastic buckling loads of rigid frames with only one member loaded can be determined from the available tabulations [12, 30, 32] in a similar manner to that described in section 6.6.2.1 for continuous beams. For example, consider a symmetrically loaded beam which is rigidly connected to two equal cross-beams as shown in Fig. 6.25a. In this case the comparatively large major axis bending stiffnesses of the cross-beams ensure that end twisting of the loaded beam is effectively prevented, and it may therefore be assumed that $R_3 = 0$. If the cross-beams are of open section so that their torsional stiffness is comparatively small, then it is not unduly conservative to assume that they do not restrain the loaded beam about its major axis, and so

$$R_1 = 0. \tag{6.66}$$

By analysing the minor axis and different flange bending of the cross-beams, it can be shown that

$$R_2 = R_4 = 1/(1 + L_1 I_{y2}/6L_2 I_{y1}). \tag{6.67}$$

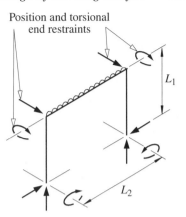

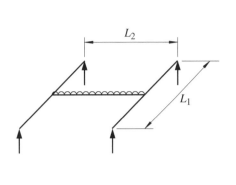

(a) Beam supported by cross-beams   (b) Symmetrical portal frame

Fig. 6.25 Simple rigid-jointed structures

This result is based on the assumption that there is no likelihood of the cross-beams themselves buckling, so that their effective minor axis rigidity can be taken as EI_{y1}. With these values for the restraint parameters, the elastic buckling load can be determined from the tabulations in [30, 32].

The buckling loads of the symmetrical portal frame shown in Fig. 6.25b can be determined in a similar manner provided there are sufficient external restraints to position-fix the joints of the frame. The columns of portal frames of this type are usually placed so that the plane of greatest bending stiffness is that of the frame. In this case, the minor axis stiffness of each column provides only a small resistance against end twisting of the beam, and this resistance is reduced by the axial force transmitted by the column, so that it may be necessary to provide additional torsional end restraints to the beam. If these additional restraints are stiff enough (and the information given in [16, 31] will give some guidance), then it may be assumed without serious error that $R_3 = 0$. The major axis stiffness of a pinned base column is $3EI_{x1}/L_1$ and of a fixed base column is $4EI_{x1}/L_1$, and it can be shown by analysing the in-plane flexure of a portal frame that for pinned base portals

$$R_1 = 1/(1 + 2L_1I_{x2}/3L_2I_{x1}),   (6.68)$$

and for fixed base portals,

$$R_1 = 1/(1 + L_1I_{x2}/2L_2I_{x1}).   (6.69)$$

If the columns are of open cross-section, then their torsional stiffness is comparatively small, and it is not unduly conservative to assume that the columns do not restrain the beam element or its flanges against minor axis flexure, and so

$$R_2 = R_4 = 0.   (6.70)$$

Using these values of the restraint parameters, the elastic buckling load can be determined from the tabulations in [30, 32].

When more than one member of a rigid frame is loaded, the buckling restraint parameters cannot be easily determined because of the interactions between members which take place during buckling. The typical member of such a frame acts as a beam-column which is subjected to a combination of axial and transverse loads and end moments. The buckling behaviour of beam-columns and rigid frames is treated in detail in Chapters 7 and 8.

6.7 Monosymmetric beams

6.7.1 ELASTIC BUCKLING RESISTANCE

When a monosymmetric I-beam (see Fig. 6.26), which is loaded in its plane of symmetry, twists during buckling, the longitudinal bending stresses $M_x y/I_x$ exert a torque (see section 6.12)

$$T_M = M_x \beta_x \frac{d\phi}{dz} \tag{6.71}$$

in which

$$\beta_x = \frac{1}{I_x} \int_A (x^2 y + y^3) dA - 2y_0 \tag{6.72}$$

is the monosymmetry property of the cross-section. An explicit expression for β_x for an I-section is given in Fig. 6.26.

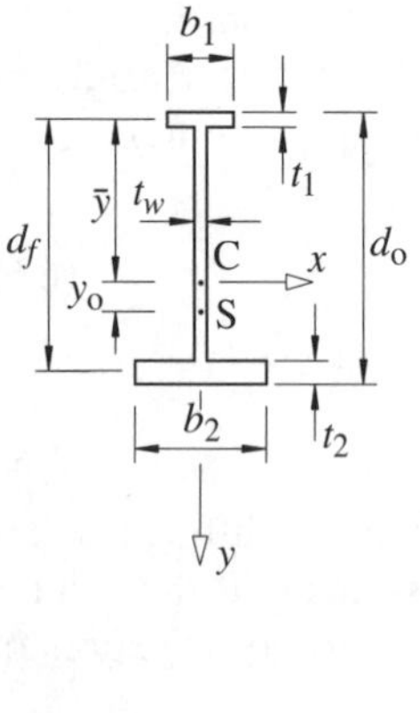

$$d_f = d_o - (t_1 + t_2)/2$$

$$\bar{y} = \frac{b_2 t_2 d_f + (d_o - t_1 - t_2)(d_o - t_2) t_w/2}{b_1 t_1 + b_2 t_2 + (d_o - t_1 - t_2) t_w}$$

$$I_x = b_1 t_1 \bar{y}^2 + b_2 t_2 (d_f - \bar{y})^2$$
$$+ (d_o - t_1 - t_2) \times t_w \times (\bar{y} - (d_o - t_2)/2)^2$$
$$+ (d_o - t_1 - t_2)^3 \times t_w/12$$

$$\alpha = \frac{1}{1 + (b_1/b_2)^3 (t_1/t_2)}$$

$$y_0 = \alpha d_f - \bar{y}$$

$$\beta_x = \frac{1}{I_x} \left\{ \begin{array}{l} (d_f - \bar{y})[b_2^3 t_2/12 + b_2 t_2 (d_f - \bar{y})^2] \\ - \bar{y}[b_1^3 t_1/12 + b_1 t_1 \bar{y}^2] \\ + [(d_f - \bar{y} - t_2/2)^4 - (\bar{y} - t_1/2)^4] t_w/4 \end{array} \right\} - 2y_0$$

$$I_w = \alpha b_1^3 t_1 d_f^2/12$$

Fig. 6.26 Properties of monosymmetric I-sections

The action of the torque T_M can be thought of as changing the effective torsional rigidity of the section from GJ to $(GJ + M_x\beta_x)$, and is related to the effect which causes some short concentrically loaded compression members to buckle torsionally (see section 3.6.5). In that case the compressive stresses exert a disturbing torque so that there is a reduction in the effective torsional rigidity. In doubly symmetric beams the disturbing torque exerted by the compressive bending stresses is exactly balanced by the restoring torque due to the tensile stresses, and β_x is zero. In monosymmetric beams, however, there is an imbalance which is dominated by the stresses in the smaller flange which is further away from the shear centre. Thus, when the smaller flange is in compression there is a reduction in the effective torsional rigidity ($M_x\beta_x$ is negative), while the reverse is true ($M_x\beta_x$ is positive) when the smaller flange is in tension. Consequently, the resistance to buckling is increased when the larger flange is in compression, and decreased when the smaller flange is in compression.

The elastic buckling moment M_{ob} of a simply supported monosymmetric beam with equal and opposite end moments can be obtained by substituting M_{ob} for M_{yz} and the effective rigidity $(GJ + M_{ob}\beta_x)$ for GJ in equation 6.3 for a doubly symmetric beam and rearranging, whence

$$M_{ob} = \sqrt{\left(\frac{\pi^2 EI_y}{L^2}\right)}\left\{\sqrt{\left[GJ + \frac{\pi^2 EI_w}{L^2} + \left\{\frac{\beta_x}{2}\sqrt{\left(\frac{\pi^2 EI_y}{L^2}\right)}\right\}^2\right]} + \frac{\beta_x}{2}\sqrt{\left(\frac{\pi^2 EI_y}{L^2}\right)}\right\}$$

(6.73)

in which the warping section constant I_w is as given in Fig. 6.26.

The evaluation of the monosymmetry property β_x is not straightforward, and it has been suggested that the more easily calculated parameter

$$\rho_m = I_{cy}/I_y$$
(6.74)

should be used instead, where I_{cy} is the section minor axis second moment of area of the compression flange. The monosymmetry property β_x may be approximated [44] by

$$\beta_x = 0.9d_f(2\rho_m - 1)(1 - I_y^2/I_x^2),$$
(6.75)

and the warping section constant I_w by

$$I_w = \rho_m(1 - \rho_m)I_y d_f^2.$$
(6.76)

The variations of the dimensionless elastic buckling moment $M_{ob}L/\sqrt{(EI_yGJ)}$ with the values of ρ_m and $K_m = \sqrt{(\pi^2 EI_y d_f^2/4GJL^2)}$ are shown in Fig. 6.27. The dimensionless buckling resistance for a T-beam with the flange in compression ($\rho_m = 1.0$) is significantly higher than for an equal flanged I-beam ($\rho_m = 0.5$) with the same value of K_m, but the resistance is greatly reduced for a T-beam with the flange in tension ($\rho_m = 0.0$).

The elastic flexural–torsional buckling of simply supported monosymmetric

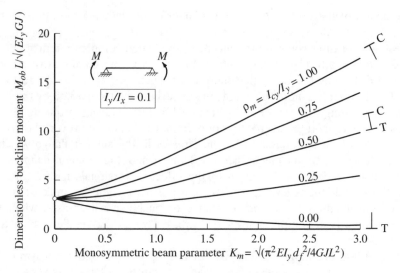

Fig. 6.27 Monosymmetric I-beams in uniform bending

beams with other loading conditions has been investigated numerically, and tabulated solutions and approximating equations are available [5, 13, 16, 19, 45, 46] for beams under moment gradient or with central concentrated loads or uniformly distributed loads. Solutions are also available [16, 19, 47] for cantilevers with concentrated end loads or uniformly distributed loads. These solutions can be used to find the maximum moment M_{ob} in the beam or cantilever at elastic buckling.

6.7.2 DESIGN RULES

For BS5950, the elastic buckling resistance given by equation 6.73 for an I-section beam with unequal flanges in uniform bending is obtained by using the approximation of equation 6.75 for ψ ($\equiv \beta_x$). With this approximation, the equivalent slenderness λ_{LT} of equation 6.28 is obtained by modifying the slenderness factor v to

$$v = \{\psi + [4\eta(1 - \eta) + 0.05(\lambda/x)^2 + \psi^2]^{1/2}\}^{-1/2} \qquad (6.77)$$

in which $\eta = \rho_m$.

For an I-section with unequal flanges, the values of m_{LT} obtained from equation 6.31 may be used in equation 6.32, provided the segment length is in single curvature bending, but modified values must be used for double curvature bending.

A worked example of checking a monosymmetric T-beam is given in section 6.13.7.

6.8 Non-uniform beams

6.8.1 ELASTIC BUCKLING RESISTANCE

Non-uniform beams are often more efficient than beams of constant section, and are frequently used in situations where the major axis bending moment varies along the length of the beam. Non-uniform beams of narrow rectangular section are usually tapered in their depth. Non-uniform I-beams may be tapered in their depth, or less commonly in their flange width, and rarely in their flange thickness, while steps in flange width or thickness are common.

Depth reductions in narrow rectangular beams produce significant reductions in their minor axis flexural rigidities EI_y and torsional rigidities GJ. Because of this, there are also significant reductions in their resistances to flexural–torsional buckling. Closed form solutions for the elastic buckling loads of many tapered beams and cantilevers are given in the papers cited in [13, 16, 48, 49].

Depth reductions in I-beams have no effect on the minor axis flexural rigidity EI_y, and little effect on the torsional rigidity GJ, although they produce significant reductions in the warping rigidity EI_w. It follows that the resistance to buckling of a beam which does not depend primarily on its warping rigidity is comparatively insensitive to depth tapering. On the other hand, reductions in the flange width cause significant reductions in GJ and even greater reductions in EI_y and EI_w, while reductions in flange thickness cause corresponding reductions in EI_y and EI_w and in GJ. Thus the resistance to buckling varies significantly with changes in the flange geometry.

General numerical methods of calculating the elastic buckling loads of tapered I-beams have been developed in [48–50], while the elastic and inelastic buckling of tapered monosymmetric I-beams is discussed in [51, 52]. Solutions for beams with constant flanges and linearly tapered depths under unequal end moments are given in [53, 54]. The buckling of I-beams with stepped flanges has also been investigated, and many solutions are tabulated in [55].

6.8.2 DESIGN RULES

Non-uniform beams can be designed according to BS5950 by using $m_{LT} = 1$ and the properties of the cross-section of maximum moment, provided the value of the equivalent slenderness λ_{LT} calculated using equation 6.28 is increased by multiplying it by an equivalent slenderness factor

$$n = 1.5 - 0.5R_f \geq 1.0 \tag{6.78}$$

in which R_f is the ratio of the flange areas at the points of minimum and maximum moment, respectively.

A slightly different method is allowed in the Australian standard AS4100 [38]. For this, non-uniform beams can be designed by using the properties of

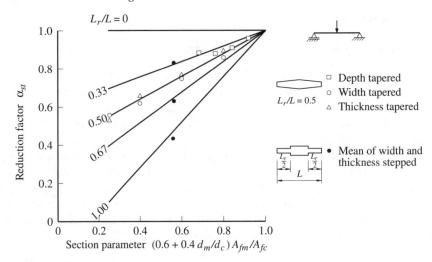

Fig. 6.28 Reduction factors for stepped and tapered I-beams

the most critical cross-section (i.e. where the ratio M^*/M_{cx} of the design bending moment to the section capacity is greatest), provided the elastic buckling moment so calculated is reduced by multiplying it by

$$\alpha_{st} = 1.0 - 1.2\left(\frac{L_r}{L}\right)\left\{1 - \left(0.6 + 0.4\frac{d_m}{d_c}\right)\frac{A_{fm}}{A_{fc}}\right\} \tag{6.79}$$

in which A_{fm}, A_{fc} are the flange areas and d_m, d_c the section depths at the minimum and critical cross-sections, and L_r is either the portion of the beam length which is reduced in section, or is $0.5L$ for a tapered beam. This equation agrees well with the buckling solutions shown in Fig. 6.28 for central concentrated loads on stepped [55] and tapered [48] beams whose minimum cross-sections are at their simply supported ends. Alternatively, the method of design by buckling analysis allows the elastic buckling solutions for stepped or tapered beams to be used directly, provided that the moments used are for the most critical cross-section.

A worked example for checking a stepped beam is given in section 6.13.8.

6.9 Design against lateral buckling

6.9.1 GENERAL

For the strength design of a beam against lateral buckling, the distribution of the bending moment is determined by an elastic analysis (if the beam is statically indeterminate), or by statics (if the beam is statically determinate). The strength design loads are obtained by summing up the loads multiplied by the appropriate partial load factors γ_f in Table 2 of BS5950 (see section 1.5.7).

The BS5950 general and simple methods of checking a uniform equal flanged

General Method			Simple Method
Rolled (R) or Welded (W)			Rolled (R)
p_y $M^* \leq M_{cx}$	T9	T9	p_y $M^* \leq M_{cx}$
L_E $\lambda = L_E / r_y$	T13	T13	L_E L_E / r_y
$u = 0.9$ (R), 1.0 (W), or B.2.3 $\eta = 0.5$ $x = D / T$ or B.2.3 $v = \mathrm{fn}\,(\eta, \lambda / x)$ $\beta_w = Z_{ex} / S_x$ $\lambda_{LT} = u\,v\,\lambda\beta_w$	T19		D / T $\beta_w = Z_{ex} / S_x$ $(\beta_w)L_E/r_y$
$p_b = \mathrm{fn}\,(p_y, \lambda_{LT})$	(R) T16 (W) T17	T20	$p_b = \mathrm{fn}(p_y, (\beta_w)L_E/r_y, D/T)$
m_{LT} $M^* \leq p_b\, Z_{ex} / m_{LT} \leq M_{cx}$	T18	T18	m_{LT} $M^* \leq p_b\, Z_{ex} / m_{LT} \leq M_{cx}$

Fig. 6.29 Flow chart for checking the buckling resistances of uniform equal flanged beams

beam are summarized in Fig. 6.29. In both methods, the section moment capacity M_{cx} is checked first, and then the effective length L_E is determined using Table 13, and then the slenderness $\lambda = L_E/r_y$. For the general method, the equivalent slenderness λ_{LT} is calculated using either simple approximations for u and x, or else the more accurate equations 6.29a,d given in Annex B.2.3 of BS5950 (accurate values for u and x are given in some tables of section properties). λ_{LT} is then used in Table 16 or 17 to determine the bending strength p_b. The member resistance $M_b = p_b Z_{ex}/m_{LT}$ is then checked. For the simple method, the values of $\sqrt{(\beta_w)}L_E/r_y$ and D/T are used in Table 20 to determine p_b.

The beam must also be checked for shear, shear and bending, and bearing as discussed in sections 5.6.1.5, 5.6.1.6, and 5.6.1.7, and for serviceability, as discussed in section 5.7.

When a beam is to be designed, the beam section is not known, and so a trial section must be chosen. An iterative process is then used, in which the trial section is evaluated and a new trial section chosen, until a satisfactory section is found. One method of finding a first trial section is to make an initial guess for p_b/m_{LT} (say $0.5p_y$), to use this to calculate a target plastic section modulus

$$S_x \geq m_{LT}M^*/p_b, \tag{6.80}$$

and then select a suitable trial section.

After the trial section has been evaluated, its value of p_b/m_{LT} can then be used to select a new trial section. The iterative process usually converges within a few cycles, but convergence can be hastened by using the mean of the previous and current values of p_b/m_{LT} in the calculation of the target section modulus.

The following subsections describe in more detail the BS5950 method of checking uniform equal flanged beams which are either supported at both ends, continuous or braced, or cantilevered. Worked examples are given in sections 6.13.1–6.13.6. The checking of monosymmetric beams was discussed in section 6.7.2, and that of non-uniform beams in section 6.8.2.

6.9.2 CHECKING BEAMS SUPPORTED AT BOTH ENDS

6.9.2.1 Section moment capacity

The classification of a specified beam cross-section as plastic, compact, semi-compact, or slender is described in sections 4.7.2 and 5.6.1.2, and the determination of the design section moment capacity M_{cx} in sections 4.7.2 and 5.6.1.3.

6.9.2.2 End restraints

The effects of end restraints are allowed for in BS5950 by adjusting the effective length L_E using Table 13. The basic case is for a beam or segment with normal (not destabilizing) loading which is fully restrained against lateral deflection and torsion and free to rotate in plan at both ends, for which the effective length L_E is the segment distance L_{LT} between the beam or segment ends.

Restraints against end rotation in plan decrease the effective length to $0.8L_{LT}$ for partial restraint, and to $0.7L_{LT}$ for full restraint. These are practical values of the theoretical predictions shown in Fig. 6.18.

Less than full end restraints against torsion increases the effective length to $(L_{LT} + 2D)$ for positive connections of the bottom flange to its supports (as in Fig. 6.19), and to $(1.2L_{LT} + 2D)$ when the restraint is only by bearing.

Many cases of practical restraint are discussed in [56], and for the Australian standard AS4100 [38] in [57].

6.9.2.3 Destabilizing loading

The effects of destabilizing loading at the top flange (see Figs 6.5 and 6.6) are allowed for in BS5950 in two ways. First of all, no account is allowed to be taken of the effects of the bending moment distribution, so that the equivalent uniform moment factor is $m_{LT} = 1$.

Secondly, the effective length L_E is increased by 20% approximately using Table 13, so that $0.7L_{LT}$ is increased to $0.85L_{LT}$, $0.75L_{LT}$ to $0.9L_{LT}$, $0.8L_{LT}$ to $0.95L_{LT}$, L_{LT} to $1.2L_{LT}$, and $1.2L_{LT}$ to $1.4L_{LT}$.

6.9.2.4 Moment distribution

BS5950 allows for the effects of the moment distribution by using an equivalent uniform moment factor m_{LT} in equation 6.32. Values of m_{LT} for non-destabilizing loads are given in Table 18, which are approximately equal to $1/\sqrt{\alpha_m}$ where α_m is the moment modification factor of equation 6.5 and Fig. 6.7. For destabilizing loads, BS5950 requires $m_{LT} = 1$.

6.9.2.5 Member moment resistance

The member moment resistance is calculated from

$$M_b/m_{LT} = p_b Z_{ex}/m_{LT} \qquad (6.80)$$

in which Z_{ex} is the effective section modulus ($Z_{ex} = S_x$ for plastic and compact sections, $S_x > Z_{ex} \geqslant Z_x$ for semi-compact sections, and $Z_{ex} < Z_x$ for slender sections). The member is satisfactory if the design moment M_x^* satisfies

$$M_x^* \leqslant M_b/m_{LT} \leqslant M_{cx} \qquad (6.81)$$

A worked example of checking the member moment resistance of a beam supported at both ends is given in section 6.13.1.

6.9.3 CHECKING A CONTINUOUS OR A BRACED BEAM

When beams supported at both ends have intermediate supports or restraints which effectively prevent lateral deflection of the compression flange, then these supports or restraints divide the beam into a series of segments.

A conservative method of checking such a beam is to check each segment individually, using its maximum moment, moment distribution and an effective length L_E which is based on there being no restraints against rotation on plan.

A less conservative method is summarized in section 6.6.2.2, in which the conservative method described above is used to identify the critical segment, and a reduced effective length ratio k_e for this segment is determined using Fig. 3.20a.

6.9.4 CHECKING A CANTILEVER

The effective length L_E of a cantilever depends on its length, its end restraints, and whether the load is destabilizing (gravity load at the top flange) or not. Table 14 of BS5950 gives expressions for L_E for cantilevers. The increases for destabilizing loads are very large when the cantilever tip is unrestrained.

6.10 Appendix – elastic beams

6.10.1 BUCKLING OF STRAIGHT BEAMS

6.10.1.1 Beams with equal end moments

The elastic buckling moment M_{ob} of the beam shown in Fig. 6.3 can be determined by finding a deflected and twisted position which is one of equilibrium. The differential equilibrium equation of bending of the beam is

$$EI_y\frac{d^2u}{dz^2} = -M_{ob}\phi, \tag{6.82}$$

which states that the internal minor axis moment of resistance $EI_y(d^2u/dz^2)$ must exactly balance the disturbing component $-M_{ob}\phi$ of the applied bending moment M_{ob} at every point along the length of the beam. The differential equation of torsion of the beam is

$$GJ\frac{d\phi}{dz} - EI_w\frac{d^3\phi}{dz^3} = M_{ob}\frac{du}{dz}, \tag{6.83}$$

which states that the sum of the internal resistance to uniform torsion $GJ(d\phi/dz)$ and the internal resistance to warping torsion $-EI_w(d^3\phi/dz^3)$ must exactly balance the disturbing torque $M_{ob}(du/dz)$ caused by the applied moment M_{ob} at every point along the length of the beam.

The derivation of the left hand side of equation 6.83 is fully discussed in sections 10.2 and 10.3. The torsional rigidity GJ in the first term determines the beam's resistance to uniform torsion, for which the rate of twist $d\phi/dz$ is constant, as shown in Fig. 10.1a. For thin-walled open sections, the torsion constant J is approximately given by the summation

$$J \approx \Sigma bt^3/3$$

in which b is the length and t the thickness of each rectangular element of the cross-section. Accurate expressions for J are given in [3: Chapter 10] from which the values for hot-rolled I-sections have been calculated [5, 6: Chapter 10]. The warping rigidity EI_w in the second term of equation 6.83 determines the additional resistance to non-uniform torsion, for which the flanges bend in opposite directions, as shown in Fig. 10.1b and c. When this flange bending varies along the length of the beam, flange shear forces are induced which exert a torque $-EI_w(d^3\phi/dz^3)$. For equal flanged I-beams,

$$I_w = \frac{I_y d_f^2}{4}$$

in which d_f is the distance between flange centroids. An expression for I_w for a monosymmetric I-beam is given in Fig. 6.26.

When equations 6.82 and 6.83 are both satisfied at all points along the beam, the deflected and twisted position is one of equilibrium. Such a position is defined by the buckled shape

$$u = \frac{M_{ob}}{\pi^2 EI_y/L^2} \phi = \delta \sin \frac{\pi z}{L} \tag{6.1}$$

in which the maximum deflection δ is indeterminate. This buckled shape satisfies the boundary conditions at the supports of lateral deflection prevented,

$$(u)_0 = (u)_L = 0, \tag{6.84}$$

twist rotation prevented,

$$(\phi)_0 = (\phi)_L = 0, \tag{6.85}$$

and ends free to warp (see section 10.8.3),

$$\left(\frac{d^2\phi}{dz^2}\right)_0 = \left(\frac{d^2\phi}{dz^2}\right)_L = 0. \tag{6.86}$$

Equation 6.1 also satisfies the differential equilibrium equations (equations 6.82 and 6.83) when $M_{ob} = M_{yz}$, where

$$M_{yz} = \sqrt{\left\{\left(\frac{\pi^2 EI_y}{L^2}\right)\left(GJ + \frac{\pi^2 EI_w}{L^2}\right)\right\}}, \tag{6.3}$$

which defines the moment at elastic flexural–torsional buckling.

6.10.1.2 Beams with unequal end moments

The major axis bending moment M_x and shear V_y in the beam with unequal end moments M and $\beta_m M$ shown in Fig. 6.4a are given by

$$M_x = M - (1 + \beta_m)Mz/L,$$
$$V_y = -(1 + \beta_m)M/L.$$

When the beam buckles, the minor axis bending equation is

$$EI_y \frac{d^2 u}{dz^2} = -M_x \phi,$$

and the torsion equation is

$$GJ \frac{d\phi}{dz} - EI_w \frac{d^3\phi}{dz^3} = M_x \frac{du}{dz} - V_y u.$$

These equations reduce to equations 6.82 and 6.83 for the case of equal and opposite end moments ($\beta_m = -1$).

Closed form solutions of these equations are not available, but numerical methods [3–11] have been used. The numerical solutions can be conveniently expressed in the form

$$M_{ob} = \alpha_m M_{yz} \tag{6.4}$$

in which the factor α_m accounts for the effect of the non-uniform distribution of the bending moment M_x on elastic flexural–torsional buckling. The variation of α_m with the end moment ratio β_m is shown in Fig. 6.4c for the two extreme values of the beam parameter

$$K = \sqrt{\left(\frac{\pi^2 EI_w}{GJL^2}\right)}$$

of 0.1 and 3. Also shown in Fig. 6.4c is the approximation

$$\alpha_m = 1.75 + 1.05\beta_m + 0.3\beta_m^2 \leq 2.56. \tag{6.5}$$

6.10.1.3 Beams with central concentrated loads

The major axis bending moment M_x and the shear V_y in the beam with a central concentrated load Q shown in Fig. 6.5 are given by

$$M_x = Qz/2 - Q\langle z - L/2\rangle,$$
$$V_y = Q/2 - Q\langle z - L/2\rangle^0$$

in which the values of the second terms are taken as zero when the values inside the Macaulay brackets $\langle\,\rangle$ are negative. When the beam buckles, the minor axis bending equation is

$$EI_y\frac{d^2u}{dz^2} = -M_x\phi,$$

and the torsion equation is

$$GJ\frac{d\phi}{dz} - EI_w\frac{d^3\phi}{dz^3} = \frac{Q}{2}(u - y_Q\phi)_{L/2}(1 - 2\langle z - L/2\rangle^0) + M_x\frac{du}{dz} - V_yu$$

in which y_Q is the distance of the point of application of the load below the centroid, and $(Q/2)(u - y_Q\phi)_{L/2}$ is the end torque.

Numerical solutions of these equations for the dimensionless buckling load $QL^2/\sqrt{(EI_yGJ)}$ are available [3, 13, 16, 19], and some of these are shown in Fig. 6.6, in which the dimensionless height ε of the point of application of the load is generally given by

$$\varepsilon = \frac{y_Q}{L}\sqrt{\left(\frac{EI_y}{GJ}\right)}, \tag{6.7}$$

or for the particular case of equal flanged I-beams, by

$$\varepsilon = \frac{K}{\pi}\frac{2y_Q}{d_f}.$$

6.10.1.4 *Cantilevers with concentrated end loads*

The elastic buckling of a cantilever with a concentrated end load Q applied at a distance y_Q below the centroid can be predicted from the solutions of the differential equations of minor axis bending

$$EI_y \frac{d^2u}{dz^2} = -M_x \phi,$$

and of torsion

$$GJ \frac{d\phi}{dz} - EI_w \frac{d^3\phi}{dz^3} = Q(u - y_Q\phi)_L + M_x \frac{du}{dz} - V_y u$$

in which

$$M_x = -Q(L - z), \quad V_y = Q.$$

These solutions must satisfy the fixed end ($z = 0$) boundary conditions of

$$(u)_0 = (\phi)_0 = (du/dz)_0 = (d\phi/dz)_0 = 0,$$

and the condition that the end $z = L$ is free to warp, whence

$$(d^2\phi/dz^2)_L = 0.$$

Numerical solutions of these equations are available [16, 19–21], and some of these are shown in Fig. 6.8.

6.10.2 DEFORMATIONS OF BEAMS WITH INITIAL CURVATURES AND TWIST

The deformations of a simply supported beam with initial curvature and twist caused by equal and opposite end moments M can be analysed by considering the minor axis bending and torsion equations

$$EI_y \frac{d^2u}{dz^2} = -M(\phi + \phi_0), \tag{6.87}$$

$$GJ \frac{d\phi}{dz} - EI_w \frac{d^3\phi}{dz^3} = M\left(\frac{du}{dz} - \frac{du_0}{dz} \right), \tag{6.88}$$

which are obtained from equations 6.85 and 6.86 by adding the additional moment $M\phi_0$ and torque $M(du_0/dz)$ induced by the initial twist and curvature.

If the initial curvature and twist are such that

$$\frac{u_0}{\delta_0} = \frac{\phi_0}{\theta_0} = \sin \frac{\pi z}{L} \tag{6.15}$$

in which the central initial lack of straightness δ_0 and twist θ_0 are related by

$$\frac{\delta_0}{\theta_0} = \frac{M_{yz}}{\pi^2 EI_y/L^2}, \tag{6.16}$$

then the solution of equations 6.87 and 6.88 which satisfies the boundary conditions (equations 6.84, 6.85) is given by

$$\frac{u}{\delta} = \frac{\phi}{\theta} = \sin \frac{\pi z}{L} \qquad (6.17)$$

in which

$$\frac{\delta}{\delta_0} = \frac{\theta}{\theta_0} = \frac{M/M_{yz}}{1 - M/M_{yz}}. \qquad (6.18)$$

The maximum longitudinal stress in the beam is the sum of the stresses due to major axis bending, minor axis bending, and warping, and is equal to

$$f_{max} = \frac{M}{Z_x} - \frac{EI_y}{Z_y}\left(\frac{d^2(u + d_f\phi/2)}{dz^2}\right)_{L/2}.$$

If the elastic limit is taken as the yield stress f_y, then the limiting nominal stress $f_L = M_L/Z_x$ for which the above elastic analysis is valid is given by

$$f_L = f_y - \frac{\delta_0 N_{oy}}{M_{yz}}\left(1 + \frac{d_f}{2}\frac{N_{oy}}{M_{yz}}\right)\frac{1}{Z_y}\frac{M_L}{1 - M_L/M_{yz}},$$

or

$$M_L = M_Y - \frac{\delta_0 N_{oy}}{M_{yz}}\left(1 + \frac{d_f}{2}\frac{N_{oy}}{M_{yz}}\right)\frac{Z_x}{Z_y}\frac{M_L}{1 - M_L/M_{yz}}$$

in which $N_{oy} = \pi^2 EI_y/L^2$, $M_L = f_L Z_x$ is the limiting moment at first yield, and $M_Y = f_y Z_x$ the nominal first yield moment. This can be solved for the dimensionless limiting moment M_L/M_Y. In the case where the central lack of straightness δ_0 is given by

$$\frac{\delta_0 N_{oy}}{M_{yz}} = \theta_0 = \frac{Z_y/Z_x}{1 + (d_f/2)(N_{oy}/M_{yz})}\frac{1}{4}\frac{M_Y}{M_{yz}}, \qquad (6.20)$$

the dimensionless limiting moment simplifies to

$$\frac{M_L}{M_Y} = \left(\frac{1.25 + M_{yz}/M_Y}{2}\right) - \sqrt{\left[\left(\frac{1.25 + M_{yz}/M_Y}{2}\right)^2 - \frac{M_{yz}}{M_Y}\right]}. \qquad (6.19)$$

6.11 Appendix – effective lengths of beams

6.11.1 BEAMS WITH ELASTIC END RESTRAINTS

The beam shown in Fig. 6.18 is restrained at its ends against minor axis rotations du/dz and against warping rotations $(d_f/2)d\phi/dz$, and the boundary conditions at the end $z = L/2$ can be expressed in the form of

$$\frac{M_B + M_T}{(du/dz)_{L/2}} = \frac{-EI_y}{L}\frac{2R_2}{1 - R_2},$$

$$\frac{M_T - M_B}{(d_f/2)(d\phi/dz)_{L/2}} = \frac{-EI_y}{L} \frac{2R_4}{1 - R_4}$$

in which M_T and M_B are the flange minor axis end restraining moments

$$M_T = \tfrac{1}{2}EI_y(d^2u/dz^2)_{L/2} + (d_f/4)EI_y(d^2\phi/dz^2)_{L/2},$$
$$M_B = \tfrac{1}{2}EI_y(d^2u/dz^2)_{L/2} - (d_f/4)EI_y(d^2\phi/dz^2)_{L/2},$$

and R_2 and R_4 are the dimensionless minor axis bending and warping end restraint parameters. If the beam is symmetrically restrained, then similar conditions apply at the end $z = -L/2$. The other support conditions are

$$(u)_{\pm L/2} = (\phi)_{\pm L/2} = 0.$$

The particular case for which the minor axis and warping end restraints are equal, so that

$$R_2 = R_4 = R \tag{6.52}$$

may be analysed. The differential equilibrium equations for a buckled position u, ϕ of the beam are

$$EI_y\frac{d^2u}{dz^2} = -M_{ob}\phi + (M_B + M_T),$$

$$GJ\frac{d\phi}{dz} - EI_w\frac{d^3\phi}{dz^3} = M_{ob}\frac{du}{dz}.$$

These differential equations and the boundary conditions are satisfied by the buckled shape

$$u = \frac{M_{ob}\phi}{\pi^2 EI_y/k_e^2 L^2} = A\left(\cos\frac{\pi z}{k_e L} - \cos\frac{\pi}{2k_e}\right)$$

in which the effective length ratio k_e satisfies

$$\frac{R}{1 - R} = -\frac{\pi}{2k_e}\cot\frac{\pi}{2k_e}. \tag{6.53}$$

The solutions of this equation are shown in Fig. 6.18c.

6.12 Appendix – monosymmetric beams

When a monosymmetric beam (see Fig. 6.26) is bent in its plane of symmetry and twisted, the longitudinal bending stresses f exert a torque which is similar to that which causes some short concentrically loaded compression members to buckle torsionally (see sections 3.6.5 and 3.11.1). The longitudinal bending force

$$f\delta A = \frac{M_x y}{I_x}\delta A$$

acting on an element δA of the cross-section rotates $a_0(d\phi/dz)$ (where a_0 is the distance to the axis of twist through the shear centre $x_0 = 0$, y_0), and so its transverse component $f\delta A a_0 (d\phi/dz)$ exerts a torque $f\delta A a_0 (d\phi/dz) a_0$ about the axis of twist. The total torque T_M exerted is

$$T_M = \frac{d\phi}{dz} \int_A a_0^2 \frac{M_x y}{I_x} dA$$

in which

$$a_0^2 = x^2 + (y - y_0)^2.$$

Thus

$$T_M = M_x \beta_x \frac{d\phi}{dz} \tag{6.71}$$

in which the monosymmetry property β_x of the cross-section is given by

$$\beta_x = \frac{1}{I_x} \left\{ \int_A x^2 y dA + \int_A y^3 dA \right\} - 2y_0. \tag{6.72}$$

An explicit expression for β_x for a monosymmetric I-section is given in Fig. 6.26, and this can also be used for tee-sections by putting the flange thickness t_1 or t_2 equal to zero. Also given in Fig. 6.26 is an explicit expression for the warping section constant I_w of a monosymmetric I-section. For a tee-section, I_w is zero.

6.13 Worked examples

6.13.1 EXAMPLE 1 – CHECKING A BEAM SUPPORTED AT BOTH ENDS

Problem.   The 7.5 m long 610 × 229 UB 125 of S275 steel shown in Fig. 6.30 is simply supported at both ends where lateral deflections u are effectively prevented and twist rotations ϕ are partially restrained. Check the adequacy of the beam for a central concentrated top flange load caused by an unfactored dead load of 60 kN (which includes an allowance for self-weight) and an unfactored imposed load of 100 kN.

Design bending moment.

$$M_x^* = \{(1.4 \times 60) + (1.6 \times 100)\} \times 7.5/4 = 457.5 \text{ kNm}.$$

Section capacity.
As in Section 5.12.15, $p_y = 265$ N/mm², the section is plastic, and the section capacity is $M_{cx} = 973.3$ kNm > 457.5 kNm $= M_x^*$ and the section capacity is adequate.

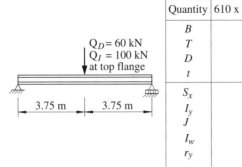

Quantity	610 x 229 UB 125	457 x 191 UB 82	Units
B	229	191.3	mm
T	19.6	16.0	mm
D	611.9	460.2	mm
t	11.9	9.9	mm
S_x	3673	1832	cm³
I_y	3932	1871	cm⁴
J	154	69.2	cm⁴
I_w	3.45	0.923	dm⁶
r_y	49.7	42.3	mm
r	12.7	10.2	mm

$Q_D = 60$ kN
$Q_I = 100$ kN
at top flange

3.75 m   3.75 m

(a) Example 1   (b) Section properties

Fig. 6.30 Examples 1 and 4

Member resistance.
For destabilizing loading at the top flange, $m_{LT} = 1$ 4.3.6.6

$$\beta_w = 1.0$$ 4.3.6.9
$$L_E = 1.2 \times 7500 + 2 \times 611.9 = 10224 \text{ mm}$$ T13
$$\surd(\beta_w)L_E/r_y = \surd(1.0) \times 10224/49.7 = 205.7$$
$$x = D/T = 611.9/19.6 = 31.2$$

Using the simple method and interpolating linearly in Table 20(2),

$$p_b = \frac{(79 \times 4.3 + 75 \times 5.7)}{10} \times \frac{3.8}{5} + \frac{(73 \times 4.3 + 69 \times 5.7)}{10} \times \frac{1.2}{5}$$

$$= 75.3 \text{ N/mm}^2$$

(Note that it is common practice to avoid this linear interpolation by determining approximate values of p_b by 'eye'.)

$$M_b/m_{LT} = 75.3 \times 3673 \times 10^3/1.0 \text{ Nmm}$$ 4.3.6.2,4
$$= 276.5 \text{ kNm} < 457.5 \text{ kNm} = M_x^*$$

and the beam is inadequate.
    Alternatively, using the general method,

$$u = 0.9$$ 4.3.6.8
$$\lambda = 10224/49.7 = 205.7$$ 4.3.6.7
$$\lambda/x = 205.7/31.2 = 6.59$$
$$\eta = 0.5$$ 4.3.6.7
$$v = (0.75 \times 0.41 + 0.73 \times 0.09)/0.5 = 0.746$$ T19
$$\beta_w = 1.0$$ 4.3.6.9
$$\lambda_{LT} = 0.9 \times 0.746 \times 205.7 \times \surd 1.0 = 138.1$$ 4.3.6.7
$$p_b = (78 \times 1.9 + 74 \times 3.1)/5 = 75.5 \text{ N/mm}^2$$ T16

$$M_b/m_{LT} = 75.5 \times 3673 \times 10^3/1.0 \text{ Nmm} \qquad\qquad 4.3.6.2,4$$
$$= 277.4 \text{ kNm} < 457.5 \text{ kNm} = M_x^*$$

and the beam is inadequate.

6.13.2 EXAMPLE 2 – CHECKING A BRACED BEAM

Problem. The 9 m long 254 × 146 UB 37 braced beam of S275 steel shown in Fig. 6.31a and b has a central concentrated design load of 70 kN (which includes an allowance for self-weight) and a design end moment of 70 kNm. Lateral deflections u and twist rotations ϕ are effectively prevented at both ends and by a brace at mid-span. Check the adequacy of the braced beam.

Design bending moment.

$$R_3^* = \{(70 \times 4.5) - 70\}/9 = 27.22 \text{ kN}, M_2^* = 27.22 \times 4.5 = 122.5 \text{ kNm}.$$

Section capacity.

$$T = 10.9 \text{ mm}, p_y = 275 \text{ N/mm}^2 \qquad\qquad \text{T9}$$
$$\varepsilon = \sqrt{(275/275)} = 1.0$$
$$b/(T\varepsilon) = (146.4/2)/(10.9 \times 1.0)$$
$$= 6.72 < 9 \text{ and the flange is plastic.} \qquad\qquad \text{T11}$$
$$d/(t\varepsilon) = (256 - 2 \times 10.9 - 2 \times 7.6)/(6.4 \times 1.0)$$
$$= 34.2 < 80 \text{ and the web is plastic.} \qquad\qquad \text{T11}$$
$$M_{cx} = 275 \times 485 \times 10^3 \text{ Nmm} \qquad\qquad 4.2.5.2$$
$$= 133.4 \text{ kNm} > 122.5 \text{ kNm} = M^*$$

and the section capacity is adequate.

Member resistance.
The beam is fully restrained at mid-span, and so consists of two equal length segments 12 and 23. By inspection, the check will be controlled by segment 23 which has the lower moment gradient, and therefore the higher value of m_{LT}.
Because lateral deflections and twist rotations are prevented at both ends of the segment 23,

$$L_E = 1.0 \times 4500 = 4500 \text{ mm} \qquad\qquad \text{T13}$$
$$\beta_w = 1.0 \qquad\qquad 4.3.6.9$$
$$\sqrt{(\beta_w)}L_E/r_y = \sqrt{(1.0)} \times 4500/34.7 = 129.7$$
$$x = D/T = 256/10.9 = 23.5$$

Using the simple method and interpolating linearly in Table 20(1),

$$p_b = \frac{(152 \times 0.3 + 147 \times 4.7)}{5} \times \frac{1.5}{5} + \frac{(140 \times 0.3 + 135 \times 4.7)}{5} \times \frac{3.5}{5}$$

$$= 138.9 \text{ N/mm}^2$$

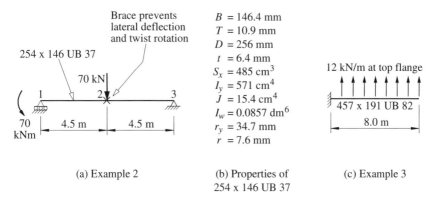

(a) Example 2

(b) Properties of
254 x 146 UB 37

(c) Example 3

Fig. 6.31 Examples 2 and 3

For an end moment ratio of $\beta = 0/122.5 = 0$, $m_{LT} = 0.60$ T18

$$M_b/m_{LT} = 138.9 \times 485 \times 10^3/0.60 \text{ Nmm} \qquad 4.3.6.2,4$$
$$= 112.3 \text{ kNm} < 122.5 \text{ kNm} = M_x^*$$

and the segment is inadequate.
Alternatively, using the general method,

$$u = 0.9 \qquad\qquad\qquad\qquad\qquad\qquad 4.3.6.8$$
$$\lambda = 4500/34.7 = 129.7 \qquad\qquad\qquad 4.3.6.7$$
$$\lambda/x = 129.7/23.5 = 5.52$$
$$\eta = 0.5 \qquad\qquad\qquad\qquad\qquad\qquad 4.3.6.7$$
$$v = 0.79 \qquad\qquad\qquad\qquad\qquad\qquad \text{T19}$$
$$\beta_w = 1.0 \qquad\qquad\qquad\qquad\qquad\qquad 4.3.6.9$$
$$\lambda_{LT} = 0.9 \times 0.79 \times 129.7 \times \sqrt{1.0} = 92.2 \qquad 4.3.6.7$$
$$p_b = (144 \times 2.8 + 134 \times 2.2)/5 = 139.6 \text{ N/mm}^2 \qquad \text{T16}$$
$$M_b/m_{LT} = 139.6 \times 485 \times 10^3/0.60 \text{ Nmm} \qquad 4.3.6.2,4$$
$$= 112.8 \text{ kNm} < 122.5 \text{ kNm} = M_x^*$$

and the segment is inadequate.

6.13.3 EXAMPLE 3 – CHECKING A CANTILEVER

Problem. The 8.0 m long 457 × 191 UB 82 cantilever of S275 steel shown in Fig. 6.31c has the section properties shown in Fig. 6.30b. The cantilever is continuous with lateral and torsional restraint at the support, free at the tip, and has a factored upwards design uniformly distributed load of 12 kN/m (which includes an allowance for self-weight) acting at the top flange. Check the adequacy of the cantilever.

Design bending moment.

$$M_x^* = 12 \times 8^2/2 = 384 \text{ kNm}$$

Section capacity.

$$T = 16.0 \text{ mm}, p_y = 275 \text{ N/mm}^2 \qquad \text{T9}$$
$$\varepsilon = \sqrt{(275/275)} = 1.0$$
$$b/(T\varepsilon) = (191.3/2)/(16.0 \times 1.0)$$
$$= 5.98 < 9 \text{ and the flange is plastic.} \qquad \text{T11}$$
$$d/(t\varepsilon) = (460.2 - 2 \times 16.0 - 2 \times 10.2)/(9.9 \times 1.0)$$
$$= 41.2 < 80 \text{ and the web is plastic.} \qquad \text{T11}$$
$$M_{cx} = 275 \times 1832 \times 10^3 \text{ Nmm} \qquad \text{4.2.5.2}$$
$$= 503.8 \text{ kNm} > 384 \text{ kNm} = M^*$$

and the section capacity is adequate.

Member resistance.
The upwards load at the top flange is equivalent to a downwards load at the
bottom flange, and so the load is not destabilizing. Thus

$$L_E = 1.0 \times 8000 = 8000 \text{ mm} \qquad \text{T14}$$
$$\beta_w = 1.0 \qquad \text{4.3.6.9}$$
$$\sqrt{(\beta_w)}L_E/r_y = \sqrt{(1.0)} \times 8000/42.3 = 189.1$$
$$D/T = 460.2/16.0 = 28.8$$

Interpolating linearly in Table 20(1),

$$p_b = \frac{(96 \times 0.9 + 93 \times 4.1)}{5} \times \frac{1.2}{5} + \frac{(87 \times 0.9 + 84 \times 4.1)}{5} \times \frac{3.8}{5}$$

$$= 86.7 \text{ N/mm}^2$$

For a cantilever, $m_{LT} = 1.0$ \qquad T18

$$M_b/m_{LT} = 86.7 \times 1832 \times 10^3/1.0 \text{ Nmm} \qquad \text{4.3.6.2,4}$$
$$= 158.8 \text{ kNm} < 384 \text{ kNm} = M_x^*$$

and the cantilever appears to be inadequate (but see section 6.13.5).

6.13.4 EXAMPLE 4 – DESIGNING A BRACED BEAM

Problem. Determine a suitable UB of S275 steel for the simply supported
beam of section 6.13.1 if twist rotations are effectively prevented at the ends
and if a brace is added which effectively prevents lateral deflection u and twist
rotation ϕ at mid-span.

Design bending moment.
Using section 6.13.1, $M_x^* = 457.5$ kNm

Selecting a trial section.
The central brace divides the beam into two segments, each of length
$L_{LT} = 3750$ mm, and changes the load from destabilizing to non-destabilizing.

The bending moment in each segment varies linearly from 0 to 457.5 kNm, so that $m_{LT} = 0.60$ (Table 18).

Guess $p_b = 150$ N/mm², so that $S_x \geqslant 457.5 \times 10^6 \times 0.60/150$ mm³ = 1830 cm³. Try a 457 × 191 UB 82 with $S_x = 1832$ cm³ > 1830 cm³.

Section capacity.
As in Section 6.13.3, $p_y = 275$ N/mm², $M_{cx} = 503.8$ kNm > 457.5 kNm = M_x^* and the section capacity is adequate.

Member resistance.
Using the general method,

$$L_E = 3750 \text{ mm} \hspace{4em} \text{T13}$$
$$\lambda = 3750/42.3 = 88.7 \hspace{4em} 4.3.6.7$$
$$u = 0.9 \hspace{4em} 4.3.6.8$$
$$x = D/T = 460.2/16.0 = 28.8 \hspace{4em} 4.3.6.8$$
$$\lambda/x = 88.7/28.8 = 3.08$$
$$\eta = 0.5 \hspace{4em} 4.3.6.7$$
$$v = (0.91 \times 0.42 + 0.89 \times 0.08)/0.5 = 0.907 \hspace{4em} \text{T19}$$
$$\beta_w = 1.0 \hspace{4em} 4.3.6.9$$
$$\lambda_{LT} = 0.9 \times 0.907 \times 88.7 \times \sqrt{1.0} = 72.4 \hspace{4em} 4.3.6.7$$
$$p_b = 182 \text{ N/mm}^2 \hspace{4em} \text{T16}$$
$$M_b/m_{LT} = 182 \times 1832 \times 10^3/0.60 \text{ Nmm} \hspace{4em} 4.3.6.2,6$$
$$= 555.7 \text{ kNm} > 503.8 \text{ kNm} = M_{cx}$$

and so the section capacity governs and the beam is adequate. Note that in this case, the initial guess of $p_b = 150$ N/mm² was only approximate. A new guess of $p_b = 180$ N/mm² will lead to a trial of a 457 × 191 UB 74, but this is just inadequate ($M_{cx} = 456$ kNm < 457.5 kNm = M_x^*).

6.13.5 EXAMPLE 5 – CHECKING A BRACED BEAM BY BUCKLING ANALYSIS

Problem. Use the method of design by buckling analysis to check the braced beam of section 6.13.2.

Elastic buckling analysis. The elastic buckling of the beam may be analysed by using the method demonstrated in section 6.6.2.2. This is summarized below, using the same step numbering as in section 6.6.2.2.

(2) The beam is fully restrained at mid-span, and so consists of two segments 12 and 23.

For segment 12 and using Table 18,

$$M_2 = -70 + (122.5 + 70)/4 = -21.88, M_3 = -70 + (122.5 + 70)/2 = 26.25,$$
$$M_4 = -70 + (122.5 + 70) \times 3/4 = 74.38, M_{max} = 122.5$$
$$m_{LT} = (0.2 \times 122.5 + 0.15 \times 21.88 + 0.5 \times 26.25 + 0.15 \times 74.38)/122.5$$
$$= 0.425 \ (<0.44).$$
$$\alpha_{m12} = 1/m_{LT} = 1/0.44 = 2.273$$

For segment 23, $\beta_{23} = 0/122.5 = 0$ and $\alpha_{m23} = 1/0.60 = 1.667$ T18

(3) $L_{E12} = L_{E23} = 4500$ mm

(4) $G = E/\{2(1 + v)\} = 205\,000/\{2 \times (1 + 0.3)\} = 78\,846$ N/mm²

$M_{yz12} = M_{yz23}$

$$= \sqrt{\left[\left[\left(\frac{\pi^2 \times 205\,000 \times 571 \times 10^4}{4500^2}\right) \times \left(78\,846 \times 15.4 \times 10^4 + \frac{\pi^2 \times 205\,000 \times 0.0857 \times 10^{12}}{4500^2}\right)\right]\right]} \text{ N mm}$$

$= 108.7$ kNm

$M_{ob12} = 2.273 \times 108.7 = 247.0$ kNm

$Q_{s12} = 70 \times 247.0/122.5 = 141.2$ kN

$M_{ob23} = 1.667 \times 108.7 = 181.1$ kNm

$Q_{s23} = 70 \times 181.1/122.5 = 103.5$ kN

(5) $Q_{s23} < Q_{s12}$ and so 23 is the critical segment.

(6) $\alpha_{r12} = (3 \times 205\,000 \times 571 \times 10^4/4500) \times (1 - 103.5/141.2)$
$= 208.4 \times 10^6$ Nmm

(7) $\alpha_{23} = 2 \times 205\,000 \times 571 \times 10^4/4500 = 520.2 \times 10^6$ Nmm

(8) $k_2 = 520.2 \times 10^6/(0.5 \times 208.4 \times 10^6 + 520.2 \times 10^6) = 0.83$
$k_3 = 1.0$ (zero restraint at 3).

(9) $k_{e23} = 0.93$, $L_{E23} = 0.93 \times 4500 = 4185$ mm

(10) $M_{ob23} =$

$$1.667 \times \sqrt{\left[\left[\left(\frac{\pi^2 \times 205\,000 \times 571 \times 10^4}{4185^2}\right) \times \left(78\,846 \times 15.4 \times 10^4 + \frac{\pi^2 \times 205\,000 \times 0.0857 \times 10^{12}}{4185^2}\right)\right]\right]} \text{ N mm}$$

$= 201.0$ kNm

Member resistance.

$$M_E = M_{ob23} = 201.0 \text{ kNm}$$
$$p_y = 275 \text{ N/mm}^2, M_{cr} = 133.4 \text{ kNm as in section 6.13.2.}$$

Using equation 6.28,

$$\lambda_{LT} = \sqrt{\left\{\left(\frac{\pi^2 \times 205\,000}{275}\right) \times \left(\frac{133.4}{201.0}\right)\right\}} = 69.9$$

$p_b = (201 \times 0.1 + 188 \times 4.9)/5 = 188.3$ N/mm²   T16

$M_b = 188.3 \times 485 \times 10^3$ Nmm
$= 91.3$ kNm < 122.5 kNm $= M_x^*$ and the beam is inadequate.

Alternatively, if M_E is calculated using M_{ob23}/α_m so that

$$M_E = 201.0/1.667 = 120.6 \text{ kNm}$$

$$\lambda_{LT} = \sqrt{\left\{\left(\frac{\pi^2 \times 205\,000}{275}\right) \times \left(\frac{133.4}{120.6}\right)\right\}} = 90.2$$

$$p_b = (144 \times 4.8 + 134 \times 0.2)/5 = 143.6 \text{ N/mm}^2 \qquad \text{T16}$$

Using $m_{LT} = 0.60$,

$$M_b = 143.6 \times 485 \times 10^3/0.60 \text{ Nmm}$$
$$= 116.1 \text{ kNm} < 122.5 \text{ kNm} = M_x^* \text{ and the beam is a little inadequate.}$$

6.13.6 EXAMPLE 6 – CHECKING A CANTILEVER BY BUCKLING ANALYSIS

Problem. Use the method of design by buckling analysis to check the cantilever of section 6.13.3.

Elastic buckling analysis. The elastic buckling of the cantilever can be analysed by using Fig. 6.8 with

$$K = \sqrt{\{\pi^2 \times 205\,000 \times 0.923 \times 10^{12}/(78\,846 \times 69.2 \times 10^4 \times 8000^2)\}} = 0.73$$

and so $q_{ob}L^3/\{2\sqrt{(EI_yGJ)}\} \approx 19$
and $M_{ob} = q_{ob}L^2/2 = 19\sqrt{(EI_yGJ)}/L$

$$= 19 \times \sqrt{(205\,000 \times 1871 \times 10^4 \times 78\,846 \times 69.2 \times 10^4)}/8000 \text{ Nmm}$$
$$= 1086 \text{ kNm}$$

Member resistance.

$$M_E = M_{ob} = 1086 \text{ kNm}$$
$$p_y = 275 \text{ N/mm}^2, M_{cr} = 503.8 \text{ kNm as in section 6.13.3.}$$

Using equation 6.28,

$$\lambda_{LT} = \sqrt{\left\{\left(\frac{\pi^2 \times 205\,000}{275}\right) \times \left(\frac{503.8}{1086}\right)\right\}} = 58.4$$

$$p_b = (226 \times 1.6 + 213 \times 3.4)/5 = 217.2 \text{ N/mm}^2 \qquad \text{T16}$$
$$M_b = 217.2 \times 1832 \times 10^3 \text{ Nmm}$$
$$= 397.8 \text{ kNm} > 384 \text{ kNm} = M^*$$

and so the cantilever is adequate after all.

6.13.7 EXAMPLE 7 – CHECKING A T-BEAM

Problem. The 5.0 m long simply supported monosymmetric 229 × 305 BT 63 T-beam of S275 steel shown in Fig. 6.32 has the section properties shown in Fig. 6.32c. Determine the uniform bending design moment capacities when the flange is in either compression or tension.

Section capacity (flange in compression).

$$T = 19.6 \text{ mm}, p_y = 265 \text{ N/mm}^2 \qquad \text{T9}$$
$$\varepsilon = \sqrt{(275/265)} = 1.019$$
$$b/(T\varepsilon) = (229/2)/(19.6 \times 1.019)$$
$$= 5.73 < 9 \text{ and the flange is plastic.} \qquad \text{T11}$$

The plastic neutral axis bisects the area, and in this case lies in the flange (at $y_p = \{229 \times 19.6 + (305.9 - 19.6) \times 11.9\}/(2 \times 229) = 17.2$ mm).

In this case the whole web is in tension, and is automatically plastic.

$$M_{cx} = 265 \times 531 \times 10^3 \text{ Nmm} = 140.7 \text{ kNm} \qquad 4.2.5.2$$

Moment resistance (flange in compression).

$$u = \left(\frac{4 \times (531 \times 10^3)^2 \times (1 - 1966/6892)}{(79.6 \times 10^2)^2 \times (305.9 - 19.6/2)^2} \right)^{0.25} = 0.617 \qquad \text{B.2.8.2}$$

$$H = 229^3 \times 19.6^3/144 + (305.9 - 19.6/2)^3 \times 11.9^3/36 \qquad \text{B.2.8.3}$$
$$= 1843 \times 10^6 \text{ mm}^6$$
$$y_0 = 7.5 \times 10 - 19.6/2 = 65.2 \text{ mm}$$

Using $\psi = \beta_x/(D - T/2)$ \hfill B.2.8.2

$$\beta_x = 2 \times 65.2 - \left\{ \begin{matrix} 65.2 \times 229^3 \times 19.6/12 + 229 \times 19.6 \times 65.2^3 + \\ (11.9/4)[(75 - 19.6)^4 - (305.9 - 75)^4] \end{matrix} \right\} \times \frac{1}{6892 \times 10^4}$$
$$= 216.1 \text{ mm}$$
$$\psi = 216.1/(305.9 - 19.6/2) = 0.730$$

(the same result is obtained by using Fig. 6.26 for β_x instead of B.2.8.2). $w = 4 \times 1843 \times 10^6/\{1966 \times 10^4 \times (305.9 - 19.6/2)^2\} = 0.0043$ (which is close to $w = 0$ obtained by using $H (\equiv I_w) = 0$).

$$x = 0.566 \times (305.9 - 19.6/2) \times (79.6 \times 10^2/76.9 \times 10^4)^{0.5} = 17.1 \quad \text{B.2.8.2}$$
$$\lambda = 5000/49.7 = 100.6 \qquad \text{B.2.8.2}$$

$$v = \frac{1}{\{[0.0043 + 0.05 \times (100.6/17.1)^2 + 0.730^2]^{0.5} + 0.730]^{0.5}} = 0.669$$
\hfill B.2.8.2

$$\beta_w = 1.0 \qquad 4.3.6.9$$
$$\lambda_{LT} = 0.617 \times 0.669 \times 100.6 \times \sqrt{1.0} = 41.5 \qquad \text{B.2.8.2}$$
$$p_b = (254 \times 3.5 + 242 \times 1.5)/5 = 250.4 \text{ N/mm}^2 \qquad \text{T16}$$
$$m_{LT} = 1.0 \qquad \text{T18}$$
$$M_b/m_{LT} = 250.4 \times 531 \times 10^3/1.0 \text{ Nmm} \qquad 4.3.6.2,4$$
$$= 133.0 \text{ kNm} < 140.7 \text{ kNm} = m_{cx}$$

Thus $M_x^* \le 133.0$ kNm.

Section capacity (flange in tension).

$$D/(t\varepsilon) = 305.9/(11.9 \times 1.019)$$
$$= 25.23 > 18 \text{ and the web is slender.} \qquad \text{T11}$$

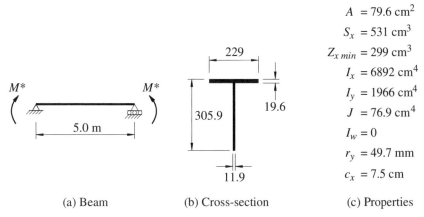

| | | $A = 79.6 \text{ cm}^2$ |

(a) Beam   (b) Cross-section   (c) Properties

Fig. 6.32 Example 7

$$p_{yr} = 265 \times (18/25.23)^2 = 134.9 \text{ N/mm}^2 \qquad 3.6.5$$
$$M_{cx} = 134.9 \times 299 \times 10^3 \text{ Nmm} = 40.3 \text{ kNm} \qquad 4.2.5.2$$

Moment resistance (flange in tension).
As above, $u = 0.617$, $w = 0.0043$, $x = 17.1$, $\lambda = 100.6$, $m_{LT} = 1.0$. With the flange in tension, $\psi = -0.730$.

$$v = \frac{1}{\{[0.0043 + 0.05 \times (100.6/17.1)^2 + 0.730^2]^{0.5} - 0.730\}^{0.5}} = 1.135 \quad \text{B.2.8.2}$$

$$\beta_w = (299 \times 10^3 \times 134.9/265)/(531 \times 10^3) = 0.287 \qquad 4.3.6.9$$
$$\lambda_{LT} = 0.617 \times 1.135 \times 100.6 \times \sqrt{0.287} = 37.7 \qquad \text{B.2.8.2}$$
$$p_b = (265 \times 2.3 + 254 \times 2.7)/5 = 259.1 \text{ N/mm}^2 \qquad \text{T16}$$

Using $Z_{x,eff}/Z_x = p_{yr}/p_y$ \hfill 3.6.5

$$Z_{x,eff} = 299 \times 10^3 \times 134.9/265 = 152.2 \times 10^3 \text{ mm}^3$$
$$M_b/m_{LT} = 259.1 \times 152.2 \times 10^3/1.0 \text{ Nmm} \qquad 4.3.6.2,4$$
$$= 39.4 \text{ kNm} < 40.3 \text{ kNm} = M_{cx}$$

Thus $M_x^* \le 39.4 \text{ kNm}$.

6.13.8 EXAMPLE 8 – CHECKING A STEPPED BEAM

Problem. The simply supported non-uniform welded beam of S275 steel shown in Fig. 6.33 has a central concentrated factored load Q^* acting at the bottom flange. The beam has a 960×16 web, and its equal flanges are 300×32 in the central 2.0 m region and 300×20 in the outer 3.0 m regions. Determine the maximum value of Q^*.

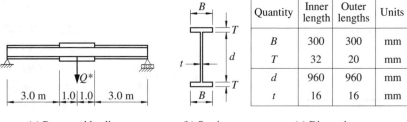

Quantity	Inner length	Outer lengths	Units
B	300	300	mm
T	32	20	mm
d	960	960	mm
t	16	16	mm

(a) Beam and loading   (b) Section   (c) Dimensions

Fig. 6.33 Example 8

Section properties.

$$S_{xc} = 300 \times 32 \times (960 + 32) + 960^2 \times 16/4 = 13.21 \times 10^6 \text{ mm}^3$$
$$S_{xo} = 300 \times 20 \times (960 + 20) + 960^2 \times 16/4 = 9.566 \times 10^6 \text{ mm}^3$$
$$I_{yc} = 2 \times 300^3 \times 32/12 = 144 \times 10^6 \text{ mm}^4$$
$$A_c = 2 \times 300 \times 32 + 960 \times 16 = 34\,560 \text{ mm}^2$$
$$r_{yc} = \sqrt{(144 \times 10^6/34\,560)} = 64.5 \text{ mm}$$

Section capacities.

$T_c = 32$ mm, $p_y = 265$ N/mm^2	T9
$T_o = 20$ mm, $p_y = 265$ N/mm^2	T9
$\varepsilon = \sqrt{(275/265)} = 1.019$	
$b/(T_o\varepsilon) = \{(300 - 16)/2\}/(20 \times 1.019)$	
$= 6.97 < 8$ and the flange is plastic.	T11
$d/(t\varepsilon) = 960/(16 \times 1.019)$	
$= 58.9 < 80$ and the web is plastic.	T11
$M_{co} = 265 \times 9.566 \times 10^6 \text{ Nmm} = 2535 \text{ kNm}$	4.2.5.2
$M_{cc} = 265 \times 13.21 \times 10^6 \text{ Nmm} = 3501 \text{ kNm}$	4.2.5.2

Moment resistance.

$R_f = (300 \times 20)/(300 \times 32) = 0.625$	B.2.5
$n = (1.5 - 0.5 \times 0.625) = 1.188 > 1.0$	B.2.5
$u = 1.0$	4.3.6.8
$\lambda = 8000/64.5 = 124.0$	4.3.6.7
$x = (960 + 2 \times 32)/32 = 32$	4.3.6.8
$\lambda/x = 124.0/32 = 3.88$	
$\eta = 0.5$	4.3.6.7
$v = (0.89 \times 0.12 + 0.86 \times 0.38)/0.5 = 0.867$	T19
$\beta_w = 1.0$	4.3.6.9
$\lambda_{LT} = 1.88 \times 1.0 \times 0.867 \times 124.0 \times \sqrt{1.0} = 127.7$	B.2.5, 4.3.6.7
$p_b = (89 \times 2.3 + 83 \times 2.7)/5 = 85.8$ N/mm^2	T17
$m_{LT} = 1.0$	B.2.5
$M_b/m_{LT} = 85.8 \times 13.21 \times 10^6/1.0 \text{ Nmm}$	4.3.6.2
$= 1133 \text{ kNm} < 3501 \text{ kNm} = M_{cc}$	

$$Q^* \leqslant 4 \times 1133/8 = 566.5 \text{ kN}.$$

Lateral deflection and twist
prevented at end and supports

Lateral deflection and twist
prevented at supports and load points

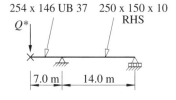

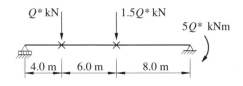

(a) Overhanging beam

(b) Braced beam

Fig. 6.34 Examples 10 and 11

6.14 Unworked examples

6.14.1 EXAMPLE 9 – CHECKING A CONTINUOUS BEAM

A continuous 533×210 UB 92 of S275 steel has two equal spans of 7.0 m. A uniformly distributed factored design load q^* is applied along the centroidal axis. Determine the maximum value of q^*.

6.14.2 EXAMPLE 10 – CHECKING AN OVERHANGING BEAM

The overhanging beam shown in Fig. 6.34a is prevented from twisting and deflecting laterally at its end and supports. The supported segment is a $250 \times 150 \times 10$ RHS of S275 steel and is rigidly connected to the overhanging segment which is a 254×146 UB 37 of S275 steel. Determine the maximum design load Q^* that can be applied at the end of the UB.

6.14.3 EXAMPLE 11 – CHECKING A BRACED BEAM

Determine the maximum moment at elastic buckling of the braced 533×210 UB 92 beam of S275 steel shown in Fig. 6.34b.

6.14.4 EXAMPLE 12 – CHECKING A MONOSYMMETRIC BEAM

A simply supported 6.0 m long 533×210 UB 92 of S275 steel has equal and opposite end moments. The moment capacity is to be increased by welding a full length 250×16 mm plate of S275 steel to one flange. Determine the increases in the design load capacity of the beam when the plate is welded to

(a) the compression flange, or
(b) the tension flange.

6.14.5 EXAMPLE 13 – CHECKING A NON-UNIFORM BEAM

A simply supported 533 × 210 UB 92 of S275 steel has a concentrated shear centre load at the centre of the 10.0 m span. The moment capacity is to be increased by welding 250 × 16 plates of S275 steel to the top and bottom flanges over a central length of 4.0 m. Determine the increase in the design load capacity of the beam.

6.15 References

1. Vacharajittiphan P., Woolcock, S.T. and Trahair, N.S. (1974) Effect of in-plane deformation on lateral buckling, *Journal of Structural Mechanics*, **3**, pp. 29–60.
2. Nethercot, D.A. and Rockey, K.C. (1971) A unified approach to the elastic lateral buckling of beams, *The Structural Engineer*, **49**, pp. 321–30.
3. Timoshenko, S.P. and Gere, J.M. (1961) *Theory of Elastic Stability*, 2nd edition, McGraw-Hill, New York.
4. Bleich, F. (1952) *Buckling Strength of Metal Structures*, McGraw-Hill, New York.
5. Structural Stability Research Council (1998) *Guide to Stability Design Criteria for Metal structures*, 5th edition, (ed. T.V. Galambos) John Wiley, New York.
6. Barsoum, R.S. and Gallagher, R.H. (1970) Finite element analysis of torsional and torsional–flexural stability problems, *International Journal of Numerical Methods in Engineering*, **2**, pp. 335–52.
7. Krajcinovic, D. (1969) A consistent discrete elements technique for thin-walled assemblages, *International Journal of Solids and Structures*, **5**, pp. 639–62.
8. Powell, G. and Klingner, R. (1970) Elastic lateral buckling of steel beams, *Journal of the Structural Division, ASCE*, **96**, No. ST9, pp. 1919–32.
9. Nethercot, D.A. and Rockey, K.C. (1971) Finite element solutions for the buckling of columns and beams; *International Journal of Mechanical Sciences*, **13**, pp. 945–9.
10. Hancock, G.J. and Trahair, N.S. (1978) Finite element analysis of the lateral buckling of continuously restrained beam-columns, *Civil Engineering Transactions*, Institution of Engineers, Australia, CE**20**, No. 2, pp. 120–7.
11. Brown, P.T. and Trahair, N.S. (1968) Finite integral solution of differential equations, *Civil Engineering Transactions*, Institution of Engineers, Australia, CE**10**, pp. 193–6.
12. Trahair, N.S. (1968) Elastic stability of propped cantilevers, *Civil Engineering Transactions*, Institution of Engineers, Australia, CE**10**, pp. 94–100.
13. Column Research Committee of Japan (1971) *Handbook of Structural Stability*, Corona, Tokyo.
14. Lee, G.C. (1960) A survey on the lateral instability of beams, *Welding Research Council Bulletin*, No. 63, August.
15. Clark, J.W. and Hill, H.N. (1960) Lateral buckling of beams, *Journal of the Structural Division, ASCE*, **86**, No. ST7, pp. 175–96.
16. Trahair, N.S. (1993) *Flexural–Torsional Buckling of Structures*, E. & F.N. Spon, London.
17. Galambos, T.V. (1968) *Structural Members and Frames*, Prentice-Hall, Englewood Cliffs, New Jersey.

18. Papangelis, J.P., Trahair, N.S. and Hancock, G.J. (1995) Elastic flexural–torsional buckling of structures by computer, *Proceedings*, Sixth International Conference on Civil and Structural Engineering Computing, Cambridge, pp. 109–19.
19. Anderson, J.M. and Trahair, N.S. (1972) Stability of monosymmetric beams and cantilevers, *Journal of the Structural Division, ASCE*, **98**, No. ST1, pp. 269–86.
20. Nethercot, D.A. (1973) The effective lengths of cantilevers as governed by lateral buckling, *The Structural Engineer*, **51**, pp. 161–8.
21. Trahair, N.S. (1983) Lateral buckling of overhanging beams, *Instability and Plastic Collapse of Steel Structures*, (ed. L.J. Morris), Granada, London, pp. 503–18.
22. Nethercot, D.A. and Trahair, N.S. (1976) Inelastic lateral buckling of determinate beams, *Journal of the Structural Division, ASCE*, **102**, No. ST4, pp. 701–17.
23. Trahair, N.S. (1983) Inelastic lateral buckling of beams, Chapter 2 in *Beams and Beam-Columns. Stability and Strength*, (ed. R. Narayanan), Applied Science Publishers, London, pp. 35–69.
24. Mutton, B.R. and Trahair, N.S. (1973) Stiffness requirements for lateral bracing, *Journal of the Structural Division, ASCE*, **99**, No. ST10, pp. 2167–82.
25. Mutton, B.R. and Trahair, N.S. (1975) Design requirements for column braces, *Civil Engineering Transactions*, Institution of Engineers, Australia, CE**17**, No. 1, pp. 30–5.
26. Nethercot, D.A. (1973) Buckling of laterally or torsionally restrained beams, *Journal of the Engineering Mechanics Division, ASCE*, **99**, No. EM4, pp. 773–91.
27. Trahair, N.S. and Nethercot, D.A. (1984) Bracing requirements in thin-walled structures, Chapter 3 of *Developments in Thin-Walled Structures – 2*, (eds J. Rhodes and A.C. Walker), Elsevier Applied Science Publishers, Barking, pp. 93–130.
28. Nethercot, D.A. and Trahair, N.S. (1975) Design of diaphragm braced I-beams, *Journal of the Structural Division, ASCE*, **101**, No. ST10, pp. 2045–61.
29. Trahair, N.S. (1979) Elastic lateral buckling of continuously restrained beam-columns, *The Profession of a Civil Engineer*, (eds D. Campbell-Allen and E.H. Davis), Sydney University Press, Sydney, pp. 61–73.
30. Austin, W.J., Yegian, S. and Tung, T.P. (1955) Lateral buckling of elastically end-restrained beams, *Proceedings of the ASCE*, **81**, No. 673, April, pp. 1–25.
31. Trahair, N.S. (1965) Stability of I-beams with elastic end restraints, *Journal of the Institution of Engineers, Australia*, **37**, pp. 157–68.
32. Trahair, N.S. (1966) Elastic stability of I-beam elements in rigid-jointed structures, *Journal of the Institution of Engineers, Australia*, **38**, pp. 171–80.
33. Nethercot, D.A. and Trahair, N.S. (1976) Lateral buckling approximations for elastic beams, *The Structural Engineer*, **54**, pp. 197–204.
34. Bradford, M.A. and Trahair, N.S. (1981) Distortional buckling of I-beams, *Journal of the Structural Division, ASCE*, **107**, No. ST2, pp. 355–70.
35. Bradford, M.A. and Trahair, N.S. (1983) Lateral stability of beams on seats, *Journal of Structural Engineering, ASCE*, **109**, No. ST9, pp. 2212–15.
36. Bradford, M.A. (1992) Lateral-distortional buckling of steel I-section members, *Journal of Constructional Steel Research*, **23**, No. 1–3, pp. 97–116.
37. Trahair, N.S. (1966) The bending stress rules of the draft ASCA1, *Journal of the Institution of Engineers, Australia*, **38**, pp. 131–41.
38. Standards Australia (1998) *AS 4100 – 1998 Steel Structures*, Standards Australia, Sydney.
39. Nethercot, D.A. (1972) Recent progress in the application of the finite element

method to problems of the lateral buckling of beams, *Proceedings of the EIC Conference on Finite Element Methods in Civil Engineering*, Montreal, June, pp. 367–91.

40. Vacharajittiphan, P. and Trahair, N.S. (1975) Analysis of lateral buckling in plane frames, *Journal of the Structural Division, ASCE*, **101**, No. ST7, pp. 1497–516.

41. Vacharajittiphan, P. and Trahair, N.S. (1974) Direct stiffness analysis of lateral buckling, *Journal of Structural Mechanics*, **3**, pp. 107–37.

42. Trahair, N.S. (1968) Interaction buckling of narrow rectangular continuous beams, *Civil Engineering Transactions*, Institution of Engineers, Australia, CE**10**, No. 2, pp. 167–72.

43. Salvadori, M.G. (1951) Lateral buckling of beams of rectangular cross-section under bending and shear, *Proceedings of the First US National Congress of Applied Mechanics*, pp. 403–5.

44. Kitipornchai, S. and Trahair, N.S. (1980) Buckling properties of monosymmetric I-beams, *Journal of the Structural Division, ASCE*, **106**, No. ST5, pp. 941–57.

45. Kitipornchai, S., Wang, C.M. and Trahair, N.S. (1986) Buckling of monosymmetric I-beams under moment gradient, *Journal of Structural Engineering, ASCE*, **112**, No. 4, pp. 781–99.

46. Wang, C.M. and Kitipornchai, S. (1986) Buckling capacities of monosymmetric I-beams, *Journal of Structural Engineering, ASCE*, **112**, No. 11, pp. 2373–91.

47. Wang, C.M. and Kitipornchai, S. (1986) On stability of monosymmetric cantilevers, *Engineering Structures*, **8**, No. 3, pp. 169–80.

48. Kitipornchai, S. and Trahair, N.S. (1972) Elastic stability of tapered I-beams, *Journal of the Structural Division, ASCE*, **98**, No. ST3, pp. 713–28.

49. Nethercot, D.A. (1973) Lateral buckling of tapered beams, *Publications, IABSE*, **33**, pp. 173–92.

50. Bradford, M.A. and Cuk, P.E. (1988) Elastic buckling of tapered monosymmetric I-beams, *Journal of Structural Engineering, ASCE*, **114**, No. 5, pp. 977–96.

51. Kitipornchai, S. and Trahair, N.S. (1975) Elastic behaviour of tapered monosymmetric I-beams, *Journal of the Structural Division, ASCE*, **101**, No. ST8, pp. 1661–78.

52. Bradford, M.A. (1989) Inelastic buckling of tapered monosymmetric I-beams, *Engineering Structures*, **11**, No. 2, pp. 119–26.

53. Lee, G.C., Morrell, M.L. and Ketter, R.L. (1972) Design of tapered members, *Bulletin* 173, Welding Research Council, June.

54. Bradford, M.A. (1988) Stability of tapered I-beams, *Journal of Constructional Steel Research*, **9**, pp. 195–216.

55. Trahair, N.S. and Kitipornchai, S. (1971) Elastic lateral buckling of stepped I-beams, *Journal of the Structural Division, ASCE*, **97**, No. ST10, pp. 2535–48.

56. Nethercot, D.A. and Lawson, R.M. (1992) Lateral stability of steel beams and columns – common cases of restraint, *SCI Publication 093*, Steel Construction Institute, Ascot.

57. Trahair, N.S., Hogan, T.J. and Syam, A.S. (1993) Design of unbraced beams, *Steel Construction*, Australian Institute of Steel Construction, **27**, No. 1, pp. 2–26.

7 Beam-columns

7.1 Introduction

Beam-columns are structural members which combine the beam function of transmitting transverse forces or moments with the compression (or tension) member function of transmitting axial forces. Theoretically, all structural members may be regarded as beam-columns, since the common classifications of tension members, compression members, and beams are merely limiting examples of beam-columns. However, the treatment of beam-columns in this chapter is generally limited to members in axial compression. The behaviour and design of members with moments and axial tension are treated in Chapter 2.

Beam-columns may act as if isolated, as in the case of eccentrically loaded compression members with simple end connections, or they may form part of a rigid frame. In this chapter, the behaviour and design of isolated beam-columns are treated. The ultimate strength and design of beam-columns in frames are discussed in Chapter 8.

It is convenient to discuss the behaviour of isolated beam-columns under the three separate headings of In-Plane Behaviour, Flexural–Torsional Buckling, and Biaxial-Bending, as is done in sections 7.2–7.4. When a beam-column is bent about its weaker principal axis, or when it is prevented from deflecting laterally while being bent about its stronger principal axis (as shown in Fig. 7.1a),

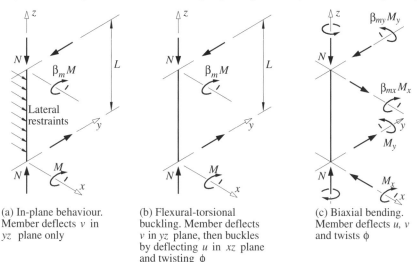

(a) In-plane behaviour. Member deflects v in yz plane only

(b) Flexural-torsional buckling. Member deflects v in yz plane, then buckles by deflecting u in xz plane and twisting ϕ

(c) Biaxial bending. Member deflects u, v and twists ϕ

Fig. 7.1 Beam-column behaviour

its action is confined to the plane of bending. This in-plane behaviour is related to the bending of beams discussed in Chapter 5 and to the buckling of compression members discussed in Chapter 3. When a beam-column which is bent about its stronger principal axis is not restrained laterally (as shown in Fig. 7.1b), it may buckle prematurely out of the plane of bending by deflecting laterally and twisting. This action is related to the flexural–torsional buckling of beams discussed in Chapter 6. More generally, however, a beam-column may be bent about both principal axes, as shown in Fig. 7.1c. This biaxial bending, which commonly occurs in three-dimensional rigid frames, involves interactions of beam bending and twisting with beam and column buckling. Sections 7.2–7.4 discuss the in-plane behaviour, flexural–torsional buckling, and biaxial bending of isolated beam-columns.

7.2 In-plane behaviour of isolated beam-columns

When the deformations of an isolated beam-column are confined to the plane of bending (see Fig. 7.1a), its behaviour shows an interaction between beam bending and compression member buckling, as indicated in Fig. 7.2. Curve 1 of this figure shows the linear behaviour of an elastic beam, while curve 6 shows the limiting behaviour of a rigid-plastic beam at the full plastic moment M_p. Curve 2 shows the transition of a real elastic–plastic beam from curve 1 to curve 6. The elastic buckling of a concentrically loaded compression member at its elastic buckling load N_{ox} is shown by curve 4. Curve 3 shows the interaction between bending and buckling in an elastic member, and allows for the additional moment $N\delta$ exerted by the axial load. Curve 7 shows the interaction

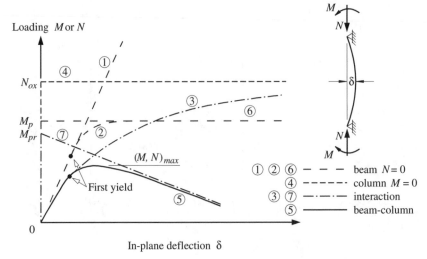

Fig. 7.2 In-plane behaviour of a beam-column

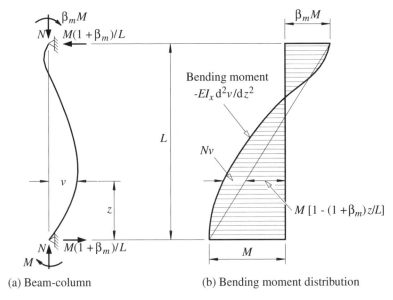

(a) Beam-column   (b) Bending moment distribution

Fig. 7.3 In-plane bending moment distribution in a beam-column

between the bending moment and axial force which causes the member to become fully plastic. This curve allows for reduction from the full plastic moment M_p to M_{pr} caused by the axial load, and for the additional moment $N\delta$. The actual behaviour of a beam-column is shown by curve 5 which provides a transition from curve 3 for an elastic member to curve 7 for full plasticity.

7.2.1 ELASTIC BEAM-COLUMNS

A beam-column is shown in Figs 7.1a and 7.3a which is acted on by axial forces N and end moments M and $\beta_m M$, where β_m can have any value between -1 (single curvature bending) and $+1$ (double curvature bending). For isolated beam-columns, these end moments are independent of the deflections, and have no effect on the elastic buckling load $N_{ox} = \pi^2 EI_x/L^2$. (They are therefore quite different from the end restraining moments discussed in section 3.5.3 which increase with the end rotations and profoundly affect the elastic buckling load.) The beam-column is prevented from bending about the minor axis out of the plane of the end moments, and is assumed to be straight before loading.

The bending moment in the beam-column is the sum of $M - M(1 + \beta_m)z/L$ due to the end moments and shears, and Nv due to the deflection v, as shown in Fig. 7.3b. It is shown in section 7.5.1 that the deflected shape of the beam-column is given by

$$v = \frac{M}{N}\left[\cos \mu z - (\beta_m \cosec \mu L + \cot \mu L) \sin \mu z - 1 + (1 + \beta_m)\frac{z}{L}\right], \quad (7.1)$$

where

$$\mu^2 = \frac{N}{EI_x} = \frac{\pi^2}{L^2}\frac{N}{N_{ox}}. \quad (7.2)$$

As the axial force N approaches the buckling load N_{ox}, the value of μL approaches π, and the values of $\cosec \mu L$, $\cot \mu L$, and therefore of v approach infinity, as indicated by curve 3 in Fig. 7.2. This behaviour is similar to that of a compression member with geometrical imperfections (see section 3.2.2). It is also shown in section 7.5.1 that the maximum moment M_m in the beam-column is given by

$$M_m = M\sqrt{[1 + \{\beta_m \cosec \pi\sqrt{(N/N_{ox})} + \cot \pi\sqrt{(N/N_{ox})}\}^2]}, \quad (7.3)$$

when $\beta_m < -\cos \pi\sqrt{(N/N_{ox})}$ and the point of maximum moment lies in the span, and is given by the end moment M, i.e.

$$M_m = M, \quad (7.4)$$

when $\beta_m \geq -\cos \pi\sqrt{(N/N_{ox})}$.

The variations of M_m/M with the axial load ratio N/N_{ox} and the end moment ratio β_m are shown in Fig. 7.4. In general, M_m remains equal to M for low values of N/N_{ox} but increases later. The value of N/N_{ox} at which M_m begins to increase above M is lowest for $\beta_m = -1$ (uniform bending), and increases with increasing β_m. Once M_m departs from M, it increases slowly at first, then more rapidly, and reaches very high values as N approaches the column buckling load N_{ox}.

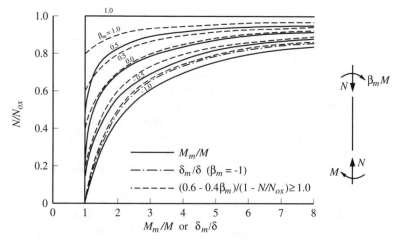

Fig. 7.4 Maximum moments and deflections in elastic beam-columns

For beam-columns which are bent in single curvature by equal and opposite end moments ($\beta_m = -1$), the maximum deflection δ_m is given by (see section 7.5.1)

$$\frac{\delta_m}{\delta} = \frac{8/\pi^2}{N/N_{ox}} \left[\sec \frac{\pi}{2} \sqrt{(N/N_{ox})} - 1 \right] \tag{7.5}$$

in which $\delta = ML^2/8EI_x$ is the value of δ_m when $N = 0$, and the maximum moment by

$$M_m = M \sec \frac{\pi}{2} \sqrt{(N/N_{ox})}. \tag{7.6}$$

These two equations are shown non-dimensionally in Fig. 7.4. It can be seen that they may be approximated by using a factor $1/(1 - N/N_{ox})$ to amplify the deflection δ and the moment M due to the applied moments alone ($N = 0$). This same approximation was used in section 3.5.3 to estimate the reduced stiffness of an axially loaded restraining member.

For beam-columns with unequal end moments ($\beta_m > -1$), the value of M_m may be approximated by using

$$\frac{M_m}{M} = \frac{c_m}{1 - N/N_{ox}} \geqslant 1.0 \tag{7.7}$$

in which

$$c_m = 0.6 - 0.4\beta_m. \tag{7.8}$$

These approximations are also shown in Fig. 7.4. It can be seen that they are generally conservative for $\beta_m > 0$, and only a little unconservative for $\beta_m < 0$.

Approximations for the maximum moments M_m in beam-columns with transverse loads can be obtained by using

$$M_m = \delta M_{max} \tag{7.9}$$

in which M_{max} is the maximum moment when $N = 0$,

$$\delta = \frac{\gamma_m(1 - \gamma_s N/N_o)}{(1 - \gamma_n N/N_o)}, \tag{7.10}$$

$$N_o = \pi^2 EI/(k_e^2 L^2), \tag{7.11}$$

and $k_e = L_E/L$ is the effective length ratio. Expressions for M_{max} and values of γ_m, γ_n, γ_s, and k_e are given in Fig. 7.5 for beam-columns with central concentrated loads Q and in Fig. 7.6 for beam-columns with uniformly distributed loads q. A worked example of the use of these approximations is given in section 7.7.1.

The maximum stress f_{max} in the beam-column is the sum of the axial stress and the maximum bending stress caused by the maximum moment M_m. It is therefore given by

$$f_{max} = f_{ac} + f_{bcx} \frac{M_m}{M} \tag{7.12}$$

Beam-column	First-order moment ($N = 0$)	k_e	γ_m	γ_n	γ_s	M_{max}
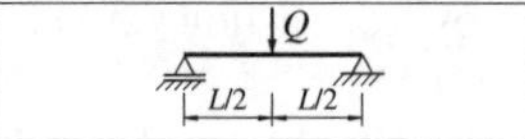	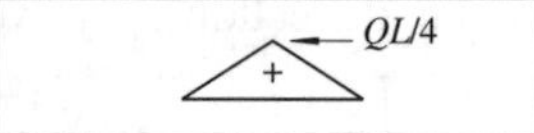 $QL/4$	1.0	1.0	1.0	0.18	$\dfrac{QL}{4}$
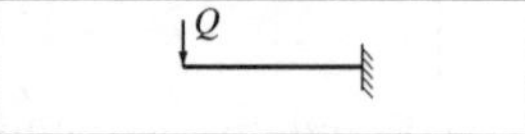	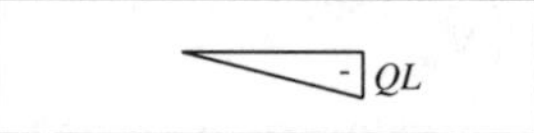 QL	2.0	1.0	1.0	0.19	QL
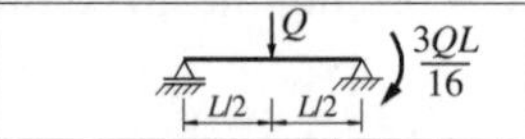 $\dfrac{3QL}{16}$	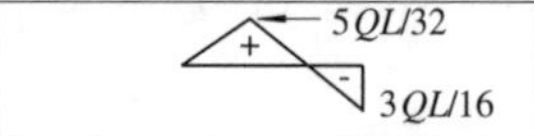 $5QL/32$ / $3QL/16$	1.0	$\dfrac{5}{6}$	1.0	0.45	$\dfrac{3QL}{16}$
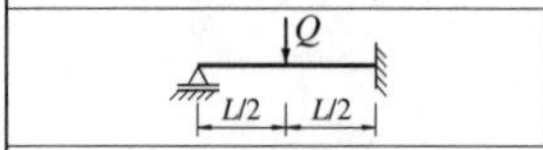	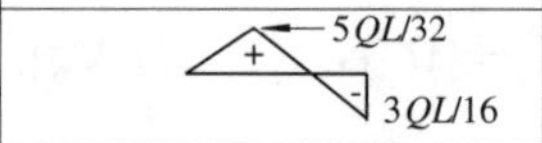 $5QL/32$ / $3QL/16$	0.7	1.0	1.0	0.28	$\dfrac{3QL}{16}$
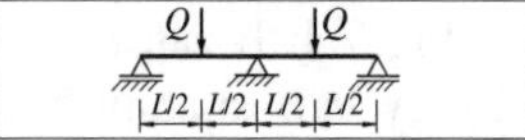	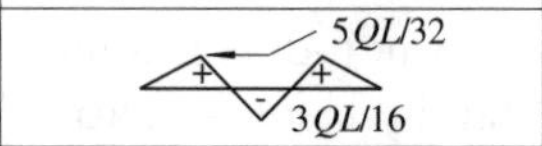 $5QL/32$ / $3QL/16$	1.0	1.0	0.49	0.14	$\dfrac{3QL}{16}$
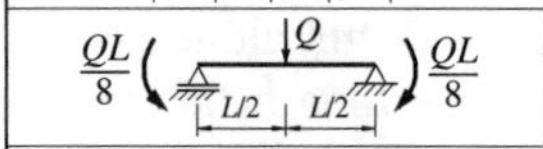 $\dfrac{QL}{8}\left(\text{}\right)\dfrac{QL}{8}$	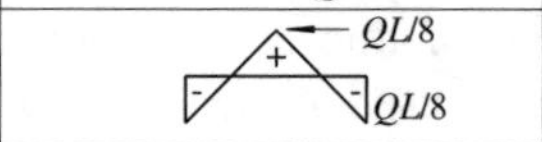 $QL/8$ / $QL/8$	1.0	1.0	1.0	0.62	$\dfrac{QL}{8}$
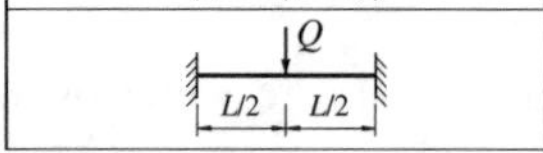	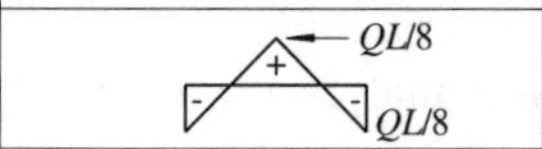 $QL/8$ / $QL/8$	0.5	1.0	1.0	0.18	$\dfrac{QL}{8}$

Fig. 7.5 Approximations for beam-columns with central concentrated loads

Beam-column	First-order moment ($N = 0$)	k_e	γ_m	γ_n	γ_s	M_{max}
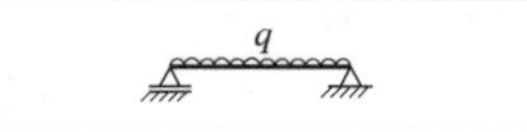	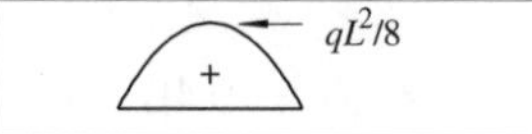 $qL^2/8$	1.0	1.0	1.0	-0.03	$\dfrac{qL^2}{8}$
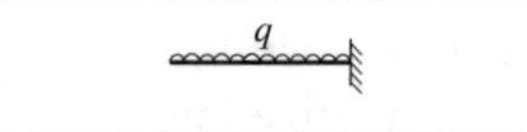	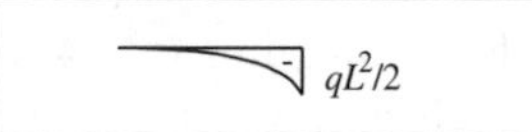 $qL^2/2$	2.0	1.0	1.0	0.4	$\dfrac{qL^2}{2}$
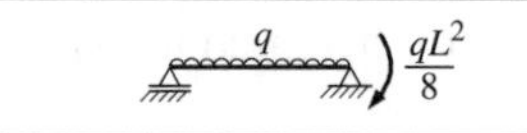 $\dfrac{qL^2}{8}$	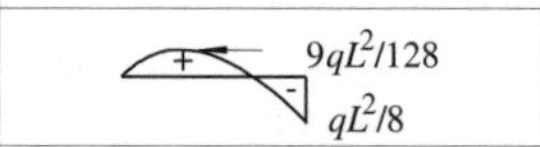 $9qL^2/128$ / $qL^2/8$	1.0	$\dfrac{9}{16}$	1.0	0.29	$\dfrac{qL^2}{8}$
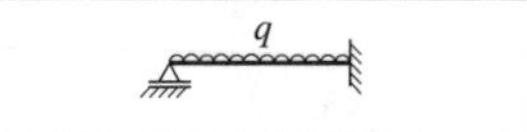	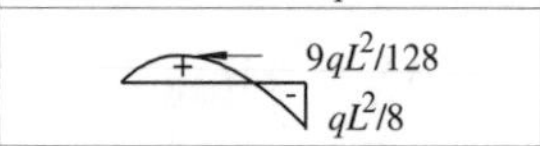 $9qL^2/128$ / $qL^2/8$	0.7	1.0	1.0	0.36	$\dfrac{qL^2}{8}$
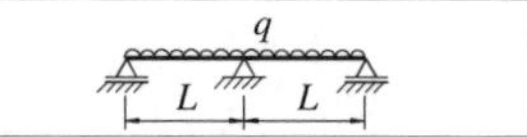	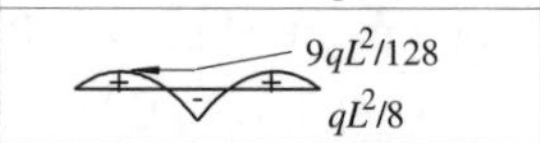 $9qL^2/128$ / $qL^2/8$	1.0	1.0	0.49	0.18	$\dfrac{qL^2}{8}$
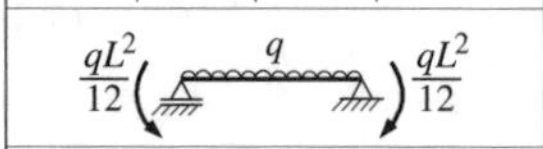 $\dfrac{qL^2}{12}\left(\text{}\right)\dfrac{qL^2}{12}$	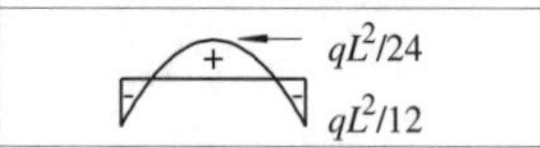 $qL^2/24$ / $qL^2/12$	1.0	0.5	1.0	0.4	$\dfrac{qL^2}{12}$
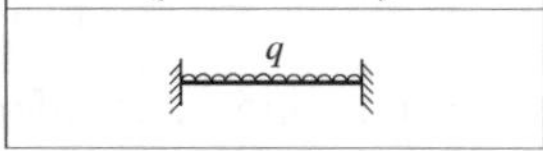	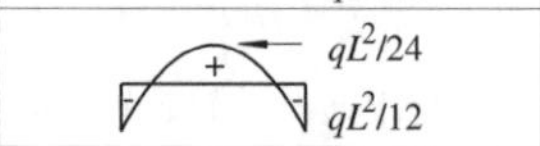 $qL^2/24$ / $qL^2/12$	0.5	1.0	1.0	0.37	$\dfrac{qL^2}{12}$

Fig. 7.6 Approximations for beam-columns with distributed loads

in which $f_{ac} = N/A$ and $f_{bcx} = M/Z_x$. If the member has no residual stresses, then it will remain elastic while f_{max} is less than the yield stress f_y, and so the above results are valid while

$$\frac{N}{N_Y} + \frac{M}{M_Y}\frac{M_m}{M} \leqslant 1.0 \tag{7.13}$$

in which $N_Y = Af_y$ is the squash load and $M_Y = f_y Z_x$ the nominal first yield moment. A typical elastic limit of this type is shown by the first yield point marked on curve 3 of Fig. 7.2. It can be seen that this limit provides a lower bound estimate of the strength of a straight beam-column, while the elastic buckling load N_{ox} provides an upper bound.

Variations of the first yield limits of N/N_Y determined from equation 7.13 with M/M_Y and β_m are shown in Fig. 7.7a for the case where $N_{ox} = N_Y/1.5$. For a beam-column in double curvature bending ($\beta_m = 1$), $M_m = M$, and so N varies linearly with M until the elastic buckling load N_{ox} is reached. For a beam-column in uniform bending ($\beta_m = -1$), the relationship between N and M is non-linear and concave due to the amplification of the bending moment from M to M_m by the axial load. Also shown in Fig. 7.7a are solutions of equation 7.13 based on the use of the approximations of equations 7.7 and 7.8. A comparison of these with the accurate solutions also shown in Fig. 7.7a again demonstrates the comparative accuracy of the approximate equations 7.7 and 7.8.

The elastic in-plane behaviour of members in which the axial forces cause tension instead of compression can also be analysed. The maximum moment in such a member never exceeds M, and so a conservative estimate of the elastic limit can be obtained from

$$\frac{N}{N_Y} + \frac{M}{M_Y} \leq 1.0. \qquad (7.14)$$

This forms the basis for the methods of designing tension members for in-plane bending, as discussed in section 2.4.

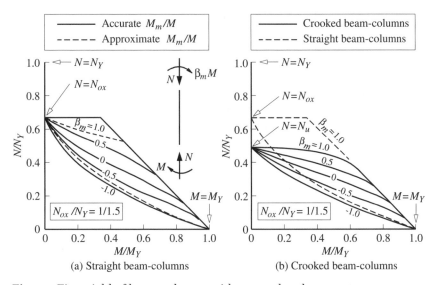

Fig. 7.7 First yield of beam-columns with unequal end moments

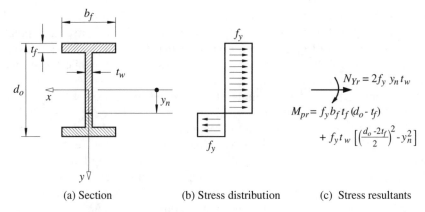

(a) Section   (b) Stress distribution   (c) Stress resultants

Fig. 7.8 Fully plastic cross-section

7.2.2 FULLY PLASTIC BEAM-COLUMNS

An upper bound estimate of the strength of an I-section beam-column bent about its major axis can be obtained from the combination of bending moment M_{pr} and axial force N_{Yr} which causes the cross-section to become fully plastic. A particular example is shown in Fig. 7.8, for which the distance y_n from the centroid to the unstrained fibre is less than $(d_o - 2t_f)/2$. This combination of moment and force lies between the two extreme combinations for members with bending moment only $(N = 0)$, which become fully plastic at

$$M_p = f_y b_f t_f (d_o - t_f) + f_y t_w \left(\frac{d_o - 2t_f}{2} \right)^2, \qquad (7.15)$$

and members with axial force only $(M = 0)$, which become fully plastic at

$$N_Y = f_y [2b_f t_f + (d_o - 2t_f)t_w]. \qquad (7.16)$$

The variations of M_{pr} and N_{Yr} are shown by the dashed interaction curve in Fig. 7.9. These combinations can be used directly for very short beam-columns for which the bending moment at the fully plastic cross-section is approximately equal to the applied end moment M.

Also shown in Fig. 7.9 is the approximation

$$\frac{M_{prx}}{M_{px}} = 1.18 \left[1 - \frac{N_{Yr}}{N_Y} \right] \leq 1.0, \qquad (7.17)$$

which is generally close to the accurate solution, except when $N_{Yr} \approx 0.15 N_Y$, when the maximum error is of the order of 5%. This small error is quite acceptable since the accurate solutions ignore the strengthening effects of strain-hardening.

A similar analysis may be made of an I-section beam-column bent about its

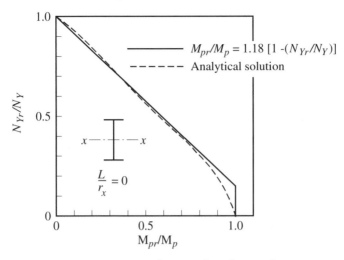

Fig. 7.9 Interaction curves for 'zero length' members

minor axis. In this case, a satisfactory approximation for the load and moment at full plasticity is given by

$$\frac{M_{pry}}{M_{py}} = 1.19\left[1 - \left(\frac{N_{Yr}}{N_Y}\right)^2\right] \leqslant 1.0. \tag{7.18}$$

Beam-columns of more general cross-section may also be analysed to determine the axial load and moment at full plasticity. In general, these may be safely approximated by the linear interaction equation

$$\frac{M_{pr}}{M_p} = 1 - \frac{N_{Yr}}{N_Y}. \tag{7.19}$$

In a longer beam-column, instability effects become important, and failure occurs before any section becomes fully plastic. A further complication arises from the fact that as the beam-column deflects by δ, its maximum moment increases from the nominal value M to $(M + N\delta)$. Thus, the value of M at which full plasticity occurs is given by

$$M = M_{pr} - N\delta, \tag{7.20}$$

and so the value of M for full plasticity decreases from M_{pr} as the deflection δ increases, as shown by curve 7 in Fig. 7.2. This curve represents an upper bound to the behaviour of beam-columns which is only approached after the maximum strength is reached.

7.2.3 ULTIMATE STRENGTH

7.2.3.1 General

An isolated beam-column reaches its ultimate strength at a load which is greater than that which causes first yield (see section 7.2.1), but is less than that which causes a cross-section to become fully plastic (see section 7.2.2), as indicated in Fig. 7.2. These two bounds are often far apart, and when a more accurate estimate of the strength is required, an elastic–plastic analysis of the imperfect beam-column must be made. Two different approximate analytical approaches to beam-column strength may be used, and these are related to the initial crookedness and residual stress methods discussed in sections 3.2.2 and 3.3.4 of allowing for the effects of imperfections on the strengths of real compression members. These two approaches are discussed below.

7.2.3.2 Elastic–plastic strengths of straight beam-columns

The strength of an initially straight beam-column with residual stresses may be found by analysing numerically its non-linear elastic–plastic behaviour and determining its maximum load capacity [1]. Some of these capacities [2, 3] are shown as interaction plots in Fig. 7.10. Figure 7.10a shows how the ultimate load N and the end moment M vary with the major axis slenderness ratio L/r_x for beam-columns with equal and opposite end moments ($\beta_m = -1$), while Fig. 7.10b shows how these vary with the end moment ratio β_m for a slenderness ratio of $L/r_x = 60$.

It has been proposed that these ultimate strengths can be simply and closely approximated by using the values of M and N which satisfy the interaction equations of both equation 7.17 and

$$\frac{N}{N_{ux}} + \frac{M}{M_p}\frac{c_m}{(1 - N/N_{ox})} \leqslant 1 \qquad (7.21)$$

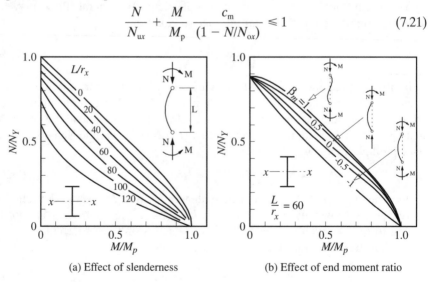

(a) Effect of slenderness   (b) Effect of end moment ratio

Fig. 7.10 Ultimate strength interaction curves

in which N_{ux} is the ultimate load of a concentrically loaded column (the beam-column with $M = 0$) which fails by deflecting about the major axis, N_{ox} is the major axis elastic buckling load of this concentrically loaded column, and c_m is given by

$$c_m = 0.6 - 0.4\beta_m \geqslant 0.4. \tag{7.22}$$

Equation 7.17 ensures that the reduced plastic moment M_{pr} is not exceeded by the end moment M, and represents the strengths of very short members $(L/r_x \to 0)$, and of some members which are not bent in single curvature (i.e. $\beta_m > 0$), as shown, for example, in Fig. 7.11b.

Equation 7.21 represents the interaction between buckling and bending which determines the strength of more slender members. If the case of equal and opposite end moments is considered first ($\beta_m = -1$ and $c_m = 1$), then it will be seen that equation 7.21 includes the same approximate amplification factor $1/(1 - N/N_{ox})$ as does equation 7.7 for the maximum moments in elastic beam-columns. It is of a similar form to the first yield condition of equation 7.13, except that a conversion to ultimate strength has been made by using the ultimate load N_{ux} and moment M_p instead of the squash load N_Y and the yield moment M_Y. These substitutions ensure that this interaction formula gives the same limit predictions for concentrically loaded columns ($M = 0$) and for beams ($N = 0$) as do the treatments given in Chapters 3 and 5. The accuracy of equation 7.21 for beam-columns with equal and opposite end moments ($\beta_m = -1$) is demonstrated in Fig. 7.11a.

The effects of unequal end moments ($\beta_m > -1$) on the ultimate strengths of beam-columns are allowed for approximately in equation 7.21 by the

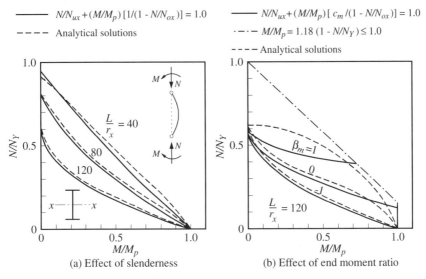

(a) Effect of slenderness   (b) Effect of end moment ratio

Fig. 7.11 Interaction formulae

coefficient c_m which converts the unequal end moments M and $\beta_m M$ into equivalent equal and opposite end moments $c_m M$. Although this conversion is for in-plane behaviour, it is remarkably similar to the conversion used for beams with unequal end moments which buckle out of the plane of bending (see section 6.2.1.2). The accuracy of equation 7.21 for beam-columns with unequal end moments is demonstrated in Fig. 7.11b, and it can be seen that it is good except for high values of β_m when it tends to be oversafe, and for high values of M when it tends to overestimate the analytical solutions for the ultimate strength (although this overestimate is reduced if strain-hardening is accounted for in the analytical solutions). The over-conservatism for high values of β_m is caused by the use of the 0.4 cut-off in equation 7.22 for c_m, but more accurate solutions can be obtained by using equation 7.8 for which the minimum value of c_m is 0.2.

Thus, the interaction equations (equations 7.17 and 7.21) provide a reasonably simple method of estimating the ultimate strengths of I-section beam-columns bent about the major axis which fail in the plane of the applied moments. The interaction equations have been successfully applied to a wide range of sections, including solid and hollow circular sections as well as I-sections bent about the minor axis. In the latter case, M_p, N_{ux} and N_{ox} must be replaced by their minor axis equivalents, while equation 7.17 should be replaced by equation 7.18.

It should be noted that the interaction equations provide a convenient way of estimating the ultimate strengths of beam-columns. An alternative method has been suggested [4] using simple design charts to show the variation of the moment ratio M/M_{pr} with the end moment ratio β_m and the length ratio L/L_c, in which L_c is the length of a column which just fails under the axial load N alone (the effects of strain-hardening must be included in the calculation of L_c).

7.2.3.3 First yield of crooked beam-columns

In the second approximate approach to the strength of a real beam-column, the first yield of an initially crooked beam-column without residual stresses is used. For this, the magnitude of the initial crookedness is increased to allow approximately for the effects of residual stresses. One logical way of doing this is to use the same crookedness as is used in the design of the corresponding compression member, since this has already been increased so as to allow for residual stresses.

The first yield of a crooked beam-column is analysed in section 7.5.2, and particular solutions are shown in Fig. 7.7b for a beam-column with $N_{ox}/N_Y = 1/1.5$ and whose initial crookedness is defined by $\eta = 0.00326(L/r - 13.5)$ (see section 3.2.2). It can be seen that as M decreases to zero, the axial load at first yield increases to the column strength N_{ux} which is reduced below the elastic buckling load N_{ox} by the effects of imperfections. As N decreases, the first yield loads for crooked beam-columns approach the corresponding loads for straight beam-columns.

For beam-columns in uniform bending ($\beta_m = -1$), the interaction between M and N is non-linear and concave, as it was for straight beam-columns (Fig. 7.7a). If this interaction is plotted using the maximum moment M_m instead of the end moment M, then the relationships between M_m and N becomes slightly convex, since the non-linear effects of the amplification of M are then included in M_m. Thus a linear relationship between M_m/M_Y and N/N_{ux} will provide a simple and conservative approximation for first yield which is of good accuracy. However, if this approximation is used for the strength, then it will become increasingly conservative as M_m increases, since it approaches the first yield moment M_Y instead of the fully plastic moment M_p. This difficulty may be overcome by modifying the approximation to a linear relationship between M_m/M_p and N/N_{ux} which passes through $M_m/M_p = 1$ when $N/N_{ux} = 0$, as shown in Fig. 7.12a. Such an approximation is also in good agreement with the uniform bending ($\beta_m = -1$) analytical strengths [2, 3] shown in Fig. 7.12b for initially straight beam-columns with residual stresses.

For beam-columns in double curvature bending ($\beta_m = 1$), the first yield interaction shown in Fig. 7.7b forms an approximately parabolic transition between the bounds of the column strength N_{ux} and the first yield condition of a straight beam-column. In this case, the effects of initial curvature are greatest near mid-span, and have little effect on the maximum moment which is generally greatest at the ends. Thus the strength may be simply approximated by a parabolic transition from the column strength N_{ux} to the full plasticity condition given by equation 7.17 for I-sections bent about the major axis, as shown in Fig. 7.12a. This approximation is in good agreement with the double curvature bending ($\beta_m = 1$) analytical strengths [2, 3] shown in Fig. 7.12b for straight beam-columns with residual stresses and zero strain-hardening.

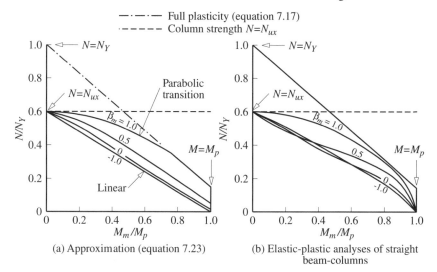

Fig. 7.12 Strengths of beam-columns

The ultimate strengths of beam-columns with unequal end moments may be approximated by suitable interpolations between the linear and parabolic approximations for beam-columns in uniform bending ($\beta_m = -1$) and double curvature bending ($\beta_m = 1$). It has been suggested [5] that this interpolation should be made according to

$$\frac{M}{M_p} = \left\{ 1 - \left(\frac{1 + \beta_m}{2} \right)^3 \right\} \left(1 - \frac{N}{N_{ux}} \right) + 1.18 \left(\frac{1 + \beta_m}{2} \right)^3 \sqrt{\left(1 - \frac{N}{N_{ux}} \right)} \leq 1.$$

(7.23)

This interpolation uses a conservative cubic weighting to shift the resulting curves downwards towards the linear curve for $\beta_m = -1$, as shown in Fig. 7.12a. This is based on the corresponding shift shown in Fig. 7.12b of the analytical results [2, 3] for initially straight beam-columns with residual stresses.

7.2.4 DESIGN RULES

7.2.4.1 Section capacity

BS5950 requires beam-columns to satisfy both section capacity and overall member resistance limitations. The section capacity limitations are intended to prevent cross-section failure due to plasticity or local buckling.

The general section capacity limitation of BS5950 is given by a modification of the first yield condition of equation 7.14 to

$$\frac{F^*}{Ap_y} + \frac{M^*}{M_c} \leq 1$$

(7.24)

in which F^* is the design axial force and M^* the design moment acting at the section under consideration, Ap_y is the section axial force capacity obtained using the effective cross-sectional area A_{eff} (Clause 3.6.2.2) when F^* causes compression or the effective area A_e (Clause 3.4.3) when F^* causes tension, and M_c is the section moment capacity.

This linear interaction equation is often conservative, and so BS5950 allows the use of the more economic

$$M^* \leq M_r$$

(7.25)

for Class 1 plastic and Class 2 compact sections, in which the reduced full plastic moment capacity

$$M_r = p_y S_r$$

(7.26)

may be obtained from expressions for the reduced plastic section modulus such as those given in Annex I.2. Less accurate approximations for the reduced full plastic moments of I-section members are given by equations 7.17 and 7.18.

BS5950 defines compact hot-rolled I-sections under combined bending in the plane of the web and compression as ones which satisfy

$$\frac{d}{t\varepsilon} \leqslant \frac{100}{1 + 1.5F_c^*/(dtp_{yw})} \tag{7.27}$$

in which dt is the web area and p_{yw} is the web design strength, so that this limit varies between 40 and 100 as F_c^*/dtp_{yw} varies between 1 and 0.

7.2.4.2 Member resistance

The general in-plane member resistance limitation of BS5950 is given by a simplification of equation 7.21 to

$$\frac{F_c^*}{P_c} + \frac{mM^*}{p_yZ} \leqslant 1 \tag{7.28}$$

in which P_c is the axial compression resistance and m is similar to but a little different from the factor m_{LT} used to allow for the effect of the bending moment distribution on the lateral buckling of beams (sections 6.4.2.2 and 6.4.2.3). Values of m are given in Table 26 of BS5950.

More exact alternatives for doubly symmetric I- or H-sections are provided by

$$\frac{F_c^*}{P_{cx}} + m_x\left(1 + 0.5\frac{F_c^*}{P_{cx}}\right)\frac{M_x^*}{M_{cx}} \leqslant 1 \tag{7.29a}$$

$$\frac{F_c^*}{P_{cy}} + m_y\left(1 + \frac{F_c^*}{P_{cy}}\right)\frac{M_y^*}{M_{cy}} \leqslant 1 \tag{7.29b}$$

in which M_{cx}, M_{cy} are the section moment capacities, and $(1 + 0.5F^*/P_{cx})$ and $(1 + F^*/P_{cy})$ are approximate amplification factors which are similar to the term $1/(1 - N/N_{ox})$ of equation 7.21. Equations 7.29 can also be expressed as

$$m_xM_x^* \leqslant M_{ax} = \frac{M_{cx}(1 - F_c^*/P_{cx})}{(1 + 0.5F_c^*/P_{cx})} \tag{7.30a}$$

$$m_yM_y^* \leqslant M_{ay} = \frac{M_{cy}(1 - F_c^*/P_{cy})}{(1 + F_c^*/P_{cy})} \tag{7.30b}$$

For Class 1 plastic and Class 2 compact double symmetric sections, further alternatives provide for the use of increased values of M_{ax}, M_{ay} in equations 7.30 obtained by linear interpolations between the values defined by equations 7.30 and M_{rx}, M_{ry} with λ_x, λ_y which vary between 85.8ε and 17.15ε, in which λ_x, λ_y are the compression member slendernesses (section 3.4.2).

7.3 Flexural–torsional buckling of isolated beam-columns

When an unrestrained beam-column is bent about its major axis, it may buckle by deflecting laterally and twisting at a load which is significantly less than the

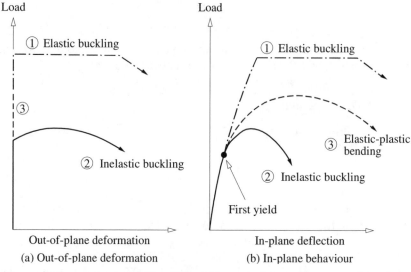

Fig. 7.13 Flexural-torsional buckling of beam-columns

maximum load predicted by an in-plane analysis. This flexural–torsional buckling may occur while the member is still elastic, or after some yielding due to
in-plane bending and compression has occurred, as indicated in Fig. 7.13.

7.3.1 ELASTIC BEAM-COLUMNS

7.3.1.1 Beam-columns with equal end moments

Consider a perfectly straight elastic beam-column bent about its major axis
(see Fig. 7.1b) by equal and opposite end moments M (so that $\beta_m = -1$), and
loaded by an axial force N. The ends of the beam-column are assumed to be
simply supported and free to warp but end twist rotations are prevented. It is
also assumed that the cross-section of the beam-column has two axes of symmetry so that the shear centre and centroid coincide.

When the applied load and moments reach the elastic buckling values N_{oc},
M_{oc}, a deflected and twisted equilibrium position is possible. It is shown in
section 7.6.1 that this position is given by

$$u = \frac{M_{oc}}{N_{oy} - N_{oc}} \quad \phi = \delta \sin \frac{\pi z}{L} \qquad (7.31)$$

in which δ is the undetermined magnitude of the central deflection, and that
the elastic buckling combination N_{oc}, M_{oc} is given by

$$\frac{M_{oc}^2}{r_0^2 N_{oy} N_{oz}} = \left(1 - \frac{N_{oc}}{N_{oy}}\right)\left(1 - \frac{N_{oc}}{N_{oz}}\right) \qquad (7.32)$$

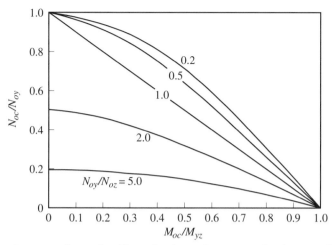

Fig. 7.14 Elastic buckling load combinations for beam-columns with equal end moments

in which $r_0 = \sqrt{[(I_x + I_y)/A]}$ is the polar radius of gyration, and

$$N_{oy} = \pi^2 EI_y/L^2, \tag{7.33}$$

$$N_{oz} = (GJ/r_0^2)(1 + \pi^2 EI_w/GJL^2) \tag{7.34}$$

are the minor axis and torsional buckling loads of an elastic axially loaded column (see sections 3.2.1 and 3.6.5). It can be seen that for the limiting case when $M_{oc} = 0$, the beam-column buckles as a compression member at the lower of N_{oy} and N_{oz}, and when $N_{oc} = 0$, it buckles as a beam at the elastic buckling moment (see equation 6.3),

$$M_{yz} = \sqrt{\left\{\left(\frac{\pi^2 EI_y}{L^2}\right)\left(GJ + \frac{\pi^2 EI_w}{L^2}\right)\right\}} = r_0\sqrt{(N_{oy}N_{oz})}. \tag{7.35}$$

Some more general solutions of equation 7.32 are shown in Fig. 7.14.

The derivation of equation 7.32 given in section 7.6.1 neglects the approximate amplification by the axial load of the in-plane moments to $M_{oc}/(1 - N_{oc}/N_{ox})$ (see section 7.2.1). When this is accounted for, equation 7.32 for the elastic buckling combination of N_{oc} and M_{oc} changes to

$$\frac{M_{oc}^2}{r_0^2 N_{oy}N_{oz}} = \left(1 - \frac{N_{oc}}{N_{ox}}\right)\left(1 - \frac{N_{oc}}{N_{oy}}\right)\left(1 - \frac{N_{oc}}{N_{oz}}\right). \tag{7.36}$$

The maximum possible value of N_{oc} is the lowest value of N_{ox}, N_{oy}, and N_{oz}. For most sections this is much less than N_{ox} and so equation 7.36 is usually very close to equation 7.32. For most hot-rolled sections, N_{oy} is less than N_{oz}, so that

$(1 - N_{oc}/N_{oz}) > (1 - N_{oc}/N_{oy})(1 - N_{oc}/N_{ox})$. In this case, equation 7.36 can be safely approximated by the interaction equation

$$\frac{N_{oc}}{N_{oy}} + \frac{1}{(1 - N_{oc}/N_{ox})} \frac{M_{oc}}{M_{yz}} = 1. \tag{7.37}$$

7.3.1.2 Beam-columns with unequal end moments

The elastic flexural–torsional buckling of simply supported beam-columns with unequal major axis end moments M and $\beta_m M$ has been investigated numerically, and many solutions are available [7–11]. The conservative interaction equation

$$\left(\frac{M/\sqrt{F}}{M_E}\right)^2 + \left(\frac{N}{N_{oy}}\right) = 1 \tag{7.38}$$

has also been proposed [9], in which

$$M_E^2 = \pi^2 EI_y GJ/L^2. \tag{7.39}$$

The factor $1/\sqrt{F}$ in equation 7.38 varies with the end moment ratio β_m as shown in Fig. 7.15, and allows the unequal end moments to be treated as equivalent equal end moments $M/\sqrt{F}$. Thus the elastic buckling of beam-columns with unequal end moments can also be approximated by modifying equation 7.32 to

$$\frac{(M_{oc}/\sqrt{F})^2}{M_{yz}^2} = \left(1 - \frac{N_{oc}}{N_{oy}}\right)\left(1 - \frac{N_{oc}}{N_{oz}}\right), \tag{7.40}$$

or by a similar modification of equation 7.37.

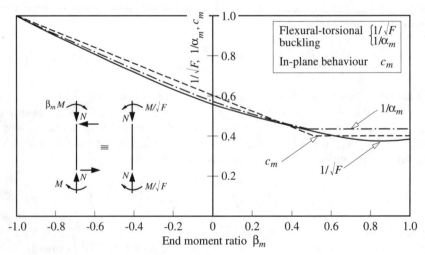

Fig. 7.15 Equivalent end moments for beam-columns

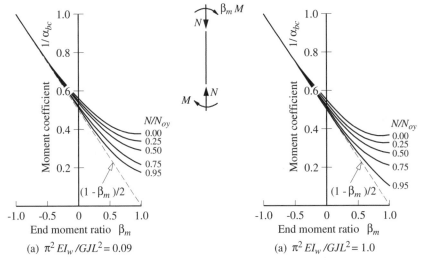

Fig. 7.16 Flexural-torsional buckling coefficients $1/\alpha_{bc}$ in equation 7.42

It is of interest to note that the variation of the factor $1/\sqrt{F}$ for the flexural–torsional buckling of beam-columns is very close to that of the coefficient $1/\alpha_m$ obtained from

$$\alpha_m = 1.75 + 1.05\beta_m + 0.3\beta_m^2 \not> 2.56 \tag{7.41}$$

used for the flexural–torsional buckling of beams (see section 6.2.1.2), and close to that of the coefficient c_m (see equation 7.22) used for the in-plane behaviour of beam-columns.

However, the results shown in Fig. 7.16 of a more recent investigation [12] of the elastic flexural–torsional buckling of beam-columns have indicated that the factor for converting unequal end moments into equivalent equal end moments should vary with N/N_{oy} as well as with β_m. More accurate predictions have been obtained using

$$\left(\frac{M_{oc}}{\alpha_{bc}M_{yz}}\right)^2 = \left(1 - \frac{N_{oc}}{N_{oy}}\right)\left(1 - \frac{N_{oc}}{N_{oz}}\right) \tag{7.42}$$

with

$$\frac{1}{\alpha_{bc}} = \left(\frac{1 - \beta_m}{2}\right) + \left(\frac{1 + \beta_m}{2}\right)^3\left(0.40 - 0.23\frac{N_{oc}}{N_{oy}}\right). \tag{7.43}$$

7.3.1.3 Beam-columns with elastic end restraints

The elastic flexural–torsional buckling of symmetrically restrained beam-columns with equal and opposite end moments ($\beta_m = -1$) is analysed in section 7.6.2. The particular example considered is one for which the minor axis bending and end warping restraints are equal. This corresponds to the situation in which both ends of both flanges have equal elastic restraints whose stiffnesses are such that

$$\frac{\text{Flange end moment}}{\text{Flange end rotation}} = \frac{-EI_y}{L} \frac{R}{1-R} \tag{7.44}$$

in which the restraint parameter R varies from 0 (flange unrestrained) to 1 (flange rigidly restrained).

It is shown in section 7.6.2 that the elastic buckling values M_{oc} and N_{oc} of the end moments and axial load for which the restrained beam-column buckles elastically are given by

$$\frac{M_{oc}^2}{r_0^2 N_{oy} N_{oz}} = \left(1 - \frac{N_{oc}}{N_{oy}}\right)\left(1 - \frac{N_{oc}}{N_{oz}}\right) \tag{7.45}$$

in which

$$N_{oy} = \pi^2 EI_y / L_E^2, \tag{7.46}$$

and

$$N_{oz} = \frac{GJ}{r_0^2}\left(1 + \frac{\pi^2 EI_w}{GJL_E^2}\right), \tag{7.47}$$

where L_E is the effective length of the beam-column

$$L_E = k_e L, \tag{7.48}$$

and the effective length ratio k_e is the solution of

$$\frac{R}{1-R} = \frac{-\pi}{2k_e}\cot\frac{\pi}{2k_e}. \tag{7.49}$$

Equation 7.45 is the same as equation 7.32 for simply supported beam-columns, except for the familiar use of the effective length L_E instead of the actual length L in equations 7.46 and 7.47 defining N_{oy} and N_{oz}. Thus the solutions shown in Fig. 7.14 can also be applied to end-restrained beam-columns.

The close relationships between the flexural–torsional buckling condition of equation 7.32 for unrestrained beam-columns with equal end moments ($\beta_m = -1$) and the buckling loads N_{oy}, N_{oz} of columns and moments M_{yz} of beams have already been discussed. It should also be noted that equation 7.49 for the effective lengths of beam-columns with equal end restraints is exactly the same as equation 3.38 for columns when $(1 - R)/R$ is substituted for γ_1 and γ_2, and equation 6.53 for beams. These suggest that the flexural–torsional buckling condition for beam-columns with unequal end restraints could well be approximated by equations 7.45–7.48 if Fig. 3.20a is used to determine the effective length ratio k_e in equation 7.48.

A further approximation may be suggested for restrained beam-columns

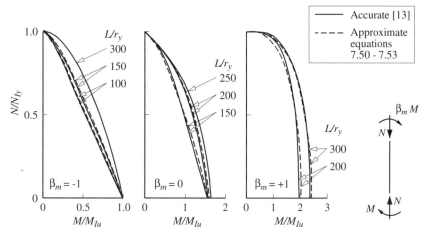

Fig. 7.17 Inelastic flexural-torsional buckling of beam-columns

with unequal end moments ($\beta_m \neq -1$), in which modifications of equations 7.42 and 7.43 (after substituting $r_0^2 N_{oy} N_{oz}$ for M_{yz}^2) are used (with equations 7.46–7.48) instead of equation 7.45.

7.3.2 INELASTIC BEAM-COLUMNS

The solutions obtained in section 7.3.1 for the flexural–torsional buckling of straight isolated beam-columns are only valid while they remain elastic. When the combination of the residual stresses with those induced by the in-plane loading causes yielding, the effective rigidities of some sections of the member are reduced, and buckling may occur at a load which is significantly less than the in-plane maximum load or the elastic buckling load, as indicated in Fig. 7.13.

A method of analysing the inelastic buckling of I-section beam-columns with residual stresses has been developed [13], and used [14] to obtain the predictions shown in Fig. 7.17 for the inelastic buckling loads of isolated beam-columns with unequal end moments. It was found that these could be closely approximated by modifying the elastic equations 7.42 and 7.43 to

$$\left(\frac{M}{\alpha_{\text{bcI}} M_{\text{Iu}}}\right)^2 = \left(1 - \frac{N}{N_{\text{Iy}}}\right)\left(1 - \frac{N}{N_{\text{oz}}}\right), \tag{7.50}$$

$$\frac{1}{\alpha_{\text{bcI}}} = \left(\frac{1 - \beta_m}{2}\right) + \left(\frac{1 + \beta_m}{2}\right)^3 \left(0.40 - 0.23\frac{N}{N_{\text{Iy}}}\right) \tag{7.51}$$

in which

$$M_{\text{Iu}}/M_{\text{p}} = 1.008 - 0.245 M_{\text{p}}/M_{\text{yz}} \leqslant 1.0 \tag{7.52}$$

is an approximation for the inelastic buckling of a beam in uniform bending (equation 7.52 is similar to equation 6.22 when $\beta_m = -1$), and

$$N_{Iy}/N_Y = 1.035 - 0.181\sqrt{(N_Y/N_{oy})} - 0.128N_Y/N_{oy} \leqslant 1.0 \tag{7.53}$$

provides an approximation for the inelastic buckling of a column.

7.3.3 ULTIMATE STRENGTH

The ultimate strengths of real beam-columns which fail by flexural–torsional buckling are reduced below their elastic and inelastic buckling loads by the presence of geometrical imperfections such as initial crookedness, just as are those of columns (section 3.4) and beams (section 6.4.1).

One method of predicting these reduced strengths is to modify the approximate interaction equation for in-plane bending (equation 7.21) to

$$\frac{N}{N_{uy}} + \frac{c_{mx}}{(1 - N/N_{ox})} \frac{M_x}{M_{uxu}} \leqslant 1, \tag{7.54}$$

where N_{uy} is the minor axis strength of a concentrically loaded column ($M_x = 0$), and M_{uxu} the strength of a beam ($N = 0$) with equal and opposite end moments ($\beta_m = -1$) which fails either by in-plane plasticity at M_{px}, or by flexural–torsional buckling. The basis for this modification is the remarkable similarity between equation 7.54 and the simple linear approximation of equation 7.37 for the elastic flexural–torsional buckling of beam-columns with equal and opposite end moments ($\beta_m = -1$). The application of equation 7.54 to beam-columns with unequal end moments is simplified by the similarity shown in Fig. 7.15 between the values of c_m for in-plane bending and $1/\sqrt{F}$ and $1/\alpha_m$ for flexural–torsional buckling.

However, the value of this simple modification is lost if the in-plane strength of a beam-column is to be predicted more accurately than by equation 7.21. If, for example, equation 7.23 is to be used for the in-plane strength as suggested in section 7.2.3.3, then it would be more appropriate to modify the flexural–torsional buckling equations 7.42 and 7.50 to obtain out-of-plane strength predictions. It has been suggested [15] that the modified equations should take the form

$$\left(\frac{M}{\alpha_{bcu}M_{uxu}}\right)^2 = \left(1 - \frac{N}{N_{uy}}\right)\left(1 - \frac{N}{N_{oz}}\right) \tag{7.55}$$

in which

$$\frac{1}{\alpha_{bcu}} = \left(\frac{1 - \beta_m}{2}\right) + \left(\frac{1 + \beta_m}{2}\right)^3 \left(0.40 - 0.23\frac{N}{N_{uy}}\right). \tag{7.56}$$

7.3.4 Design rules

Both the general and the more exact alternative rules of BS5950 for designing beam-columns against out-of-plane buckling are given by

$$\frac{F_c^*}{P_{cy}} + \frac{m_{LT}M_x^*}{M_b} \leq 1 \tag{7.57}$$

in which M_b is the beam lateral buckling resistance (section 6.4.2). This is an adaptation of equation 7.28 for in-plane bending, and is similar to the approximate equation 7.54 for the out-of-plane strength, except that m_{LT} is used instead of c_m to allow for the effect of non-uniform bending and the amplification factor $1/(1 - N/N_{ox})$ is omitted. Equation 7.57 can also be expressed as

$$m_{LT}M_x^* \leq M_{ab} = M_b(1 - F_c^*/P_{cy}) \tag{7.58}$$

For Class 1 plastic and Class 2 compact doubly symmetric members, a further alternative provides for the use of an increased value of M_{ab} obtained by a linear interpolation between $M_b(1 - F_c^*/P_{cy})$ and M_{rx} with λ_r as it varies between 85.8ε and λ_{ro}, in which the slenderness λ_r is related to the beam and column slendernesses λ_{LT} and λ_y, and λ_{ro} is given by

$$\lambda_{ro} = 17.15\varepsilon \frac{(2m_{LT}M_x^*/M_b + F_c^*/P_{cy})}{(m_{LT}M_x^*/M_b + F_c^*/P_{cy})} \tag{7.59}$$

BS5950 also provides an out-of-plane rule for beam-columns which are bent only about the minor axis. This takes the form of

$$\frac{F_c^*}{P_{cx}} + 0.5 \frac{m_{yx}M_y^*}{M_{cy}} \leq 1 \tag{7.60}$$

in which m_{yx} is the minor axis equivalent uniform moment factor obtained from Table 26 of BS5950. This equation allows for a reduction in the column resistance P_{cx} to buckling about the x axis caused by the minor axis moment M_y^*.

The design of members subjected to bending and tension is discussed in section 2.4.

A worked example of checking the out-of-plane resistance of a beam-column is given in section 7.7.4.

7.4 Biaxial bending of isolated beam-columns

7.4.1 BEHAVIOUR

The geometry and loading of most framed structures are three-dimensional, and the typical member of such a structure is compressed, bent about both principal axes and twisted by the other members connected to it, as shown in Fig. 7.1c. The structure is usually arranged so as to produce significant bending about the major axis of the member, but the minor axis deflections and twists are often significant as well, because the minor axis bending and torsional stiff-nesses are small. In addition, these deformations are amplified by the compo-nents of the axial load and the major axis moment which are induced by the deformations of the member.

The elastic biaxial bending of isolated beam-columns with equal and oppos-ite end moments acting about each principal axis has been analysed [16–22], but the solutions obtained cannot be simplified without approximation. These analyses have shown that the elastic biaxial bending of a beam-column is similar to its in-plane behaviour (see curve 3 of Fig. 7.2), in that the major and minor axis deflections and the twist all begin at the commencement of loading, and increase rapidly as the elastic buckling load (which is the load which causes elastic flexural–torsional buckling when there are no minor axis moments and torques acting) is approached. First yield predictions based on these analyses have given conservative estimates of the member strengths. The elastic biaxial bending of beam-columns with unequal end moments has also been investi-gated. In one of these investigations [16], approximate solutions were obtained by changing both sets of unequal end moments into equivalent equal end moments by multiplying them by the conversion factor $1/\sqrt{F}$ (see Fig. 7.15) for flexural–torsional buckling.

Although the first yield predictions for the strengths of slender beam-columns are of reasonable accuracy, they are rather conservative for stocky members in which considerable yielding occurs before failure. Sophisticated numerical analyses have been made [18, 20–24] of the biaxial bending of inelas-tic beam-columns, and good agreement with test results has been obtained. However such an analysis requires a specialized computer program, and is only useful as a research tool.

A number of attempts have been made to develop approximate methods of predicting the strengths of inelastic beam-columns. One of the simplest of these uses a linear extension

$$\frac{N}{N_u} + \frac{M_x}{M_{uxu}} \frac{c_{mx}}{(1 - N/N_{ox})} + \frac{M_y}{M_{uyu}} \frac{c_{my}}{(1 - N/N_{oy})} \leq 1 \qquad (7.61)$$

of the linear interaction equations for in-plane bending and flexural–torsional buckling. In this extension, M_{uxu} is the ultimate moment which the beam-column can support when $N = M_y = 0$ for the case of equal end moments ($\beta_m = -1$), while M_{uyu} is similarly defined. Thus equation 7.61 reduces to an equation similar to equation 7.21 for in-plane behaviour when $M_x = 0$, and to equation 7.54 for flexural–torsional buckling when $M_y = 0$. Equation 7.61 is used in conjunction with

$$\frac{M_x}{M_{prx}} + \frac{M_y}{M_{pry}} \leq 1, \qquad (7.62)$$

where M_{prx} is given by equation 7.17 and M_{pry} by equation 7.18. Equation 7.62 represents a linear approximation to the biaxial bending full plasticity limit for

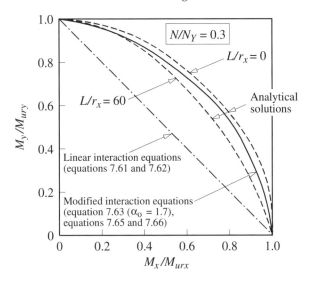

Fig. 7.18 Interaction curves for biaxial bending

the cross-section of the beam-column, and reduces to equation 7.18 or 7.17 which provides the uniaxial bending full plasticity limit when $M_x = 0$ or $M_y = 0$.

A number of criticisms may be made of these extensions of the interaction equations. Firstly there is no allowance in equation 7.61 for the amplification of the minor axis moment M_y by the major axis moment M_x (the term $(1 - N/N_{oy})$ only allows for the amplification caused by axial load N). A simple method of allowing for this would be to replace the term $(1 - N/N_{oy})$ by either $(1 - N/N_{oc})$ or by $(1 - M_x/M_{oc})$ where the flexural–torsional buckling load N_{oc} and moment M_{oc} are given by equation 7.42.

Secondly, studies [22–26] have shown that the linear additions of the moment terms in equations 7.61 and 7.62 generally lead to predictions which are too conservative, as indicated in Fig. 7.18. It has been proposed that for column type hot-rolled I-sections, the section full plasticity limit of equation 7.62 should be replaced by

$$\left(\frac{M_x}{M_{prx}} \right)^{\alpha_0} + \left(\frac{M_y}{M_{pry}} \right)^{\alpha_0} \leqslant 1, \qquad (7.63)$$

where M_{prx} and M_{pry} are given in equations 7.17 and 7.18 as before, and the index α_0 by

$$\alpha_0 = 1.60 - \frac{N/N_Y}{2\ln(N/N_Y)}. \qquad (7.64)$$

At the same time, the linear member interaction equation (equation 7.61) should be replaced by

$$\left(\frac{M_x}{M_{\text{urx}}}\right)^{\alpha_L} + \left(\frac{M_y}{M_{\text{ury}}}\right)^{\alpha_L} \leq 1, \tag{7.65}$$

where M_{urx}, the maximum value of M_x when N acts but $M_y = 0$, is obtained from equation 7.54, M_{ury} is similarly obtained from an equation similar to equation 7.21, and the index α_L is given by

$$\alpha_L = 1.40 + \frac{N}{N_Y}. \tag{7.66}$$

The approximations of equations 7.63–7.66 have been shown to be of reasonable accuracy when compared with test results [27]. This conclusion is reinforced by the comparison of some analytical solutions with the approximations of equations 7.63–7.66 (but with $\alpha_0 = 1.70$ instead of 1.725 approximately for $N/N_Y = 0.3$) shown in Fig. 7.18. Unfortunately, these approximations are of limited application, although it seems to be possible that their use may be extended.

7.4.2 DESIGN RULES

The general biaxial bending section capacity limitation of BS5950 is given by

$$\frac{F^*}{Ap_y} + \frac{M_x^*}{M_{\text{cx}}} + \frac{M_y^*}{M_{\text{cy}}} \leq 1 \tag{7.67}$$

which is a simple linear extension of the uniaxial section capacity limitation of equation 7.24. For Class 1 plastic and Class 2 compact sections, a more economic alternative is provided by a modification of equation 7.62 to

$$\left(\frac{M_x^*}{M_{\text{rx}}}\right)^{z1} + \left(\frac{M_y^*}{M_{\text{ry}}}\right)^{z2} \leq 1 \tag{7.68}$$

in which $z_1 = 2.0$ and $z_2 = 1.0$ for equal flanged I-sections, $z_1 = z_2 = 2.0$ for circular sections, $z_1 = z_2 = 5/3$ for rectangular sections, and $z_1 = z_2 = 1.0$ otherwise.

The general biaxial bending member resistance limitations are given by extensions of equations 7.28 and 7.57 for uniaxial bending to

$$\frac{F_c^*}{P_c} + \frac{m_x M_x^*}{p_y Z_x} + \frac{m_y M_y^*}{p_y Z_y} \leq 1 \tag{7.69a}$$

in which P_c is the lesser of P_{cx} and P_{cy}, and

$$\frac{F_c^*}{P_{\text{cy}}} + \frac{m_{\text{LT}} M_x^*}{M_b} + \frac{m_y M_y^*}{p_y Z_y} \leq 1 \tag{7.69b}$$

The first of these is more likely to govern the design of members which do not buckle laterally, such as square or circular hollow sections, while the second is more likely to govern the design of members susceptible to lateral buckling.

More exact alternatives for doubly symmetric I- or H-sections are provided by

$$\frac{F_c^*}{P_{cx}} + m_x \left(1 + 0.5 \frac{F_c^*}{P_{cx}}\right) \frac{M_c^*}{M_{cx}} + 0.5 \frac{m_{yx}M_y^*}{M_{cy}} \leqslant 1 \qquad (7.70a)$$

$$\frac{F_c^*}{P_{cy}} + m_{LT} \frac{M_x^*}{M_b} + m_y \left(1 + \frac{F_c^*}{P_{cy}}\right) \frac{M_y^*}{M_{cy}} \leqslant 1 \qquad (7.70b)$$

$$m_x \frac{(1 + 0.5F_c^*/P_{cx})}{(1 - F_c^*/P_{cx})} \frac{M_x^*}{M_{cx}} + m_y \frac{(1 + F_c^*/P_{cy})}{(1 - F_c^*/P_{cy})} \frac{M_y^*}{M_{cy}} \leqslant 1 \qquad (7.70c)$$

which are related to equations 7.29, 7.57, and 7.60.

For Class 1 plastic and Class 2 compact doubly symmetric sections, more economic further alternatives are given in Annex I of BS5950.

A worked example for checking the biaxial bending of a beam-column is given in section 7.7.5.

7.5 Appendix – in-plane behaviour of elastic beam-columns

7.5.1 STRAIGHT BEAM-COLUMNS

The bending moment in the beam-column shown in Fig. 7.3 is the sum of the moment $[M - M(1 + \beta_m)z/L]$ due to the end moments and reactions and the moment Nv due to the deflection v. Thus the differential equation of bending is obtained by equating this bending moment to the internal moment of resistance $-EI_x d^2v/dz^2$, whence

$$-EI_x \frac{d^2v}{dz^2} = M - M(1 + \beta_m)\frac{z}{L} + Nv. \qquad (7.71)$$

The solution of this equation which satisfies the support conditions $(v)_0 = (v)_L = 0$ is

$$v = \frac{M}{N}\left[\cos \mu z - (\beta_m \operatorname{cosec} \mu L + \cot \mu L)\sin \mu z - 1 + (1 + \beta_m)\frac{z}{L}\right], \quad (7.1)$$

where

$$\mu^2 = \frac{N}{EI_x} = \frac{\pi^2}{L^2}\frac{N}{N_{ox}}, \qquad (7.2)$$

and the bending moment distribution can be obtained by substituting equation 7.1 into equation 7.71. The maximum bending moment can be determined by solving the condition

$$\frac{d(-EI_x d^2v/dz^2)}{dz} = 0$$

for the position z_m of the maximum moment, whence

$$\tan \mu z_m = -\beta_m \operatorname{cosec} \mu L - \cot \mu L. \tag{7.72}$$

The value of z_m is positive while

$$\beta_m < -\cos \pi \sqrt{(N/N_{ox})},$$

and the maximum moment obtained by substituting equation 7.72 into equation 7.71 is

$$M_m = M\sqrt{[1 + \{\beta_m \operatorname{cosec} \pi\sqrt{(N/N_{ox})} + \cot \pi\sqrt{(N/N_{ox})}\}^2]}. \tag{7.3}$$

When $\beta_m > -\cos \pi\sqrt{(N/N_{ox})}$, z_m is negative, in which case the maximum moment in the beam-column is the end moment M at $z = 0$, whence

$$M_m = M. \tag{7.4}$$

For beam-columns which are bent in single curvature by equal and opposite end moments ($\beta_m = -1$), equation 7.1 for the deflected shape simplifies, and the central deflection δ_m can be expressed non-dimensionally as

$$\frac{\delta_m}{\delta} = \frac{8/\pi^2}{(N/N_{ox})}\left[\sec \frac{\pi}{2} \sqrt{(N/N_{ox})} - 1\right] \tag{7.5}$$

in which $\delta = ML^2/8EI_x$ is the value of δ_m when $N = 0$. The maximum moment occurs at the centre of the beam-column, and is given by

$$\frac{M_m}{M} = \sec \frac{\pi}{2} \sqrt{(N/N_{ox})}. \tag{7.6}$$

Equations 7.5 and 7.6 are plotted in Fig. 7.4.

7.5.2 BEAM-COLUMNS WITH INITIAL CURVATURE

If the beam-column shown in Fig. 7.3 is not straight, but has an initial crooked-ness given by

$$v_0 = \delta_0 \sin \pi z/L,$$

then the differential equation of bending becomes

$$-EI_x \frac{d^2v}{dz^2} = M - M(1 + \beta_m)\frac{z}{L} + N(v + v_0). \tag{7.73}$$

The solution of this equation which satisfies the support conditions of $(v)_0 = (v)_L = 0$ is

$$v = (M/N)[\cos \mu z - (\beta_m \operatorname{cosec} \mu L + \cot \mu L) \sin \mu z - 1 \\ + (1 + \beta_m)z/L] + [(\mu L/\pi)^2/\{1 - (\mu L/\pi)^2\}]\delta_0 \sin \pi z/L, \tag{7.74}$$

where

$$(\mu L/\pi)^2 = N/N_{ox}. \tag{7.75}$$

The bending moment distribution can be obtained by substituting equation 7.74 into equation 7.73. The position of the maximum moment can then be determined by solving the condition $d(-EI_x d^2 v/dz^2)dz = 0$, whence

$$\frac{N\delta_0}{M} = \frac{\mu L}{\pi}\left(1 - \frac{\mu^2 L^2}{\pi^2}\right)\frac{\{(\beta_m \operatorname{cosec} \mu L + \cot \mu L)\cos \mu z_m + \sin \mu z_m\}}{\cos \pi z_m/L}.$$

$$(7.76)$$

The value of the maximum moment M_m can be obtained from equations 7.73 and 7.74 (with $z = z_m$) and from equation 7.76, and is given by

$$\frac{M_m}{M} = \cos \mu z_m - (\beta_m \operatorname{cosec} \mu L + \cot \mu L)\sin \mu z_m + \frac{(N\delta_0/M)}{(1 - \mu^2 L^2/\pi^2)}\sin\frac{\pi z_m}{L}.$$

$$(7.77)$$

First yield occurs when the maximum stress is equal to the yield stress f_y, in which case

$$\frac{N}{N_Y} + \frac{M_m}{M_Y} = 1,$$

or

$$\frac{N}{N_Y}\left\{1 + \frac{M_m}{M}\frac{N_Y\delta_0}{M_Y}\frac{M}{N\delta_0}\right\} = 1.$$

$$(7.78)$$

In the special case of $M = 0$, the first yield solution is the same as that given in section 3.2.2 for a crooked column. Equation 7.76 is replaced by $z_m = L/2$, and equation 7.77 by

$$M_m = N\delta_0/(1 - \mu^2 L^2/\pi^2).$$

The value N_0 of N which then satisfies equation 7.78 is given by the solution of

$$\frac{N_0}{N_Y} + \frac{N_0\delta_0/M_Y}{\{1 - (N_0/N_Y)/(N_{ox}/N_Y)\}} = 1.$$

$$(7.79)$$

The value of M/M_Y at first yield can be determined for any specified set of values of N_{ox}/N_Y, β_m, $\delta_0 N_Y/M_Y$ and N/N_Y. An initial guess for z_m/L allows $N\delta_0/M$ to be determined from equation 7.76, and M_m/M from equation 7.77, and these can then be used to evaluate the left hand side of equation 7.78. This process can be repeated until the values are found which satisfy equation 7.78. Solutions obtained in this way are plotted in Fig. 7.7b.

7.6 Appendix – flexural–torsional buckling of elastic beam-columns

7.6.1 SIMPLY SUPPORTED BEAM-COLUMNS

The elastic beam-column shown in Fig. 7.1b is simply supported and prevented from twisting at its ends so that

$$(u)_{0,L} = (\phi)_{0,L} = 0, \tag{7.80}$$

and is free to warp (see section 10.8.3) so that

$$\left(\frac{d^2\phi}{dz^2}\right)_{0,L} = 0. \tag{7.81}$$

The combination at elastic buckling of the axial load N_{oc} and equal and opposite end moments M_{oc} (i.e. $\beta_m = -1$) can be determined by finding a deflected and twisted position which is one of equilibrium. The differential equilibrium equations for such a position are

$$EI_y \frac{d^2u}{dz^2} + N_{oc}u = -M_{oc}\phi \tag{7.82}$$

for minor axis bending, and

$$(GJ - N_{oc}r_0^2)\frac{d\phi}{dz} - EI_w \frac{d^3\phi}{dz^3} = M_{oc}\frac{du}{dz} \tag{7.83}$$

for torsion, where $r_0 = \sqrt{[(I_x + I_y)/A]}$ is the polar radius of gyration. Equation 7.82 reduces to equation 3.57 for compression member buckling when $M_{oc} = 0$, and to equation 6.82 for beam buckling when $N_{oc} = 0$. Equation 7.83 can be used to derive equation 3.70 for torsional buckling of a compression member when $M_{oc} = 0$, and reduces to equation 6.83 for beam buckling when $N_{oc} = 0$.

The buckled shape of the beam-column is given by

$$u = \frac{M_{oc}}{N_{oy} - N_{oc}} \phi = \delta \sin \frac{\pi z}{L}, \tag{7.31}$$

where $N_{oy} = \pi^2 EI_y/L^2$ is the minor axis elastic buckling load of an axially loaded column, and δ the undetermined magnitude of the central deflection. The boundary conditions of equations 7.80 and 7.81 are satisfied by this buckled shape, as is equation 7.82, while equation 7.83 is satisfied when

$$\frac{M_{oc}^2}{r_0^2 N_{oy} N_{oz}} = \left(1 - \frac{N_{oc}}{N_{oy}}\right)\left(1 - \frac{N_{oc}}{N_{oz}}\right), \tag{7.32}$$

where

$$N_{oz} = \frac{GJ}{r_0^2}\left(1 + \frac{\pi^2 EI_w}{GJL^2}\right)$$

is the elastic torsional buckling load of an axially loaded column (see section 3.6.5). Equation 7.32 defines the combinations of axial load N_{oc} and end moments M_{oc} which cause the beam-column to buckle elastically.

7.6.2 ELASTICALLY RESTRAINED BEAM-COLUMNS

If the elastic beam-column shown in Fig. 7.1b is not simply supported but is elastically restrained at its ends against minor axis rotations du/dz and against warping rotations $(d_f/2)d\phi/dz$, then the boundary conditions at the end $z = L/2$ of the beam-column can be expressed in the form

$$\left.\begin{aligned}\frac{M_B + M_T}{(du/dz)_{L/2}} &= \frac{-EI_y}{L}\frac{2R_2}{1 - R^2} \\[2mm] \frac{M_T - M_B}{(d_f/2)(d\phi/dz)_{L/2}} &= \frac{-EI_y}{L}\frac{2R_4}{1 - R_4}\end{aligned}\right\} , \tag{7.84}$$

where M_T and M_B are the flange minor axis end restraining moments

$$M_T = \tfrac{1}{2}EI_y(d^2u/dz^2)_{L/2} + (d_f/4)EI_y(d^2\phi/dz^2)_{L/2},$$
$$M_B = \tfrac{1}{2}EI_y(d^2u/dz^2)_{L/2} - (d_f/4)EI_y(d^2\phi/dz^2)_{L/2},$$

and R_2 and R_4 are dimensionless minor axis bending and warping end restraint parameters, which vary between zero (no restraint) and one (rigid restraint). If the beam-column is symmetrically restrained, similar conditions apply at the end $z = -L/2$. The other support conditions are

$$(u)_{\pm L/2} = (\phi)_{\pm L/2} = 0 \tag{7.85}$$

The particular case for which the minor axis and warping end restraints are equal, so that

$$R_2 = R_4 = R \tag{7.86}$$

can be analysed. When the restrained beam-column has equal and opposite end moments $(\beta_m = -1)$, the differential equilibrium equations for a buckled position u, ϕ are

$$EI_y \frac{d^2u}{dz^2} + N_{oc}u = -M_{oc}\phi + (M_B + M_T)$$

$$(GJ - N_{oc}r_0^2)\frac{d\phi}{dz} - EI_w\frac{d^3\phi}{dz^3} = M_{oc}\frac{du}{dz}. \tag{7.87}$$

These differential equations and the boundary conditions of equations 7.84–7.86 are satisfied by the buckled shape

$$u = \frac{M_{oc}\phi}{N_{oy} - N_{oc}} = A\left(\cos\frac{\pi z}{k_e L_E} - \cos\frac{\pi}{2k_e}\right),$$

where

$$N_{oy} = \pi^2 EI_y / L_E^2, \tag{7.46}$$

$$L_E = k_e L, \tag{7.48}$$

when the effective length ratio k_e satisfies

$$\frac{R}{1 - R} = \frac{-\pi}{2k_e} \cot \frac{\pi}{2k_e}. \tag{7.49}$$

This is the same as equation 6.53 for the buckling of restrained beams, whose solutions are shown in Fig. 6.18c. The values M_{oc} and N_{oc} for flexural–torsional buckling of the restrained beam-column are obtained by substituting the buckled shape u, ϕ into equations 7.87, and are related by

$$\frac{M_{oc}^2}{r_0^2 N_{oy} N_{oz}} = \left(1 - \frac{N_{oc}}{N_{oy}}\right)\left(1 - \frac{N_{oc}}{N_{oz}}\right), \tag{7.45}$$

where

$$N_{oz} = \frac{GJ}{r_0^2}\left(1 + \frac{\pi^2 EI_w}{GJL_e^2}\right). \tag{7.47}$$

7.7 Worked examples

7.7.1 EXAMPLE 1 – APPROXIMATING THE MAXIMUM ELASTIC MOMENT

Problem. Use the approximations of Figs 7.5 and 7.6 to determine the maximum moments M_{xm}, M_{ym} for the 254 × 146 UB 37 beam-columns shown in Fig. 7.19d.

Solution for M_{xm} for the beam-column of Fig. 7.19d. Using Fig. 7.5, $k_e = 1.0$, $\gamma_m = 1.0$, $\gamma_n = 1.0$, $\gamma_s = 0.18$

$$M_x = 20 \times 9/4 = 45 \text{ kNm}$$

$$N_{ox} = \pi^2 \times 205\,000 \times 5547 \times 10^4/(1.0 \times 9000)^2 \text{ N} = 1386 \text{ kN}.$$

Using equation 7.10, $\delta_x = \dfrac{1.0 \times (1 - 0.18 \times 200/1386)}{(1 - 1.0 \times 200/1386)} = 1.138.$

Using equation 7.9, $M_{xm} = 1.138 \times 45 = 51.2 \text{ kNm}$.

Solution for M_{ym} for the beam-column of Fig. 7.19c. Using Fig. 7.6, $k_e = 1.0$, $\gamma_m = 1.0$, $\gamma_n = 0.49$, $\gamma_s = 0.18$

$$M_y = 3.2 \times 4.5^2/8 = 8.1 \text{ kNm}$$

$$N_{oy} = \pi^2 \times 205\,000 \times 571 \times 10^4/(1.0 \times 4500)^2 \text{ N} = 570.5 \text{ kN}.$$

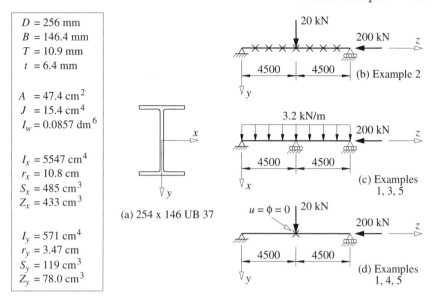

$D = 256$ mm
$B = 146.4$ mm
$T = 10.9$ mm
$t = 6.4$ mm

$A = 47.4$ cm^2
$J = 15.4$ cm^4
$I_w = 0.0857$ dm^6

$I_x = 5547$ cm^4
$r_x = 10.8$ cm
$S_x = 485$ cm^3
$Z_x = 433$ cm^3

(a) 254 x 146 UB 37

$I_y = 571$ cm^4
$r_y = 3.47$ cm
$S_y = 119$ cm^3
$Z_y = 78.0$ cm^3

20 kN

200 kN

4500 | 4500

(b) Example 2

3.2 kN/m

200 kN

4500 | 4500

(c) Examples 1, 3, 5

$u = \phi = 0$ | 20 kN

200 kN

4500 | 4500

(d) Examples 1, 4, 5

Fig. 7.19 Examples 1–5

Using equation 7.10,

$$\delta_y = \frac{1.0 \times (1 - 0.18 \times 200/570.5)}{(1 - 0.49 \times 200/570.5)} = 1.131.$$

Using equation 7.9, $M_{ym} = 1.131 \times 8.1 = 9.2$ kNm.

7.7.2 EXAMPLE 2 – CHECKING THE MAJOR AXIS IN-PLANE CAPACITY (ELASTIC ANALYSIS)

Problem.   The 9 m long simply supported beam-column shown in Fig. 7.19b has a factored design axial compression force of 200 kN and a design concentrated load of 20 kN (which includes an allowance for self-weight) acting in the major principal plane at mid-span. The beam-column is the 254 × 146 UB 37 of S275 steel shown in Fig. 7.19a. The beam-column is continuously braced against lateral deflections u and twist rotations ϕ. Check the adequacy of the beam-column.

Simplified approach for section capacity.

$$M_x^* = 20 \times 9/4 = 45.0 \text{ kNm}$$
$$T = 10.9 \text{ mm}, p_y = 275 \text{ N/mm}^2 \qquad \text{T9}$$
$$\varepsilon = \sqrt{(275/275)} = 1.0$$
$$b/(t\varepsilon) = (146.4/2)/(10.9 \times 1.0) = 6.72 < 9$$

and the flange is plastic.   T11

$$r_1 = \frac{200 \times 10^3}{(256 - 2 \times 10.9 - 2 \times 7.6) \times 6.4 \times 275} = 0.519 \qquad 3.5.5$$

$$(1 + r_1)d/(t\varepsilon) = (1 + 0.519) \times (256 - 2 \times 10.9 - 2 \times 7.6)/(6.4 \times 1.0)$$
$$= 52.0 < 80 \text{ and the web is plastic.} \qquad \text{T11}$$

Thus the section is Class 1 plastic.

$$M_{cx} = 275 \times 485 \times 10^3 \text{ Nmm} = 133.4 \text{ kNm} \qquad 4.2.5.2$$

$$\frac{200 \times 10^3}{47.4 \times 10^2 \times 275} + \frac{45.0}{133.4} = 0.491 < 1.0 \qquad 4.8.3.2$$

and the section capacity is adequate.

Alternative approach for section capacity.
Because the section is Class 1 plastic, Clause 4.8.2.3 (and Annex I.2) can be
used. $\qquad$ 4.8.3.2(b)

$$n = (200 \times 10^3)/(47.4 \times 10^2 \times 275) = 0.153 \qquad \text{I.2.1}$$
$$t(D - 2T)/A = 6.4 \times (256 - 2 \times 10.9)/(47.4 \times 10^2) \qquad \text{I.2.1}$$
$$= 0.316 > 0.153 = n$$
$$S_{rx} = 485 \times 10^3 - (47.4 \times 10^2)^2 \times 0.153^2/(4 \times 6.4) \text{ mm}^3 \qquad \text{I.2.1}$$
$$= 464.5 \text{ cm}^3$$
$$M_{rx} = 275 \times 464.5 \times 10^3 \text{ Nmm} \qquad \text{I.2.1}$$
$$= 127.7 \text{ kNm} > 45.0 \text{ kNm} = M_x^*$$

and the section capacity is adequate. $\qquad$ 4.8.2.3

Simplified approach for member resistance.
Because the member is continuously braced, beam lateral buckling and column
minor axis buckling need not be considered.

$$\lambda_x = 9000/(10.8 \times 10) = 83.3$$

For x-axis buckling of a rolled I-section with $T < 40$ mm, use Strut Curve (a).
$\qquad$ T23

$$p_{cx} = (199 \times 0.7 + 194 \times 1.3)/2 = 195.7 \text{ N/mm}^2 \qquad \text{T24(1)}$$
$$P_{cx} = 47.4 \times 10^2 \times 195.7 \text{ N} = 927.9 \text{ kN} \qquad 4.7.4$$
$$m_x = 0.90 \qquad \text{T26}$$

$$\frac{200 \times 10^3}{927.9 \times 10^3} + \frac{0.90 \times 45.0 \times 10^6}{275 \times 433 \times 10^3} = 0.556 < 1.0 \qquad 4.8.3.3.1$$

and the member resistance is adequate.

More exact approach for member resistance.

$$\frac{200}{927.9} + \frac{0.90 \times 45.0}{133.4}\left(1 + \frac{0.5 \times 200}{927.9}\right) = 0.552 < 1.0 \qquad 4.8.3.3.2(a)$$

and the member resistance is adequate.

Further alternative approach for member resistance.

$$M_{ox} = \frac{133.4 \times (1 - 200/927.9)}{(1 + 0.5 \times 200/927.9)} = 94.5 \text{ kNm} \qquad \text{I.1}$$

$$\lambda_x/\varepsilon = 83.3/1.0 = 83.3 < 85.8 \qquad \text{I.1}$$

$$M_{ax} = 94.5 + \frac{(85.8 \times 1.0 - 83.3)}{68.6 \times 1.0} \times (127.7 - 94.5) = 95.7 \text{ kNm} \qquad \text{I.1}$$

$$m_x M_x^* = 0.90 \times 45.0 = 40.5 \text{ kNm} < 95.7 \text{ kNm} = M_{ax} \qquad \text{I.1(a)}$$

and the member resistance is adequate.

7.7.3 EXAMPLE 3 – CHECKING THE MINOR AXIS IN-PLANE CAPACITY

Problem. The 9 m long two span beam-column shown in Fig. 7.19c has a factored design axial compression force of 200 kN and a factored design uniformly distributed load of 3.2 kN m acting in the minor principal plane. The beam-column is the 254 × 146 UB 37 of S275 steel shown in Fig. 7.19a. Check the adequacy of the beam-column.

Simplified approach for section capacity.

$$M_y^* = 3.2 \times 4.5^2/8 = 8.1 \text{ kNm}$$
$$M_{cy} = 275 \times 119 \times 10^3 \text{ Nmm} = 32.7 \text{ kNm} \qquad 4.2.5.2$$

$$\frac{200 \times 10^3}{47.4 \times 10^2 \times 275} + \frac{8.1}{32.7} = 0.401 < 1.0 \qquad 4.8.3.2$$

and the section capacity is adequate.

Alternative approach for section capacity.

$$n = 0.153, \text{ as in section 7.7.2}$$
$$tD/A = 6.4 \times 256/(47.4 \times 10^2) = 0.346 > 0.153 = n \qquad \text{I.2.1}$$
$$S_{ry} = 119 \times 10^3 - \{(47.4 \times 10^2)^2/(4 \times 256)\} \times 0.153^2 \text{ mm}^3 \qquad \text{I.2.1}$$
$$= 118.5 \text{ cm}^3$$
$$M_{ry} = 275 \times 118.5 \times 10^3 \text{ Nmm} \qquad \text{I.2.1}$$
$$= 32.6 \text{ kNm} > 8.1 \text{ kNm} = M_y^*$$

and the section capacity is adequate. $\qquad 4.8.2.3$

Simplified approach for member resistance.
Because the member is bent about the minor axis, beam lateral buckling need not be considered.

$$p_{cy} = 195.7 \text{ N/mm}^2, \text{ as in section 7.7.2.}$$
$$\lambda_y = 4500/(3.47 \times 10) = 129.7$$

For y-axis buckling of a rolled I-section with $T < 40$ mm, use Strut Curve (b)

T23

$$p_{cy} = (97 \times 0.3 + 95 \times 1.7)/2 = 95.3 \text{ N/mm}^2 \qquad \text{T24(4)}$$
$$P_{cy} = 47.4 \times 10^2 \times 95.3 \text{ N} = 451.7 \text{ kN} \qquad 4.7.4$$
$$M_{max} = -8.1 \text{ kNm}, M_2 = 0.5 \times 8.1 \text{ kNm}, M_3 = 0.5 \times 8.1 \text{ kNm}, M_4 = 0 \text{ kNm}.$$

$$m_y = \frac{\{0.2 + 0.1 \times 0.5 + 0.6 \times 0.5 + 0.1 \times 0\} \times 8.1}{8.1} = 0.55 > 0.40 = 0.8 \times 0.5$$

T26

$$\frac{200 \times 10^3}{451.7 \times 10^3} + \frac{0.55 \times 8.1 \times 10^6}{275 \times 78.0 \times 10^3} = 0.650 < 1.0 \qquad 4.8.3.3.1$$

and the member resistance is adequate.

More exact approach for member resistance.

$$\frac{200}{451.7} + \frac{0.55 \times 8.1}{32.7}\left(1 + \frac{200}{451.7}\right) = 0.639 < 1.0 \qquad 4.8.3.3.2\text{(b)}$$

$$m_{yx} = \frac{\{0.2 + 0.1 \times 0.5 + 0.6 \times 1.0 + 0.1 \times 0.5\} \times 8.1}{8.1} = 0.90 > 0.80 = 0.8 \times 1.0$$

T26

$$\frac{200}{927.9} + \frac{0.50 \times 0.90 \times 8.1}{32.7} = 0.327 < 1.0 \qquad 4.8.3.3.2\text{(b)}$$

and the member resistance is adequate.

Further alternative approach for member resistance.

$$\lambda_y/\varepsilon = 129.7/1.0 = 129.7 > 85.8 \qquad \text{I.1}$$

and so $M_{ay} = M_{oy} = M_{cy}(1 - F_c^*/P_{cy})/(1 + F_c^*/P_{cy})$, and there is no advantage to be gained in this case by using the further alternative approach.

7.7.4 EXAMPLE 4 – CHECKING THE OUT-OF-PLANE CAPACITY

Problem. The 9 m long simply supported beam-column shown in Fig. 7.19d has a factored design axial compression force of 200 kN and a factored design concentrated load of 20 kN (which includes an allowance for self-weight) acting in the major principal plane at mid-span. Lateral deflections u and twist rotations ϕ are prevented at the ends and at mid-span. The beam-column is the 254×146 UB 37 of S275 steel shown in Fig. 7.19a. Check the adequacy of the beam-column.

Design moment.

$$M_x^* = 45.0 \text{ kNm (section 7.7.1).}$$

Simplified approach for section capacity.
The section capacity was checked in section 7.7.2.

Simplified approach for member resistance.
The member in-plane resistance was checked in section 7.7.2.
For the member out-of-plane resistance,

$$P_{cy} = 451.7 \text{ kN as in section 7.7.3.}$$
$$M_b = 67.4 \text{ kNm}, m_{LT} = 0.60, \text{ as in section 6.13.2.}$$

$$\frac{200}{451.7} + \frac{0.60 \times 45.0}{67.4} = 0.843 < 1.0 \qquad 4.8.3.3.1$$

and the member out-of-plane resistance is adequate.

More exact approach for member resistance.
The member in-plane resistance was checked in section 7.7.2.
For the member out-of-plane resistance,

$$\frac{200}{451.7} + \frac{0.60 \times 45.0}{67.4} = 0.843 < 1.0 \qquad 4.8.3.3.1$$

as above, and the member out-of-plane resistance is adequate.

Further alternative approach for member resistance.

$$r_b = 0.60 \times 45.0/67.4 = 0.401 \qquad \text{I.1}$$
$$r_c = 200/451.7 = 0.443 \qquad \text{I.1}$$
$$\lambda_{LT} = 92.2 \text{ as in section 6.13.2.}$$
$$\lambda_y = 129.7 \text{ as in section 7.7.3.}$$
$$\lambda_r = (0.401 \times 92.2 + 0.443 \times 129.7)/(0.401 + 0.443) \qquad \text{I.1}$$
$$= 111.9 > 85.8 = 85.8\varepsilon \qquad \text{I.1}$$

and so $M_{ab} = M_{ob} = M_b(1 - F_c^*/P_{cy})$, and there is no advantage to be gained in this case by using the further alternative approach.

7.7.5 EXAMPLE 5 – CHECKING THE BIAXIAL BENDING CAPACITY

Problem. The 9 m long simply supported beam-column shown in Fig. 7.19c and d has a factored design axial compression force of 200 kN, a factored design concentrated load of 20 kN (which includes an allowance for self-weight) acting in the major principal plane, and a factored design uniformly distributed load of 3.2 kN m acting in the minor principal plane. Lateral deflections u and twist rotations ϕ are prevented at the ends and at mid-span. The beam-column is the 254 × 146 UB 37 of S275 steel shown in Fig. 7.19a. Check the adequacy of the beam-column.

Design moments.

$$M_x^* = 45.0 \text{ kNm (section 7.7.1)}, M_y^* = 8.1 \text{ kNm (section 7.7.1)}$$

Simplified approach for section capacity.
Combining the calculations of sections 7.7.2 and 7.7.3,

$$\frac{200 \times 10^3}{47.4 \times 10^2 \times 275} + \frac{45.0}{133.4} + \frac{8.1}{32.7} = 0.738 < 1.0 \qquad 4.8.3.2$$

and the section capacity is adequate.

Alternative approach for section capacity.
Combining the calculations of sections 7.7.2 and 7.7.3,

$$\left(\frac{45.0}{127.7}\right)^{2.0} + \left(\frac{8.1}{32.6}\right)^{1.0} = 0.373 < 1.0 \qquad 4.8.2.3$$

and the section capacity is adequate.

Simplified approach for member resistance.
Combining the appropriate calculations of sections 7.7.2, 7.7.3, and 7.7.4,

$$\frac{200}{451.7} + \frac{0.90 \times 45.0 \times 10^6}{275 \times 433 \times 10^3} + \frac{0.55 \times 8.1 \times 10^6}{275 \times 78.0 \times 10^3} = 0.991 < 1.0 \qquad 4.8.3.3.1$$

$$\frac{200}{451.7} + \frac{0.6 \times 45.0}{67.4} + \frac{0.55 \times 8.1 \times 10^6}{275 \times 78.0 \times 10^3} = 1.051 > 1.0 \qquad 4.8.3.3.1$$

and the member is inadequate.

More exact approach for member resistance.
Combining the appropriate calculations of sections 7.7.2, 7.7.3, and 7.7.4,

$$\frac{200}{927.9} + \frac{0.90 \times 45.0}{133.4}\left(1 + \frac{0.5 \times 200}{927.9}\right) + \frac{0.5 \times 0.90 \times 8.1}{32.7} = 0.663 < 1.0$$

$$4.8.3.3.2(c)$$

$$\frac{200}{451.7} + \frac{0.60 \times 45.0}{67.4} + \frac{0.55 \times 8.1}{32.7}\left(1 + \frac{200}{451.7}\right) = 1.040 > 1.0$$

$$4.8.3.3.2(c)$$

and the member is inadequate.

Further alternative approach for member resistance.
Combining the appropriate calculations of sections 7.7.2, 7.7.3, and 7.7.4,

$$\frac{0.90 \times 45.0}{95.7} + \frac{0.5 \times 0.90 \times 8.1}{32.7(1 - 200/927.9)} = 0.565 < 1.0 \qquad I.1(c)$$

$$M_{ab} = M_{ob} = 67.4 \times (1 - 200/451.7) = 37.6 \text{ kNm} \qquad I.1$$
$$M_{ay} = M_{oy} = 32.7 \times (1 - 200/451.7)/(1 + 200/451.7) = 12.6 \text{ kNm} \qquad I.1$$

$$\frac{0.60 \times 45.0}{37.6} + \frac{0.55 \times 8.1}{12.6} = 1.072 > 1.0 \qquad 4.8.3.3.2$$

and the member is inadequate.

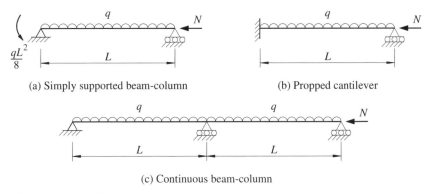

(a) Simply supported beam-column   (b) Propped cantilever

(c) Continuous beam-column

Fig. 7.20 Examples 6–8

7.8 Unworked examples

7.8.1 EXAMPLE 6 – NON-LINEAR ANALYSIS

Analyse the non-linear elastic in-plane bending of the simply supported beam-column shown in Fig. 7.20a, and show that the maximum moment M_m may be closely approximated by using the greater of $M_m = qL^2/8$ and

$$M_m = \frac{9qL^2}{128}\frac{(1 - 0.29N/N_{ox})}{(1 - N/N_{ox})}$$

in which $N_{ox} = \pi^2 EI_x/L^2$.

7.8.2 EXAMPLE 7 – NON-LINEAR ANALYSIS

Analyse the non-linear elastic in-plane bending of the propped cantilever shown in Fig. 7.20b, and show that the maximum moment M_m may be closely approximated by using

$$\frac{M_m}{qL^2/8} \approx \frac{(1 - 0.18N/N_{ox})}{(1 - 0.49N/N_{ox})}$$

in which $N_{ox} = \pi^2 EI_x/L^2$.

7.8.3 EXAMPLE 8 – NON-LINEAR ANALYSIS

Analyse the non-linear elastic in-plane bending of the two span beam-column shown in Fig. 7.20c, and show that the maximum moment M_m may be closely approximated by using

$$\frac{M_m}{qL^2/8} \approx \frac{(1 - 0.18N/N_{ox})}{(1 - 0.49N/N_{ox})}$$

while $N \le N_{ox}$, in which $N_{ox} = \pi^2 EI_x/L^2$.

7.8.4 EXAMPLE 9 – IN-PLANE DESIGN

A 7.0 m long simply supported beam-column, which is prevented from swaying and from deflecting out of the plane of bending, is required to support a factored design axial load of 1200 kN and factored design end moments of 300 and 150 kNm which bend the beam-column in single curvature about the major axis. Design a suitable UB or UC beam-column of S275 steel.

7.8.5 EXAMPLE 10 – OUT-OF-PLANE DESIGN

The intermediate restraints of the beam-column of example 9 which prevent deflection out of the plane of bending are removed. Design a suitable UB or UC beam-column of S275 steel.

7.8.6 EXAMPLE 11 – BIAXIAL BENDING DESIGN

The beam-column of example 10 has to carry additional factored design end moments of 75 and 75 kNm which cause double curvature bending about the minor axis. Design a suitable UB or UC beam-column of S275 steel.

7.9 References

1. Chen, W-F. and Atsuta, T. (1976) *Theory of Beam-Columns, Vol. 1 In-Plane Behavior and Design*, McGraw-Hill, New York.
2. Galambos, T.V. and Ketter, R.L. (1959) Columns under combined bending and thrust, *Journal of the Engineering Mechanics Division, ASCE*, **85**, No. EM2, April, pp. 1–30.
3. Ketter, R.L. (1961) Further studies of the strength of beam-columns, *Journal of the Structural Division, ASCE*, **87**, No. ST6, August, pp. 135–52.
4. Young, B.W. (1973) The in-plane failure of steel beam-columns, *The Structural Engineer*, **51**, pp. 27–35.
5. Trahair, N.S. (1986) Design strengths of steel beam-columns, *Canadian Journal of Civil Engineering*, **13**, No. 6, December, pp. 639–46.
6. Bridge, R.Q. and Trahair, N.S. (1987) Limit state design rules for steel beam-columns, *Steel Construction*, Australian Institute of Steel Construction, **21**, No. 2, September, pp. 2–11.
7. Salvadori, M.G. (1955) Lateral buckling of I-beams, *Transactions, ASCE*, **120**, pp. 1165–77.
8. Salvadori, M.G. (1955) Lateral buckling of eccentrically loaded I-columns, *Transactions, ASCE*, **121**, pp. 1163–78.
9. Horne, M.R. (1954) The flexural–torsional buckling of members of symmetrical I-section under combined thrust and unequal terminal moments, *Quarterly Journal of Mechanics and Applied Mathematics*, **7**, pp. 410–26.
10. Column Research Committee of Japan (1971) *Handbook of Structural Stability*, Corona, Tokyo.

11. Trahair, N.S. (1993) *Flexural–Torsional Buckling of Structures*, E. & F.N. Spon, London.
12. Cuk, P.E. and Trahair, N.S. (1981) Elastic buckling of beam-columns with unequal end moments, *Civil Engineering Transactions*, Institution of Engineers, Australia, **CE 23**, No. 3, August, pp. 166–71.
13. Bradford, M.A., Cuk, P.E., Gizejowski, M.A. and Trahair, N.S. (1987) Inelastic lateral buckling of beam-columns, *Journal of Structural Engineering, ASCE*, **113**, No. 11, November, pp. 2259–77.
14. Bradford, M.A. and Trahair, N.S. (1985) Inelastic buckling of beam-columns with unequal end moments, *Journal of Constructional Steel Research*, **5**, No. 3, pp. 195–212.
15. Cuk, P.E., Bradford, M.A. and Trahair, N.S. (1986) Inelastic lateral buckling of steel beam-columns, *Canadian Journal of Civil Engineering*, **13**, No. 6, December, pp. 693–9.
16. Horne, M.R. (1956) The stanchion problem in frame structures designed according to ultimate load carrying capacity, *Proceedings of the Institution of Civil Engineers*, Part III, **5**, pp. 105–46.
17. Culver, C.G. (1966) Exact solution of the biaxial bending equations, *Journal of the Structural Division, ASCE*, **92**, No. ST2, pp. 63–83.
18. Harstead, G.A., Birnsteil, C. and Leu, K-C. (1968) Inelastic H-columns under biaxial bending, *Journal of the Structural Division, ASCE*, **94**, No. ST10, pp. 2371–98.
19. Trahair, N.S. (1969) Restrained elastic beam-columns, *Journal of the Structural Division, ASCE*, **95**, No. ST12, pp. 2641–64.
20. Vinnakota, S. and Aoshima, Y. (1974) Inelastic behaviour of rotationally restrained columns under biaxial bending, *The Structural Engineer*, **52**, pp. 245–55.
21. Vinnakota, S. and Aysto, P. (1974) Inelastic spatial stability of restrained beam-columns, *Journal of the Structural Division, ASCE*, **100**, No. ST11, pp. 2235–54.
22. Chen, W-F. and Atsuta, T. (1977) *Theory of Beam-Columns, Vol. 2 Space Behavior and Design*, McGraw-Hill, New York.
23. Pi, Y-L. and Trahair, N.S. (1994) Nonlinear inelastic analysis of steel beam-columns. I: Theory, *Journal of Structural Engineering, ASCE*, **120**, No. 7, pp. 2041–61.
24. Pi, Y-L. and Trahair, N.S. (1994) Nonlinear inelastic analysis of steel beam-columns. II: Applications, *Journal of Structural Engineering, ASCE*, **120**, No. 7, pp. 2062–85.
25. Tebedge, N. and Chen, W-F. (1974) Design criteria for H-columns under biaxial loading, *Journal of the Structural Division, ASCE*, **100**, No. ST3, pp. 579–98.
26. Young, B.W. (1973) Steel column design, *The Structural Engineer*, **51**, pp. 323–36.
27. Bradford, M.A. (1995) Evaluation of design rules of the biaxial bending of beam-columns, *Civil Engineering Transactions*, Institution of Engineers, Australia, **CE 37**, No. 3, pp. 241–45.

8 Frames

8.1 Introduction

Structural frames are composed of one-dimensional members connected together in skeletal arrangements which transfer the applied loads to the supports. While most frames are three-dimensional, they may often be considered as a series of parallel two-dimensional frames, or as two perpendicular series of two-dimensional frames. The behaviour of a structural frame depends on its arrangement and loading, and on the type of connections used.

Triangulated frames with joint loading only have no primary bending actions, and the members act in simple axial tension or compression. The behaviour and design of these members have already been discussed in Chapters 2 and 3.

Frames which are not triangulated include rectangular frames, which may be multi-storey, or multi-bay, or both, and pitched-roof portal frames. The members usually have substantial bending actions (Chapters 5 and 6), and if they also have significant axial forces, then they must be designed as beam-ties (Chapter 2), or beam-columns (Chapter 7). In frames with simple connections (see Fig. 9.3a), the moments transmitted by the connections are small, and often can be neglected, and the members can be treated as isolated beams, or as eccentrically loaded beam-ties or beam-columns. However, when the connections are semi-rigid or rigid, there are important moment interactions between the members.

When a frame is or can be considered as two-dimensional then its behaviour is similar to that of the beam-columns of which it is composed. With in-plane loading only, it will fail either by in-plane bending, or by flexural–torsional buckling out of its plane. If, however, the frame or its loading is three-dimensional, then it will fail in a mode in which the individual members are subjected to primary biaxial bending and torsion actions.

In this chapter, the in-plane, out-of-plane, and biaxial behaviour, analysis, and design of two- and three-dimensional frames are treated, and related to the behaviour and design of isolated tension members, compression members, beams, and beam-columns discussed in Chapters 2, 3, and 5–7.

8.2 Triangulated frames

8.2.1 STATICALLY DETERMINATE FRAMES

The primary actions in triangulated frames whose members are concentrically connected and whose loads act concentrically through the joints are those of axial compression or tension in the members, and any bending actions are secondary only. These bending actions are usually ignored, in which case the member forces may be determined by a simple analysis for which the member connections are assumed to be made through frictionless pin-joints.

If the assumed pin-jointed frame is statically determinate, then each member force can be determined by statics alone, and is independent of the behaviour of the remaining members. Due to this, the frame may be assumed to fail when its weakest member fails, as indicated in Fig. 8.1a and b. Thus, each member may be designed independently of the others by using the procedures discussed in Chapter 2 (for tension members) or 3 (for compression members).

If all the members of a frame are designed to fail simultaneously (either by tension yielding or by compression buckling), then there will be no significant moment interactions between the members at failure, even if the connections are not pin-jointed. Hence, the effective lengths of the compression members should be taken as equal to the lengths between their ends. If, however, the members of a rigid-jointed frame are not designed to fail simultaneously, then there will be moment interactions between them at failure. While it is a common and conservative practice to ignore these interactions and design each member as if pin-ended, some account may be taken of these by using one of the procedures discussed in section 8.3.5.2 to estimate the effective lengths of the compression members, and exemplified in section 8.5.1.

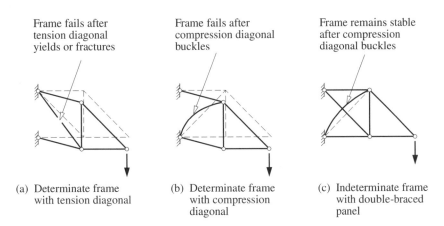

Frame fails after tension diagonal yields or fractures

Frame fails after compression diagonal buckles

Frame remains stable after compression diagonal buckles

(a) Determinate frame with tension diagonal

(b) Determinate frame with compression diagonal

(c) Indeterminate frame with double-braced panel

Fig. 8.1 Determinate and indeterminate triangulated frames

8.2.2 STATICALLY INDETERMINATE FRAMES

When an assumed pin-jointed triangulated frame without primary bending actions is statically indeterminate, the forces transferred by the members will depend on their axial stiffnesses. Since these decrease with tension yielding or compression buckling, there is usually some force redistribution as failure is approached. To design such a frame, it is necessary to estimate the member forces at failure, and then design the individual tension and compression members for these forces, as in Chapters 2 and 3. Four different methods may be used to estimate the member forces.

In the first method, a sufficient number of members are ignored so that the frame becomes statically determinate. For example, it is common to ignore completely the compression diagonal in the double-braced panel of the indeterminate frame of Fig. 8.1c, in which case the strength of the now determinate frame is controlled by the strength of one of the other members. This method is simple and conservative.

A less conservative method can be used when each indeterminate member can be assumed to be ductile, so that it can maintain its maximum strength over a considerable range of axial deformations. In this method, the frame is converted to a statically determinate frame by replacing a sufficient number of early failing members by sets of external forces equal to their maximum load capacities. For the frame shown in Fig. 8.1c, this might be done for the diagonal compression member, by replacing it by forces equal to its buckling strength. This simple method is very similar in principle to the plastic method of analysing the collapse of flexural members (section 5.5) and frames (section 8.3.5.5). However, the method can only be used when the ductilities of the members being replaced are assured.

In the third method, each member is assumed to behave linearly and elastically, which allows the use of a linear elastic method to analyse the frame and determine its member force distribution. The frame strength is then controlled by the strength of the most severely loaded member. For the frame shown in Fig. 8.1c, this would be the compression diagonal if this reaches its buckling strength before any of the other members fail. This method ignores redistribution, and is conservative when the members are ductile, or when there are no lacks-of-fit at member connections.

In the fourth method, account is taken of the effects of changes in the member stiffnesses on the force distribution in the frame. As a result of its complexity, this method is rarely used.

When the members of a triangulated frame are eccentrically connected, or when the loads act eccentrically or are applied to the members between the joints, the flexural effects are no longer secondary, and the frame should be treated as a flexural frame. Flexural frames are discussed in sections 8.3 and 8.4.

8.3 Two-dimensional flexural frames

8.3.1 GENERAL

The structural behaviour of a frame may be classified as two-dimensional when there are a number of independent two-dimensional frames with in-plane loading, or when this is approximately so. The primary actions in a two-dimensional frame in which the members support transverse loads are usually flexural, and are often accompanied by significant axial actions. The structural behaviour of a flexural frame is influenced by the behaviour of the member joints, which are usually considered to be either simple, semi-rigid, or rigid, according to their ability to transmit moment.

In the following subsections, the behaviour, analysis and design of two-dimensional flexural frames with simple, semi-rigid, or rigid joints are discussed.

8.3.2 FRAMES WITH SIMPLE JOINTS

Simple joints may be defined as being those that will not develop restraint moments which adversely affect the members and the structure as a whole, and which therefore allow the structure to be designed as if pin-jointed. Thus it is usually assumed that no moment is transmitted through a simple joint, and that the members connected to the joint may rotate. The joint should therefore have sufficient rotation capacity to permit member end rotations to occur without causing failure of the joint or its elements. An example of a typical simple joint between a beam and column is shown in Fig. 9.3a.

If there is a sufficient number of pin-joints to make the structure statically determinate, then each member will act independently of the others, and may be designed as an isolated tension member (Chapter 2), compression member (Chapter 3), beam (Chapters 5 and 6), or beam-column (Chapter 7). However, if the pin-jointed structure is indeterminate, then some part of it may act as a rigid-jointed frame. The behaviour, analysis, and design of rigid-jointed frames are discussed in sections 8.3.4–8.3.6.

One of the most common methods of designing frames with simple joints is often used for rectangular frames with vertical (column) and horizontal (beam) members under the action of vertical loads only. The columns in such a frame are assumed to act as if eccentrically loaded. A minimum eccentricity of 100 mm from the face of the column is specified in BS5950, which gives approximate procedures for distributing the moment caused by the eccentric beam reaction between the column lengths above and below the connection. It should be noted that such a pin-jointed frame is usually incapable of resisting transverse forces, and must therefore be provided with an independent bracing or shear wall system.

8.3.3 FRAMES WITH SEMI-RIGID JOINTS

Semi-rigid joints are those which have dependable moment capacities and which partially restrain the relative rotations of the members at the joints. The action of these joints in rectangular frames is to reduce the maximum moments in the beams, and so the semi-rigid design method offers potential economies over the simple design method [1–5]. An example of a typical semi-rigid joint between a beam and column is shown in Fig. 9.3b.

A method of semi-rigid design is permitted by BS5950. In this method, the stiffness, strength, and rotation capacities of the joints are based on experimental evidence, and used to assess the moments and forces in the members. However, this method has not found great favour with designers, and will therefore not be discussed further.

8.3.4 IN-PLANE BEHAVIOUR OF FRAMES WITH RIGID JOINTS

A rigid joint may be defined as a joint which has sufficient rigidity to virtually prevent relative rotation between the members connected. Properly arranged welded and high-strength friction grip bolted joints are usually assumed to be rigid. An example of a typical rigid joint between a beam and a column is shown in Fig. 9.3c. There are important interactions between the members of frames with rigid joints, which are generally stiffer and stronger than frames with simple or semi-rigid joints. Due to this, rigid frames offer significant economies, while many difficulties associated with their analysis have been greatly reduced by the widespread availability of standard computer programs.

The in-plane behaviour of rigid frames is discussed in general terms in section 1.4.2, where it is pointed out that, although a rigid frame may behave in an approximately linear fashion while its service loads are not exceeded, especially when the axial forces are small, it becomes non-linear near its in-plane ultimate load because of yielding and buckling effects. When the axial compression forces are small, failure occurs when a sufficient number of plastic hinges have developed to cause the frame to form a collapse mechanism, in which case the load capacity of the frame can be determined by plastic analysis of the collapse mechanism. More generally, however, the in-plane stability effects associated with the axial compression forces significantly modify the behaviour of the frame near its ultimate loads. The in-plane analysis of rigid-jointed frames is discussed in section 8.3.5, and their design in 8.3.6.

8.3.5 IN-PLANE ANALYSIS OF FRAMES WITH RIGID JOINTS

8.3.5.1 First-order elastic analysis

A first-order elastic analysis of a rigid-jointed frame is based on the assumptions that:

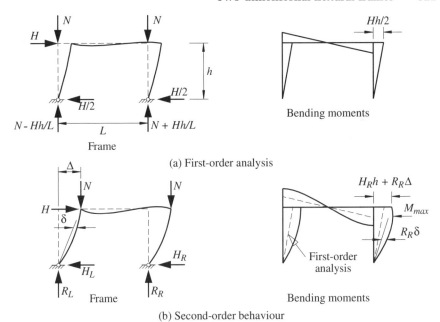

Bending moments

Frame

(a) First-order analysis

(b) Second-order behaviour

Fig. 8.2 First-order analysis and second-order behaviour

(a) the material behaves linearly, so that all yielding effects are ignored;
(b) the members behave linearly, with no member instability effects such as those caused by axial compressions which reduce the members' flexural stiffnesses (these are often called the P–δ effects);
(c) the frame behaves linearly, with no frame instability effects such as those caused by the moments of the vertical forces and the horizontal frame deflections (these are often called the P–Δ effects).

For example, for the portal frame of Fig. 8.2, a first-order elastic analysis ignores all second-order moments such as $R_R(\delta + \Delta z/h)$ in the right hand column, so that the bending moment distribution is linear in this case. First-order analyses predict linear behaviour in elastic frames, as shown in Fig. 1.15.

Rigid-jointed frames are invariably statically indeterminate, and while there are many manual methods of first-order elastic analysis available [6–8], these are labour-intensive and error prone for all but the simplest frames. In the past, designers were often forced to rely on approximate methods or available solutions for specific frames [9, 10]. However computer methods of first-order elastic analysis [11, 12] have formed the basis of computer programs such as [13–15] which are now used extensively. These first-order elastic analysis programs require the geometry of the frame and its members to have been established (usually by a preliminary design), and then compute the member moments and forces as well as the joint deflections for each specified set of

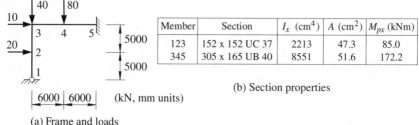

(a) Frame and loads

(b) Section properties

Member	Section	I_x (cm^4)	A (cm^2)	M_{px} (kNm)
123	152 x 152 UC 37	2213	47.3	85.0
345	305 x 165 UB 40	8551	51.6	172.2

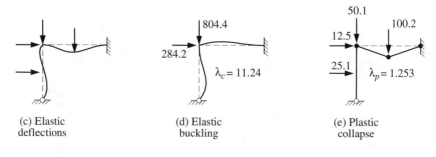

(c) Elastic deflections

(d) Elastic buckling

(e) Plastic collapse

Quantity	Units	First-order elastic	Elastic buckling	Second-order elastic	First-order plastic
M_2	kNm	23.5	0	24.7	20.2
M_3	kNm	-53.0	0	-53.6	-85.0
M_4	kNm	136.6	0	137.4	172.2
M_5	kNm	-153.8	0	-154.6	-172.2
N_{123}	kN	-71.6	-804.4	-71.6	-93.0
N_{345}	kN	-25.3	-284.2	-25.1	-33.6
v_4	mm	58.5	(0.155)	58.9	–
u_2	mm	19.0	(1.000)	21.2	–

(f) Analysis results

Fig. 8.3 Analysis of a braced frame

loads. As these are proportional to the loads, the results of individual analyses may be combined by linear superposition. The results of computer first-order elastic analyses [13] of a braced and an unbraced frame are given in Figs 8.3 and 8.4.

8.3.5.2 Elastic buckling of braced frames

The results of an elastic buckling analysis may be used to approximate any second-order effects (Fig. 8.2). The elastic buckling analysis of a rigid-jointed braced frame is carried out by replacing the initial set of frame loads by a set which produces the same set of member axial forces without any

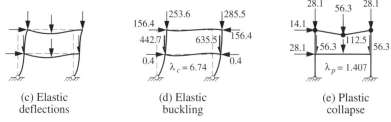

	20	40	20	
10	3	5	8	
	40	80 40		5000
20	2	4	7	
	1		6	5000
	6000	6000	(kN, mm units)	

(a) Frame and loads

(b) Section properties

Member	Section	I_x (cm^4)	A (cm^2)	M_{px} (kNm)
12, 67	203 x 203 UC 52	5254	66.4	155.9
23, 78	152 x 152 UC 37	2213	47.3	85.0
247	356 x 171 UB 51	14160	64.6	246.1
358	254 x 102 UB 25	3420	32.3	84.4

(c) Elastic
deflections

(d) Elastic
buckling

$\lambda_c = 6.74$

(e) Plastic
collapse

$\lambda_p = 1.407$

Quantity	Units	First-order elastic	Elastic buckling	Second-order elastic	First-order plastic
M_4	kNm	153.1	0	153.5	237.5
M_5	kNm	71.6	0	72.9	84.4
M_{76}	kNm	-119.2	0	-130.8	-155.9
M_{74}	kNm	-172.7	0	-185.0	-240.9
M_{78}	kNm	53.5	0	54.3	85.0
M_8	kNm	-62.5	0	-64.5	-84.4
N_{12}	kN	-103.3	-696.3	-100.8	-145.4
N_{23}	kN	-37.6	-253.6	-37.6	-56.3
N_{67}	kN	-136.7	-921.0	-138.9	-192.3
N_{78}	kN	-42.4	-285.5	-42.5	-56.3
v_4	mm	45.8	(0.004)	47.2	–
v_5	mm	81.9	(0.001)	85.2	–
u_2	mm	87.6	(0.826)	103.2	–
u_3	mm	124.3	(1.000)	144.3	–

(f) Analysis results

Fig. 8.4 Analysis of an unbraced frame

bending, as indicated in Fig. 8.5b. The set of member forces N_{om} which causes buckling depends on the distribution of the axial forces in the frame, and is often expressed in terms of a load factor λ_c by which the initial set of axial forces N_{im} must be multiplied to obtain the member forces N_{om} at frame buckling, so that

$$N_{om} = \lambda_c N_{im}. \tag{8.1}$$

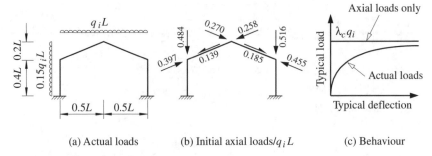

(a) Actual loads   (b) Initial axial loads/$q_i L$ (c) Behaviour

Fig. 8.5 In-plane behaviour of a rigid frame

Alternatively, it may be expressed by a set of effective length ratios $k_m = L_{Em}/L_m$ which define the member forces at frame buckling by

$$N_{om} = \pi^2 EI_m/(k_m L_m)^2. \tag{8.2}$$

For all but isolated members and very simple frames, the determination of the frame buckling factor λ_c is best carried out numerically, using a suitable computer program. The bases of frame elastic buckling programs are discussed in [11, 12]. The computer program PRFSA [13] first finds the initial member axial forces $\{N_i\}$ by carrying out a first-order elastic analysis of the frame under the initial loads. It then uses a finite element method to determine the elastic buckling load factors λ_c for which the total frame stiffness vanishes [16], so that

$$[K]\{D\} - \lambda_c[G]\{D\} = \{0\} \tag{8.3}$$

in which $[K]$ is the elastic stiffness matrix, $[G]$ the stability matrix associated with the initial axial forces $\{N_i\}$, and $\{D\}$ the vector of nodal deformations which defines the buckled shape of the frame. The results of a computer elastic buckling analysis [13] of a braced frame are given in Fig. 8.3.

For isolated braced members or very simple braced frames, a buckling analysis may also be made from first principles, as demonstrated in section 3.10. Alternatively, the effective length ratio k_m of each compression member may be obtained by using estimates of the member end stiffness factors k_1, k_2 in a braced member chart such as that of Fig. 3.20a. The direct application of this chart is limited to the vertical columns of regular rectangular frames with regular loading patterns in which each horizontal beam has zero axial force, and all the columns buckle simultaneously in the same mode. In this case, the values of the column end stiffness factors can be obtained from

$$k = \frac{\Sigma(I_c/L_c)}{\{\Sigma(I_c/L_c + \Sigma(\beta_e I_b/L_b)\}}, \tag{8.4}$$

where β_e is a factor which varies with the restraint conditions at the far end of the beam ($\beta_e = 0.5$ if the restraint condition at the far end is the same as that at the column, $\beta_e = 0.75$ if the far end is pinned, and $\beta_e = 1.0$ if the far end is fixed

– note that the proportions $0.5 : 0.75 : 1.0$ of these are the same as $2 : 3 : 4$ of the appropriate stiffness multipliers of Fig. 3.18), and where the summations are carried out for the columns (c) and beams (b) at the column end.

The effective length chart of Fig. 3.20a may also be used to approximate the buckling forces in other braced frames. In the simplest application, an effective length factor k_m is determined for each compression member of the frame by approximating its member end stiffness factors (see equation 3.40) by

$$k = \frac{I_m/L_m}{\{I_m/L_m + \sum(\beta_e \alpha_r I_r/L_r)\}}, \tag{8.5}$$

where α_r is a factor which allows for the effect of axial force on the flexural stiffness of each restraining member (Fig. 3.18), and the summation is carried out for all of the members connected to that end of the compression member. Each effective length ratio so determined is used with equations 8.1 and 8.2 to obtain an estimate of the frame buckling load factor as

$$\lambda_{cm} = \frac{\pi^2 EI_m/(k_m L_m)^2}{N_{im}}. \tag{8.6}$$

The lowest of these provides a conservative approximation of the actual frame buckling load factor λ_c. An equivalent method is allowed by BS5950.

The accuracy of this method of calculating effective lengths using the stiffness approximations of Fig. 3.18 is indicated in Fig. 8.6c for the rectangular frame shown in Fig. 8.6a and b. For the buckling mode shown in Fig. 8.6a, the buckling of the vertical members is restrained by the horizontal members. These horizontal members are braced restraining members which bend in symmetrical single curvature, so that their stiffnesses are $(2EI_1/L_1)(1 - N_1/N_{oL1})$ in which $N_{oL1} = \pi^2 EI_1/L_1^2$, as indicated in Fig. 3.18. It can be seen from Fig. 8.6c that the approximate buckling loads are in very close agreement with the

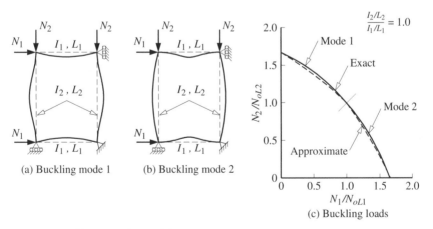

(a) Buckling mode 1   (b) Buckling mode 2   (c) Buckling loads

Fig. 8.6 Buckling of a braced frame

accurate values. Worked examples of the application of this method are given in sections 8.5.1 and 8.5.2.

A more accurate iterative procedure [17] may also be used, in which the accuracies of the approximations for the member end stiffness factors increase with each iteration.

8.3.5.3 Elastic buckling of unbraced frames

The determination of the frame buckling load factor λ_c of a rigid-jointed unbraced frame may also be carried out using a suitable computer program such as that described in [13].

For isolated unbraced members or very simple unbraced frames, a buckling analysis may also be made from first principles, as demonstrated in section 3.10. Alternatively the effective length ratio k_m of each compression member may be obtained by using estimates of the end stiffness factors k_1, k_2 in a chart such as that of Fig. 3.20b. The application of this chart is limited to the vertical columns of regular rectangular frames with regular loading patterns in which each horizontal beam has zero axial force, and all the columns buckle simultaneously in the same mode and with the same effective length ratio. The member end stiffness factors are then given by equation 8.4.

The effects of axial forces in the beams of an unbraced regular rectangular frame can be approximated by using equation 8.5 to approximate the end stiffness factors k_1, k_2 (but with $\beta_e = 1.5$ if the restraint conditions at the far end are the same as at the column, $\beta_e = 0.75$ if the far end is pinned, and $\beta_e = 1.0$ if the far end is fixed – note that the proportions 1.5 : 0.75 : 1 of these are the same as 6 : 3 : 4 of the appropriate stiffness multipliers of Fig. 3.18) and Fig. 3.20b.

The accuracy of this method of using the stiffness approximations of Fig. 3.18 is indicated in Fig. 8.7c for the rectangular frame shown in Fig. 8.7a

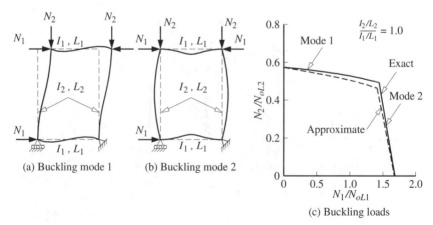

(a) Buckling mode 1   (b) Buckling mode 2   (c) Buckling loads

Fig. 8.7 Buckling of an unbraced frame

and b. For the sway buckling mode shown in Fig. 8.7a, the sway buckling of the vertical members is restrained by the horizontal members. These horizontal members are braced restraining members which bend in antisymmetrical double curvature, so that their stiffnesses are $(6EI_1/L_1)(1 - N_1/4N_{oL1})$ as indicated in Fig. 3.18. It can be seen from Fig. 8.7c that the approximate buckling loads are in close agreement with the accurate values. A worked example of the application of this method is given in section 8.5.4.

The effective length ratio chart of Fig. 3.20b may also be used to approximate the storey buckling load factor λ_{ms} for each storey of an unbraced rectangular frame with negligible axial forces in the beams as

$$\lambda_{ms} = \frac{\Sigma(N_{om}/L_m)}{\Sigma(N_{im}/L_m)} \tag{8.7}$$

in which N_{om} is the buckling load of a column in a storey obtained by using equation 8.4, and the summations are made for each column in the storey. For a storey in which all the columns are of the same length, this equation implies that the approximate storey buckling load factor depends on the total stiffness of the columns and the total load on the storey, and is independent of the distribution of stiffness or load. The frame buckling load factor λ_c may be approximated by the lowest of the values of λ_{ms} calculated for all the storeys of the frame. A worked example of this application of this method is given in section 8.5.3.

This method gives close approximations when the buckling pattern is the same in each storey, and is conservative when the horizontal members have zero axial forces and the buckling pattern varies from storey to storey.

Another method of analysing the buckling of an unbraced rectangular frame is to carry out a series of first-order elastic analyses of the frame under its actual loading (if this does not cause lateral deflection, then it should be augmented by a small set of horizontal forces), together with a series of fictitious horizontal forces which correspond to the P–Δ moments caused by the vertical forces and the horizontal deflections Δ [11, 18, 19]. A series of analyses must be carried out at each load level, since the fictitious horizontal forces increase with the deflections Δ. If the analysis series converges at a particular load level, then the frame is stable. The load level is then increased, and the series repeated. This process is continued until the analysis series fails to converge, indicating that the frame buckling loads have been exceeded. An approximate method of anticipating convergence is to examine the value of $(\Delta_{n+1} - \Delta_n)/(\Delta_n - \Delta_{n-1})$ computed from the values of Δ at the steps $(n - 1)$, n, and $(n + 1)$. If this is less than $n/(n + 1)$, then the analysis series probably converges. This method of analysing frame buckling is slow and clumsy but it does allow the use of the widely available first-order elastic computer programs. The method ignores the decreases in the member stiffnesses caused by axial compressions (the P–δ effects), and therefore tends to overestimate the buckling loads. A simpler related method is given in Annex F of BS5950.

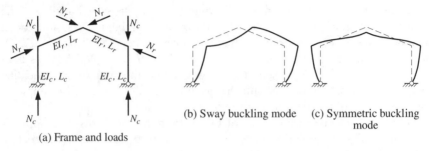

(b) Sway buckling mode   (c) Symmetric buckling mode

(a) Frame and loads

Fig. 8.8 Pinned-base portal frame

The approximate methods described above for analysing the elastic buckling of unbraced frames are limited to rectangular frames. While the buckling of non-rectangular unbraced frames may be analysed by using a suitable computer program such as that described in [13], there are many published solutions for specific frames [20].

For the symmetrical pinned-base portal frame shown in Fig. 8.8a, the buckling load factor for the sway mode shown in Fig. 8.8b may be approximated by [21, 22]

$$\lambda_{sp} = \frac{3EI_r/(N_rL_r^2)}{[(1 + 1.2/R)N_cL_c/(N_rL_r) + 0.3]} \tag{8.8}$$

in which

$$R = (I_c/L_c)/(I_r/L_r), \tag{8.9}$$

and N_c, N_r are the average compression forces in the columns and rafters. In some cases, the symmetric buckling mode shown in Fig. 8.8c may occur. The buckling load factor for this mode may be approximated by treating the two rafters as a single member of length $2L_r$ which is elastically restrained by the columns, so that

$$\lambda_r = \frac{\pi^2 EI_r}{(2k_eL_r)^2N_r} \tag{8.10}$$

in which the rafter effective length ratio k_e is obtained from Fig. 3.20a for braced members by using

$$k_{1p} = k_{2p} = \frac{I_r/2L_r}{(I_r/2L_r + 0.75I_c/L_c)}. \tag{8.11}$$

A worked example of the application of this method is given in section 8.5.5.

For a symmetrical fixed-base portal frame, the sway buckling load factor may be approximated by [21]

$$\lambda_{sf} = \frac{(10 + R)5EI_r/(N_rL_r^2)}{2N_cL_c/(N_rL_r) + 5} \tag{8.12}$$

in which R is given by equation 8.9. The symmetric buckling load factor may be approximated by using equation 8.10 in which the rafter effective length ratio k_e is obtained from Fig. 3.20a for braced members by using

$$k_{1f} = k_{2f} = \frac{I_r/2L_r}{(I_r/2L_r + I_c/L_c)}. \tag{8.13}$$

Approximate methods for multi-bay frames are given in [22].

Similar approximate methods for single- and multi-bay portal frames are given in Annex F of BS5950.

8.3.5.4 Second-order elastic analysis

Second-order effects in elastic frames include additional moments such as $R_R(\delta + \Delta z/h)$ in the right hand column of Fig. 8.2 which results from the finite deflections δ and Δ of the frame. The second-order moments arising from the member deflections from the straight line joining the member ends are often called the P–δ effects, while the second-order moments arising from the joint displacements Δ are often called the P–Δ effects. In braced frames, the joint displacements Δ are small, and only the P–δ effects are important. In unbraced frames, the P–Δ effects are important, and often much more so than the P–δ effects. Other second-order effects include those arising from the end-to-end shortening of the members due to stress ($= NL/EA$), due to bowing ($= \frac{1}{2}\int_0^L (d\delta/dz)^2 dz$), and from finite deflections. Second-order effects cause non-linear behaviour in elastic frames, as shown in Fig. 1.14.

The second-order P–δ effects in a number of elastic beam-columns have been analysed, and approximations for the maximum moments are given in section 7.2.1. A worked example for the use of these approximations is given in section 7.7.1.

The P–δ and P–Δ second-order effects in elastic frames are most easily and accurately accounted for by using a computer second-order elastic analysis program such as those of [13–15]. The results of second-order elastic analyses [13] of braced and unbraced frames are given in Figs 8.3 and 8.4. For the braced frame of Fig. 8.3, the second-order results are only slightly higher than the first-order results. This is usually true for well-designed braced frames that have substantial bending effects and small axial compressions. For the unbraced frame of Fig. 8.4, the second-order sway deflections are about 18% larger than those of the first-order analysis, while the moment at the top of the right-hand lower storey column is about 10% larger than that of the first-order analysis. Second-order effects are often significant in unbraced frames.

There are a number of methods of approximating second-order effects which allow a general second-order analysis to be avoided. In the simplest of these, the results of a first-order elastic analysis are amplified by using the results of an elastic buckling analysis (sections 8.3.5.2 and 8.3.5.3). For braced frames, the maximum member moment M_m^* determined by first-order analysis may be used to approximate the maximum second-order design moment M^* as

$$M^* = \delta M_m^*$$ (8.14)

in which

$$\delta = \delta_b = \frac{c_m}{1 - N^*/N_{omb}} \geq 1,$$ (8.15)

N_{omb} is the elastic buckling load calculated for the braced restrained member, and

$$c_m = 0.6 - 0.4\beta_m \leq 1.0,$$ (8.16)

where β_m is the ratio of the smaller to the larger end moment (equations 8.15 and 8.16 are related to equations 7.7 and 7.8 used for isolated beam-columns). A worked example of the application of this method is given in section 8.5.6.

A procedure for approximating the value of β_m to be used for members with transverse forces is available [23], while more accurate solutions are obtained by using equations 7.9–7.11 and Figs 7.5 and 7.6.

For unbraced frames, BS5950 allows the maximum moment M_m^* determined by first-order analysis to be amplified using

$$\delta = \delta_s = \frac{1}{1 - 1/\lambda_c}$$ (8.17)

in which λ_c is the elastic sway buckling load factor calculated for the unbraced frame. A worked example of the application of this method is given in section 8.5.7.

For unbraced rectangular frames with negligible axial forces in the beams, a more accurate approximation can be obtained by amplifying the column moments of each storey by using the sway buckling load factor λ_{ms} for that storey obtained using equation 8.7 in

$$\delta = \delta_s = \frac{1}{1 - 1/\lambda_{ms}}.$$ (8.18)

A worked example of the application of this method is given in section 8.5.8.

An alternative approximate method is permitted by BS5950 for each storey of an unbraced rectangular frame which uses the storey amplification factor

$$\delta = \delta_s = \frac{1}{1 - (\Delta_s/h_s)(\sum N^*/\sum V^*)}$$ (8.19)

in which Δ_s is the relative horizontal displacement of the top of the storey of height h_s from the bottom, and the storey design shears V^* and design axial forces N^* are summed up for the storey. The values of Δ_s may be obtained by first-order analysis. A worked example of the calculation of these storey amplification factors is given in section 8.5.9.

Alternatively, the first-order end moment

$$M_f^* = M_{fb}^* + M_{fs}^* \tag{8.20}$$

may be separated into a braced frame component M_{fb}^* and a complementary sway component M_{fs}^* [24]. The second-order maximum moment M^* may be approximated by using these components in

$$M^* = M_{fb}^* + \frac{M_{fs}^*}{(1 - 1/\lambda_c)} \tag{8.21}$$

in which λ_c is the load factor at frame elastic sway buckling (section 8.3.5.3). A similar method is permitted by BS5950.

A more accurate second-order analysis can be made by including the P–Δ effects of sway displacement directly in the analysis. An approximate method of doing this is to include fictitious horizontal forces equivalent to the P–Δ effects in an iterative series of first-order analyses [18, 19], which is similar to the approximate method of elastic buckling analysis discussed in section 8.3.5.3. The greater second-order end moment M_m^* determined in this way may then be used in equations 8.14 and 8.15 to determine the design second-order moment M^*. As the P–Δ effects of sway have already been included in the determination of M_m^*, it is appropriate to use

$$N_{omb} = \pi^2 EI/L^2 \tag{8.22}$$

in equation 8.15.

8.3.5.5 First-order plastic analysis

In the first-order method of plastic analysis, all instability effects are ignored, and the collapse strength of the frame is determined by using the rigid-plastic assumption (section 5.5.2) and finding the plastic hinge locations which first convert the frame to a collapse mechanism. All members must be ductile so that the plastic moment capacity can be maintained at each hinge over a range of hinge rotations sufficient to allow the plastic collapse mechanism to develop. Ductility is usually ensured by restricting the steel type to one which has a substantial yield plateau and significant strain-hardening (section 1.3.1) and by preventing reductions in rotation capacity by local buckling effects (by restricting the cross-section dimensions), by out-of-plane buckling effects (by restricting the unbraced lengths), and by in-plane buckling effects (by restricting the member in-plane slendernesses and axial compression forces).

The methods used for the plastic analysis of frames are extensions of those discussed in section 5.5.5 which incorporate reductions in the plastic moment capacity to account for the presence of axial force (section 7.2.2). These methods are discussed in many texts, such as those [18–24: Chapter 5] referred to in section 5.14. The manual application of these methods is not as simple as it is for beams, but computer programs such as that in [13] are available. The results of computer first-order plastic analyses [13] of a braced and an unbraced frame are given in Figs 8.3 and 8.4.

8.3.5.6 Advanced analysis

Present methods of designing statically indeterminate steel structures rely on predictions of the stress resultants acting on the individual members and models of the strengths of these members. These member strength models include second-order effects and inelastic behaviour, as well as residual stresses and geometrical imperfections. It is therefore possible that the common methods of elastic second-order analysis used to determine the member stress resultants, which do not account for inelastic behaviour, residual stresses and geometrical imperfections, may lead to inaccurate predictions.

Second-order effects and inelastic behaviour cause redistributions of the member stress resultants in indeterminate structures as some of the more critical members lose stiffness and transfer moments to their neighbours. Non-proportional loading also causes redistribution. For the example frame shown in Fig. 8.9, the column end moments induced initially by the beam loads decrease as the column load increases and the column becomes less stiff, due to second-order effects at first, and then due to inelastic behaviour. The end moments may reduce to zero, or even reverse in direction. In this case the end moments change from initially disturbing the column to stabilizing it by resisting further end rotations. This change in the sense of the column end moments, which contributes significantly to the column strength, cannot be predicted unless an analysis is made which more closely models the behaviour of the real structure than does a simple plastic analysis (which ignores second-order effects) or a second-order elastic analysis (which ignores inelastic behaviour).

Ideally, the member stress resultants should be determined by a method of

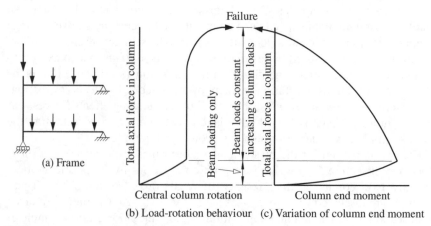

(a) Frame

(b) Load-rotation behaviour (c) Variation of column end moment

Fig. 8.9 Reversal of column end moments in frames

frame analysis which accounts for both second-order effects (P–δ and P–Δ), inelastic behaviour, residual stresses and geometrical imperfections, and any local or out-of-plane buckling effects. Such a method has been described as an advanced analysis. Not only can an advanced analysis be expected to lead to predictions that are of high accuracy, but it also has the further advantage that the design process can be greatly simplified, since the structure can be considered to be satisfactory if the advanced analysis can show that it can reach an equilibrium position under the design loads.

Past research on advanced analysis and the prediction of the strengths of real structures has concentrated on frames for which local and lateral buckling are prevented (by using compact sections and full lateral bracing). One of the earliest, and perhaps the simplest, methods of approximating the strengths of rigid-jointed frames was suggested in [25] where it was proposed that the ultimate load factor λ_u of a frame should be calculated from the load factor λ_p at which plastic collapse occurs (all stability effects being ignored) and the load factor λ_c at which elastic in-lane buckling takes place (all plasticity effects being ignored) by using

$$\frac{1}{\lambda_u} = \frac{1}{\lambda_p} + \frac{1}{\lambda_c},$$ (8.23)

or

$$\lambda_u/\lambda_p = 1 - \lambda_u/\lambda_c.$$ (8.24)

Equation 8.24 is shown in Fig. 8.10 as a linear relationship between λ_u/λ_c and λ_u/λ_p. This is similar to the linear approximation of equation 7.27 used for

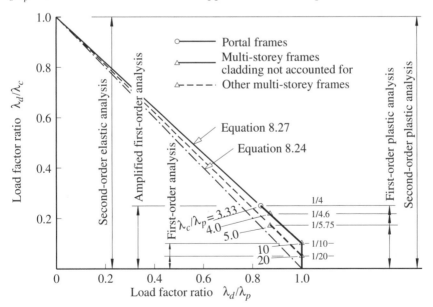

Fig. 8.10 Interaction diagram for plastic design of sway frames

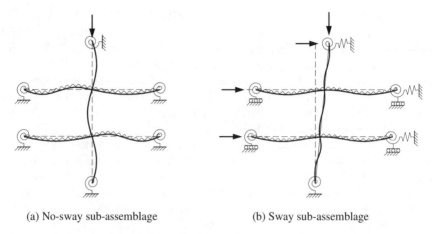

(a) No-sway sub-assemblage (b) Sway sub-assemblage

Fig. 8.11 Sub-assemblages for multi-storey frames

beam-column strength. However, it can be said that this method represents a considerable simplification of a complex relationship between the ultimate strength and the plastic collapse and elastic buckling strengths of a frame.

More accurate predictions of the strengths of two-dimensional rigid frames can be made using a subassemblage method of analysis. In this method, the braced or unbraced frame is considered as a number of subassemblies [26–28] such as those shown in Fig. 8.11. The conditions at the ends of the subassemblage are approximated according to the structural experience and intuition of the designer (it is believed that these approximations are not critical) and the subassemblage is analysed by adding together the load–deformation characteristics [29] and curve 5 in Fig. 7.2) of the individual members. Although the application of this manual technique to the analysis of braced frames was developed to an advanced stage [30, 31], it does not appear to have achieved widespread popularity.

Further improvements in the prediction of frame strength are provided by computer methods of second-order plastic analysis, such as the deteriorated stiffness method of analysis, in which progressive modifications are made to a second-order elastic analysis to account for the formation of plastic hinges [11, 32, 33]. The use of such an analysis is limited to structures with appropriate material properties, for which local and out-of-plane buckling effects are prevented. However, the use of such an analysis has in the past been largely limited to research studies.

Recent research [34] has concentrated on extending advanced computer methods of analysis so that they allow for residual stresses and geometrical imperfections as well as second-order effects and inelastic behaviour. Such methods can duplicate design code predictions of the strengths of individual members (through the incorporation of allowances for residual stresses and geometrical imperfections) as well as the behaviour of complete frames. Two general types of analyses have been developed, called plastic zone analysis [35, 36] and concentrated plasticity analysis [37]. While plastic zone analysis is the more accurate, concentrated plasticity is the more economical of computer memory and time. It seems to be likely that advanced methods of analysis will play important roles in the future design of two-dimensional rigid-jointed frames in which local and lateral buckling are prevented.

8.3.6 IN-PLANE DESIGN OF FRAMES WITH RIGID JOINTS

8.3.6.1 Analysis for strength design

The BS5950 methods of strength design may be classified according to the analysis method used as elastic, plastic, or advanced. The range of application of each of these methods varies approximately with the ratio λ_d/λ_c of the design load to the elastic buckling load, as shown in Fig. 8.10.

When there are no compression forces, or when they are so small that they have no effect on the design of the members, rigid-jointed frames may be analysed by first-order elastic analysis alone (section 8.3.5.1). This is likely to be the case when the members are braced and have low compression forces, or low slendernesses, or high moment gradients, so that their amplification factors δ_b (section 8.3.5.4) are equal to 1.0. BS5950 permits braced frames to be analysed by first-order elastic analysis alone.

When the compression forces are not so small, then either the results of a first-order elastic analysis (section 8.3.5.1) should be amplified using equation 8.14 for braced members or the results of an elastic buckling analysis (section 8.3.5.2 or 8.3.5.3), or a second-order elastic analysis should be used (section 8.3.5.4). Specifically, a second-order analysis is suggested when the amplification factor δ exceeds 1.4, which for unbraced frames corresponds to a ratio λ_d/λ_c of the design load set to the elastic buckling load set of more than 0.286. It is also suggested that a full second-order analysis incorporating the stability stiffness reduction ($P-\delta$) and sway buckling ($P-\Delta$) effects should be used when λ_d/λ_c is greater than 0.2, but that an approximate second-order analysis for the sway ($P-\Delta$) effects is sufficient when λ_d/λ_c is less than 0.2. These suggestions are shown diagramatically in Fig. 8.10.

The use of plastic analysis is permitted by BS5950 when the structure and its members satisfy restrictions which are intended to ensure that premature failure does not occur before the plastic collapse mechanism develops. These restrictions include requirements that prevent local buckling and lateral buckling, and limit the effects of in-plane buckling. A second-order plastic analysis must be used when the ratio λ_d/λ_c of the design load set to the elastic buckling load set is greater than 0.25, as indicated in Fig. 8.10.

8.3.6.2 Strength design using elastic analysis

When an elastic method of analysis has been used to determine the member design actions, BS5950 requires that the section and member design capacities of the individual members are not less than the corresponding design actions. The section and member design capacities are determined by using the appropriate clauses of BS5950, as explained in section 2.6 for tension members, in section 3.7 for compression members, in sections 5.6.1 and 6.9 for beams, and in sections 7.2.4 and 7.3.4 for beam-columns.

For a beam-column in an unbraced frame, BS5950 permits the use of the actual member length L instead of the effective length L_E when determining the compression capacity to be used in the design of the beam-column under combined actions, provided second-order effects are allowed for. The justification for this is based on the assumption that the analysis accurately models the frame behaviour, so that the calculated member design actions can be assumed to act as if on an isolated beam-column, as in Chapter 7.

However, this procedure will overestimate the member capacity when compression effects dominate. As a result of this, the use of L instead of L_E in beam-column design is only permitted by BS5950 if the member is also designed as a compression member ($M^* = 0$) using the effective length L_E.

8.3.6.3 Strength design using plastic analysis

A first-order plastic analysis may be used when λ_d/λ_c is not greater than 0.25. When $\lambda_d/\lambda_c \leqslant 0.1$, then the design load effects need not be amplified, which is equivalent to requiring that

$$\lambda_d \leqslant \lambda_p \tag{8.25}$$

in which λ_p is the plastic collapse load set factor. When $0.1 \leqslant \lambda_d/\lambda_c \leqslant 0.25$, the design load effects must be amplified by using

$$\delta_p = \frac{0.9}{1 - \lambda_d/\lambda_c}, \tag{8.26}$$

which is equivalent to requiring that

$$\lambda_d \leqslant \lambda_p(1 - \lambda_d/\lambda_c)/0.9, \tag{8.27}$$

where λ_d is the unamplified design load set factor. Equations 8.25 and 8.27 are shown in Fig. 8.10.

8.3.6.4 Strength design using advanced analysis

BS5950 does not make any recommendations for advanced analysis. It is suggested that an advanced analysis may be used when local and lateral buckling are prevented, provided the analysis accounts for second-order effects, inelastic behaviour, residual stresses, and geometrical imperfections. The structure is then satisfactory when the section capacity requirements are met. As these requirements are automatically satisfied by the prevention of local buckling and the accounting for inelastic behaviour, the structure can be regarded as satisfactory if the analysis shows that it can reach an equilibrium position under the factored loads represented by λ_d.

8.3.6.5 Serviceability design

As the service loads are usually substantially less than the factored loads used for strength design, the behaviour of a frame under its service loads is usually closely approximated by the predictions of a first-order elastic analysis (section 8.3.5.1). For this reason, the serviceability design of a frame is usually carried out using the results of a first-order elastic analysis. The serviceability design of a frame is often based on the requirement that the service load deflections must not exceed specified values. Thus the member sizes are systematically changed until this requirement is met.

8.3.7 OUT-OF-PLANE BEHAVIOUR OF FRAMES WITH RIGID JOINTS

Most two-dimensional rigid frames which have in-plane loading are arranged so that the stiffer planes of their members coincide with that of the frame. Such a frame deforms only in its plane until the out-of-plane (flexural–torsional) buckling loads are reached, and if these are less than the in-plane ultimate loads, then the members and the frame buckle by deflecting out of the plane and twisting.

   In some cases, the action of one loaded member dominates, and its elastic buckling load can be determined by evaluating the restraining effects of the remainder of the frame. For example, when the frame shown in Fig. 8.12 has a zero beam load Q_2, the beam remains straight and does not induce moments in the columns, which buckle elastically out of the plane of the frame at a load given by $Q_1 \approx \pi^2 EI_y/(0.7L_1)^2$ (see Fig. 3.14c). On the other hand, if the effects of the compressive forces in the columns are small enough to be neglected, then the elastic stiffnesses and restraining effects of the columns on the beam can be estimated, and the elastic flexural–torsional buckling load Q_2 of the beam can be evaluated. The solution of a related problem is discussed in section 6.6.3.

   In general, however, both the columns and the beams of a rigid frame buckle, and there is an interaction between them during out-of-plane buckling. This interaction is related to that which occurs during the in-plane buckling of rigid frames (see sections 8.3.5.2 and 8.3.5.3 and Figs 8.6 and 8.7) and also to that which occurs during the elastic flexural–torsional buckling of continuous beams (see section 6.6.2.2 and Fig. 6.20). Due to the similarities between the elastic flexural–torsional buckling of restrained beam-columns (section 7.3.1.3) and restrained columns (section 3.5.4), it may prove possible to extend further

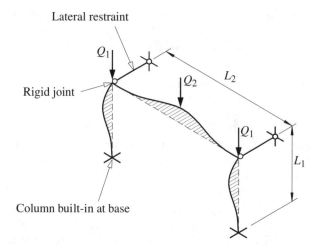

Fig. 8.12 Flexural–torsional buckling of a portal frame

the approximate method given in section 8.3.5.2 for estimating the in-plane elastic buckling loads of rigid frames (which was applied to the elastic flexural–torsional buckling of some beams in section 6.6.2.2). Thus, the elastic flexural–torsional buckling loads of some rigid frames might be calculated approximately by using the column effective length chart of Fig. 3.20a. However, this development has not yet been investigated.

On the other hand, there have been developments [38–40] in the analytical techniques used to determine the elastic flexural–torsional buckling loads of general rigid plane frames with general in-plane loading systems, and a computer program [41] has been prepared which requires only simple data specifying the geometry of the frame, its supports and restraints, and the arrangement of the loads. The use of this program to calculate the elastic flexural–torsional buckling load of a frame will simplify the design problem considerably.

Interaction diagrams for the elastic flexural–torsional buckling loads of a number of frames have been determined [38], and two of these (for the portal frames shown in Fig. 8.12) are given in Fig. 8.13. These diagrams show that the region of stability is convex, as it is for the in-plane buckling of rigid frames (Figs 8.6 and 8.7) and for the flexural–torsional buckling of continuous beams (Fig. 6.22b). As a result of this convexity, linear interpolations (as in Fig. 6.23) made between known buckling load sets are conservative.

Although the elastic flexural–torsional buckling of rigid frames has been investigated, the related problems of inelastic buckling and its influence on the strength of rigid frames have not yet been systematically studied. However, advanced computer programs for the inelastic out-of-plane behaviour of beam-columns [42–44] have been developed, and it seems to be likely that these will be extended in the near future to rigid-jointed frames.

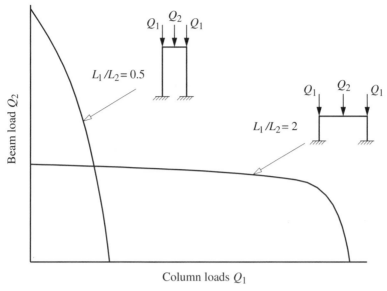

Fig. 8.13 Elastic flexural–torsional buckling loads of portal frames

8.3.8 OUT-OF-PLANE DESIGN OF FRAMES

While there is no general method yet available for designing for flexural–torsional buckling in rigid frames, design codes imply that the strength formulations for isolated beam-columns (see section 7.3.4) can be used in conjunction with the moments and forces determined from an appropriate elastic in-plane analysis of the frame.

A rational design method has been developed [45] for the columns of two-dimensional building frames in which laterally braced horizontal beams form plastic hinges. Thus, when plastic analysis is used, columns which do not participate in the collapse mechanism may be designed as isolated beam-columns against flexural–torsional buckling. In later publications [46, 47] the method was extended so that the occurrence of plastic hinges at one or both ends of the column (instead of in the beams) could be allowed for. As the assumption of plastic hinges at all the beam-column joints is only valid for frames with vertical loads, the analysis for lateral loads must be carried out independently. A frame analysed by this method must therefore have an independent bracing or shear wall system to resist the lateral loads.

A number of current research efforts are being directed towards the extension of advanced methods of analysing the in-plane behaviour of frames (section 8.3.5.6) to include the flexural–torsional out-of-plane buckling effects. The availability of a suitable computer program will greatly simplify the design of two-dimensional frames in which local buckling is prevented, since a frame can be considered to be satisfactory if it can be shown that it can reach an equilibrium position under a set of loads represented by the factored design loads.

8.4 Three-dimensional flexural frames

The design methods commonly used when either the frame or its loading is three-dimensional are very similar to the methods used for two-dimensional frames. The actions caused by the design loads are usually calculated by allowing for second-order effects in the elastic analysis of the frame, and then compared with the design capacities determined for the individual members acting as isolated beam-columns. In the case of three-dimensional frames or loading, the elastic frame analysis is three-dimensional, while the design capacities are based on the biaxial bending capacities of isolated beam-columns (see section 7.4).

A more rational method was developed [45, 46] for designing three-dimensional braced rigid frames in which it could be assumed that plastic hinges form at all major and minor axis beam ends. This method is an extension of the method of designing braced two-dimensional frames ([45–47] and section 8.3.8). For such frames, all the beams can be designed plastically, while the columns can again be designed as if independent of the rest of the frame. The method of designing these columns is based on a second-order elastic analysis of the biaxial bending of an isolated beam-column. Once again, the assumption of plastic hinges at all the beam ends is valid only for frames with vertical loads, and so an independent design must be made for the effects of lateral loads.

In many three-dimensional rigid frames, the vertical loads are carried principally by the major axis beams, while the minor axis beams are lightly loaded and do not develop plastic hinges at collapse, but restrain and strengthen the columns. In this case, an appropriate design method for vertical loading is one in which the major axis beams are designed plastically and the minor axis beams elastically. The chief difficulty in using such a method is in determining the minor axis column moments to be used in the design. In the method presented in [48], this is done by first making a first-order elastic analysis of the minor axis frame system to determine the column end moments. The larger of these is then increased to allow for the second-order effects of the axial forces, the amount of the increase depending on the column load ratio N/N_{oy} (in which N_{oy} is the minor axis buckling load of the elastically restrained column) and the end moment ratio β_m. The increased minor axis end moment is then used in a nominal first yield analysis of the column which includes the effects of initial curvature.

However, it appears that this design method is erratically conservative [49], and that the small savings achieved do not justify the difficulty of using it.

Another design method was proposed [50] which avoids the intractable problem of inelastic biaxial bending in frame structures by omitting altogether the calculation of the minor axis column moments. This omission is based on the observation that the minor axis moments induced in the columns by the working loads reduce to zero as the ultimate loads are approached, and even reverse in sign so that they finally restrain the column, as shown in Fig. 8.9. With this simplification, the moments for which the column is to be designed are the same as those in a two-dimensional frame. However, the beneficial restraining effects of the minor axis beams must be allowed for, and a simple way of doing this has been proposed [50]. For this, the elastic minor axis stiffness of the column is reduced to allow for plasticity caused by the axial load and the major axis moments, and this reduced stiffness is used to calculate the effect of the minor axis restraining beams on the effective length L_E of the column. The reduced stiffness is also used to calculate the effective minor axis flexural rigidity EI_e of the column, and the ultimate strength N_u of the column is taken as

$$N_u = \pi^2 EI_e/L_E^2. \tag{8.28}$$

The results obtained from two series of full scale three-storey tests indicate that the predicted ultimate loads obtained by this method vary between 84% and 104% of the actual ultimate loads.

8.5 Worked examples

8.5.1 EXAMPLE 1 – BUCKLING OF A TRUSS

Problem.   All the members of the rigid-jointed triangulated frame shown in Fig. 8.14a have the same in-plane flexural rigidity which is such that $\pi^2 EI/L^2 = 1$. Determine the effective length ratio of member 12 and the elastic in-plane buckling loads of the frame.

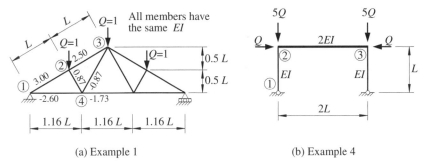

(a) Example 1   (b) Example 4

Fig. 8.14 Worked examples 1 and 4

Solution.   The member 12 is one of the longest compression members and one of the most heavily loaded, and so at buckling it will be restrained by its adjacent members. An initial estimate is required of the buckling load, and this can be obtained after observing that end 1 will be heavily restrained by the tension member 14 and end 2 moderately restrained by the short lightly loaded compression member 24. Thus the initial estimate of the effective length ratio of member 12 in this braced frame should be closer to the rigidly restrained value of 0.5 than to the unrestrained value of 1.0.

Accordingly, assume $k_{m12} = 0.70$. Then, by using equation 8.2

$$N_{om12} = \pi^2 EI/(0.70L)^2 = 2.04,$$
$$Q_{oc} = N_{om12}/3.00 = 0.68.$$

Thus

$$N_{23} = 1.70, \quad N_{24} = 0.59, \quad N_{14} = -1.77.$$

$$K_{b23} = \frac{0.75I}{L}\left[1 - \frac{1.0 \times 1.70}{\pi^2 EI/L^2}\right] = -0.525\ \frac{I}{L} \qquad\qquad \text{TE.3}$$

$$K_{b24} = \frac{I}{0.58L}\left[1 - \frac{0.4 \times 0.59}{\pi^2 EI/(0.58L)^2}\right] = 1.587\ \frac{I}{L} \qquad\qquad \text{TE.3}$$

$$K_{b14} = \frac{I}{1.16L} = 0.862\ \frac{I}{L} \qquad\qquad \text{TE.3}$$

$$k_1 = \frac{I/L}{(I/L + 0.862I/L)} = 0.537 \qquad\qquad \text{E.2.1}$$

$$k_2 = \frac{I/L}{(I/L - 0.525I/L + 1.587I/L)} = 0.485 \qquad\qquad \text{E.2.1}$$

$$k_{e12} = (L_E/L)_{12} = 0.69 \qquad\qquad \text{Fig. E.1}$$

The calculation can be repeated using $k_{m12} = 0.69$ instead of the initial estimate of 0.70, in which case the solution $k_{m12} = 0.69$ will be obtained. The corresponding frame buckling loads are $Q_{oc} = \{\pi^2 EI/(0.69L)^2\}/3 = 0.70$.

8.5.2 EXAMPLE 2 – BUCKLING OF A BRACED FRAME

Problem.   Determine the effective length ratios of the members of the braced frame shown in Fig. 8.3a, and estimate the frame buckling load factor λ_c.

Solution.   For vertical member 13, using equation 8.4

$k_1 = 1.0$ (theoretical value for a frictionless hinge).

$$k_3 = \frac{2213 \times 10^4/10\,000}{(2213 \times 10^4/10\,000 + 1.0 \times 8551 \times 10^4/12\,000)} = 0.237 \qquad \text{TE.2}$$

$$k_{e13} = (L_E/L)_{13} = 0.74 \hspace{2cm} \text{Fig. E.1}$$
$$(k_{e13} = (L_E/L)_{13} = 0.73 \text{ if a base stiffness ratio of 0.1 is used} \hspace{0.5cm} 5.1.3.3$$

so that

$$k_1 = (2213 \times 10^4/10\ 000)/(1.1 \times 2213 \times 10^4/10\ 000) = 0.909)$$
$$N_{om13} = \pi^2 \times 205\ 000 \times 2213 \times 10^4/(0.74 \times 10\ 000)^2 \text{ N} = 817.7 \text{ kN}$$
$$\lambda_{c13} = 817.7/71.6 = 11.4 \hspace{4cm} \text{E.6}$$

For the horizontal member 35, using equation 8.4

$$k_3 = \frac{8551 \times 10^4/12\ 000}{(8551 \times 10^4/12\ 000 + 0.75 \times 2213 \times 10^4/10\ 000)} = 0.811 \hspace{0.5cm} \text{TE.2}$$

$k_5 = 0$ (theoretical value for a fixed end).
$$k_{e35} = (L_E/L)_{35} = 0.66 \hspace{4cm} \text{Fig. E1}$$
$$(k_{e35} = (L_E/L)_{35} = 0.77 \text{ if a base stiffness ratio of 1.0 is used so that } k_{35} = 0.5)$$
$$\hspace{11cm} 5.1.3.2$$
$$N_{om35} = \pi^2 \times 205\ 000 \times 8551 \times 10^4/(0.66 \times 12\ 000)^2 \text{ N} = 2758.2 \text{ kN}$$
$$\lambda_{c35} = 2758.2/25.3 = 109 > 11.4 = \lambda_{c13} \hspace{3cm} \text{E.6}$$

$\therefore \lambda_c = 11.4$ (compare with the computer analysis value of $\lambda_c = 11.24$ given in Fig. 8.3d).

A slightly more accurate estimate might be obtained by using equation 8.5 with α_r obtained from Fig. 3.18 as

$$\alpha_r = 1 - 25.3 \times 10^3/(2 \times \pi^2 \times 205\ 000 \times 8551 \times 10^4/12\ 000^2) = 0.989,$$

so that

$$k_3 = \frac{2213 \times 10^4/10\ 000}{(2213 \times 10^4/10\ 000 + 0.989 \times 8551 \times 10^4/12\ 000)} = 0.239 \hspace{0.5cm} \text{TE.3}$$

but k_{e13} is virtually unchanged, and so therefore is λ_c.

8.5.3 EXAMPLE 3 – BUCKLING OF AN UNBRACED TWO-STOREY FRAME

Problem. Determine the storey buckling load factors of the unbraced two-storey frame shown in Fig. 8.4a, and estimate the frame buckling load factor.

Solution. For the upper storey columns, using equation 8.4

$$k_T = \frac{2213 \times 10^4/5000}{(2213 \times 10^4/5000 + 1.5 \times 3420 \times 10^4/12\ 000)} = 0.509 \hspace{0.5cm} \text{TE.2}$$

$$k_B = \frac{(2213 \times 10^4/5000 + 0.75 \times 5254 \times 10^4/5000)}{(2213 \times 10^4/5000 + 0.75 \times 5254 \times 10^4/5000 + 1.5 \times 14\ 160 \times 10^4/12\ 000)}$$

$$= 0.410 \hspace{10cm} \text{TE.2}$$
$$k_{eu} = (L_E/L)_u = 1.40 \hspace{7cm} \text{Fig. E.2}$$
$$N_{omnu} = \pi^2 \times 205\ 000 \times 2213 \times 10^4/(1.40 \times 5000)^2 \text{ N} = 913.8 \text{ kN}.$$

For the lower storey columns, using equation 8.4

$$k_T = \frac{(2213 \times 10^4/5000 + 5254 \times 10^4/5000)}{(2213 \times 10^4/5000 + 5254 \times 10^4/5000 + 1.5 \times 14\,160 \times 10^4/12\,000)}$$

$$= 0.458 \hspace{6cm} \text{TE.2}$$

$k_B = 1.0$ (theoretical value for a frictionless hinge).

$$k_{el} = (L_E/L)_1 = 2.4 \hspace{5cm} \text{Fig. E.2}$$

$$N_{oml} = \pi^2 \times 205\,000 \times 5254 \times 10^4/(2.4 \times 5000)^2\,\text{N} = 738.2\,\text{kN}$$

Using equation 8.7 and the axial forces of Fig. 8.4f,

$$\lambda_{msu} = \frac{2 \times 913.8/5.0}{(37.6 + 42.4)/5.0} = 22.8$$

$$\lambda_{msl} = \frac{2 \times 738.2/5.0}{(103.3 + 136.7)/5.0} = 6.15 < 22.8 = \lambda_{msu}$$

$\therefore \lambda_c = 6.15$ (compare with the computer analysis value of $\lambda_c = 6.74$ given in Fig. 8.4d).

8.5.4 EXAMPLE 4 – BUCKLING OF AN UNBRACED SINGLE STOREY FRAME

Problem. Determine the effective length ratio of member 12 of the unbraced frame shown in Fig. 8.14b (and for which $\pi^2 EI/L^2 = 1$), and the elastic in-plane buckling loads.

Solution. The sway member 12 is unrestrained at end 1 and moderately restrained at end 2 by the lightly loaded member 23. Its effective length ratio is therefore greater than 2, the value for a rigidly restrained sway member (see Fig. 3.14e). Accordingly, assume $k_{12} = 2.5$. Then, by using equations 3.36 and 3.37,

$$N_{om12} = \pi^2 EI(2.5L)^2 = 0.16,$$
$$Q_{oc} = N_{om12}/5 = 0.032,$$
$$N_{23} = Q_{oc} = 0.032.$$
$$K_{b23} = (1.5 \times 2I/2L)(1 - 0.2 \times 0.032/0.5) = 1.48I/L \hspace{2cm} \text{TE.3}$$

$$k_2 = \frac{I/L}{(I/L + 1.48I/L)} = 0.40 \hspace{3cm} \text{E.2.1}$$

$k_1 = 1.0$ (theoretical value for a frictionless hinge).
$$k_{e12} = (L_E/L)_{12} = 2.3 \hspace{5cm} \text{Fig. E.2}$$

The calculation can be repeated using $k_{12} = 2.3$ instead of the initial estimate of 2.5, in which case the solution $k_{12} = 2.3$ will be obtained. The corresponding frame buckling loads are given by

$$Q_{oc} = \frac{\pi^2 EI/(2.3L)^2}{5} = 0.038.$$

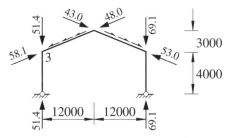

356 x 171 UB 45, I = 12080 cm^4

Fig. 8.15 Worked example 5

8.5.5 EXAMPLE 5 – BUCKLING OF A PORTAL FRAME

Problem. A uniform section pinned-base portal frame is shown in Fig. 8.15 with the member axial compression forces. Determine the elastic buckling load factor of the frame.

Solution.

$$L_r = \sqrt{(12\ 000^2 + 3000^2)} = 12\ 369 \text{ mm.}$$

The average axial forces are

$$N_c = (51.4 + 69.1)/2 = 60.3 \text{ kN,}$$
$$N_r = (58.1 + 48.0 + 43.0 + 53.0)/4 = 50.5 \text{ kN.}$$

For sway buckling,

$$R = (12\ 080 \times 10^4/4000)/(12\ 080 \times 10^4/12\ 369) = 3.09,$$

$$\frac{N_c L_c}{N_r L_r} = \frac{60.3 \times 10^3 \times 4000}{50.5 \times 10^3 \times 12\ 369} = 0.386,$$

$$\lambda_{sp} = \frac{3 \times 205\ 000 \times 12\ 080 \times 10^4/(50.5 \times 10^3 \times 12\ 369^2)}{[(1 + 1.2/3.09) \times 0.386 + 0.3]} = 11.5.$$

For symmetric buckling,

$$k_{1p} = k_{2p} \approx \frac{\{12\ 080 \times 10^4/(2 \times 12\ 369)\}}{\{12\ 080 \times 10^4/(2 \times 12\ 369) + 0.75 \times 12\ 080 \times 10^4/4000\}} = 0.177$$

TE.2

$$k_{ep} = (L_E/L)_p \approx 0.55$$

Fig. E.1

$$\lambda_r = \frac{\pi^2 \times 205\ 000 \times 12\ 080 \times 10^4}{(2 \times 0.55 \times 12\ 369)^2 \times 50.5 \times 10^3} = 26.1 > 11.5 = \lambda_{sp}$$

E.6

and so the frame buckling factor is $\lambda_s = 11.5$. This is reasonably close to the value of $\lambda_s = 13.6$ predicted by a computer elastic frame buckling analysis program [13].

8.5.6 EXAMPLE 6 – MOMENT AMPLIFICATION IN A BRACED FRAME

Problem. Determine the amplified design moments for the braced frame shown in Fig. 8.3.

Second-order effects are rarely important in braced members which have significant moment gradients. It can be inferred from Section 5.6 of BS5950 that second-order effects need not be considered in braced frames. Nevertheless, an approximate method of allowing for second-order effects by estimating the amplified design moments is illustrated below.

Solution. For vertical member 123, the first-order moments are $M_1 = 0$, $M_2 = 23.5$ kNm, $M_3 = -53.0$ kNm, and for these, the method of [23] leads to $\beta_m = 0.585$, and so

$$c_m = 0.6 - 0.4 \times 0.585 = 0.366$$

Using $N_{om13} = 817.7$ kN from section 8.5.2, and using equations 8.14 and 8.15,

$$M_3^* = \frac{0.366 \times 53.0}{1 - 71.6/817.7} = 21.3 \text{ kNm} < 53.0 \text{ kNm},$$

and so $M_3^* = 53.0$ kNm. This is close to the second-order moment of 53.6 kNm determined using the computer program of [13].

For horizontal member 345, the first-order moments are $M_3 = -53.0$ kNm, $M_4 = 136.6$ kNm, $M_5 = -153.8$ kNm, and for these, the method of [23] leads to $\beta_m = 0.292$, and so $c_m = 0.6 - 0.4 \times 0.292 = 0.483$.

Using $N_{om35} = 2758.2$ kN from section 8.5.2, and using equations 8.14 and 8.15,

$$M_5^* = \frac{0.483 \times 153.8}{1 - 25.3/2758.2} = 75.0 \text{ kNm} < 153.8 \text{ kNm},$$

and so $M_5^* = 153.8$ kNm. This is close to the second-order moment of 154.6 kNm determined using the computer program of [13].

8.5.7 EXAMPLE 7 – MOMENT AMPLIFICATION IN A PORTAL FRAME

Problem. The maximum first-order elastic moment in the portal frame whose buckling load factor was determined in section 8.5.5 is 151.8 kNm. Determine the amplified design moment.

Solution. Using the elastic frame buckling load factor determined in section 8.5.5 of $\lambda_s = 11.5$ and using equations 8.14 and 8.17,

$$M^* = \frac{151.8}{1 - 1/11.5} = 166.3 \text{ kNm},$$

which is about 7% higher than the value of 155.4 kNm obtained using the computer program of [13].

8.5.8 EXAMPLE 8 – MOMENT AMPLIFICATION IN AN UNBRACED FRAME

Problem.   Determine the amplified design moments for the unbraced frame shown in Fig. 8.4.

Solution.   For the upper storey columns, using the value of $\lambda_{msu} = 22.8$ determined in section 8.5.3, and using equation 8.18,

$$\delta_{msu} = 1/(1 - 1/22.8) = 1.046$$

and so

$$M_8^* = 1.046 \times 62.5 = 65.4 \text{ kNm}$$

which is close to the second-order moment of 64.5 kNm determined using the computer program of [13].

For the lower storey columns, using the value of $\lambda_{msl} = 6.15$ determined in section 8.5.3, and using equation 8.18,

$$\delta_{msl} = 1/(1 - 1/6.15) = 1.194$$

and so

$$M_{76}^* = 1.194 \times 119.2 = 142.3 \text{ kNm}$$

which is 9% higher than the value of 130.8 kNm determined using the computer program of [13].

For the lower beam, the second-order end moment M_{74}^* may be approximated by adding the second-order end moments M_{76}^* and M_{78}^*, so that

$$M_{74}^* = 142.3 + 53.5/(1 - 1/22.8) = 198.3 \text{ kNm,}$$

which is 7% higher than the value of 185.0 kNm determined using the computer program of [13].

8.5.9 EXAMPLE 9 – AMPLIFICATION FACTORS FOR AN UNBRACED FRAME

Problem.   Determine the storey amplification factors δ_s for the unbraced frame shown in Fig. 8.4

Solution.   For the upper storey, and using the first-order analysis results shown in Fig. 8.4 in equation 8.19,

$$\delta_{su} = \frac{1}{1 - \{(124.3 - 87.6)/5000\}(37.6 + 42.4)/10} = 1.062,$$

which is slightly higher than the value of $\delta_{msu} = 1.046$ obtained in section 8.5.8.

For the lower storey, and using the first-order analysis results shown in Fig. 8.4 in equation 8.19,

$$\delta_{sl} = \frac{1}{1 - \{(87.6 - 0)/5000\}(103.3 + 136.7)/(10 + 20)} = 1.163,$$

which is a little lower than the value of $\delta_{msl} = 1.194$ obtained in section 8.5.8.

8.5.10 EXAMPLE 10 – PLASTIC ANALYSIS OF A BRACED FRAME

Problem. Determine the plastic collapse load factor for the braced frame shown in Fig. 8.3, if the members are all of S275 steel.

Solution. For the horizontal member plastic collapse mechanism shown in Fig. 8.3e,

$$\delta W = 80 \, \lambda_{ph} \times (\delta\theta_h \times 6.0) = 480 \, \lambda_{ph}\delta\theta_h$$

for a virtual rotation $\delta\theta_h$ of 34, and

$$\delta U = (85.0 \times \delta\theta_h) + (172.2 \times 2\delta\theta_h) + (172.2 \times \delta\theta_h) = 601.6\delta\theta_h$$

so that

$$\lambda_{ph} = 601.6/480 = 1.253.$$

Checking for reductions in M_p using equation 7.17,

$$M_{pr13} = 1.18 \times 85.0 \times \{1 - 93.0/(275 \times 47.3 \times 10^2/10^3)\}$$
$$= 93.1 \text{ kNm} > 85.0 \text{ kNm} = M_{p13}$$

and so there is no reduction in M_p for member 13.

$$M_{pr35} = 1.18 \times 172.2 \times \{1 - 33.6/(275 \times 51.6 \times 10^2/10^3)\}$$
$$= 198.4 \text{ kNm} > 172.2 \text{ kNm} = M_{p35}$$

and so there is no reduction in M_p for member 35. Therefore, $\lambda_{ph} = 1.253$.

For a plastic collapse mechanism in the vertical member with a frictionless hinge at 1 and plastic hinges at 2 and 3,

$$\lambda_{pv} = \frac{85.0 \times 2 \times \delta\theta_v + 85.0 \times \delta\theta_v}{20 \times 5.0 \times \delta\theta_v} = 2.55 > 1.253 = \lambda_{ph}$$

and so $\lambda_p = 1.253$.

8.5.11 EXAMPLE 11 – PLASTIC ANALYSIS OF AN UNBRACED FRAME

Problem. Determine the plastic collapse load factor for the unbraced frame shown in Fig. 8.4, if the members are all of S275 steel.

Solution. For the beam plastic collapse mechanism shown in Fig. 8.4e,

$$\delta W = 40\lambda_p \times 6\delta\theta = 240\lambda_p\theta$$

for virtual rotations $\delta\theta$ of the half beams 35 and 58, and

$$\delta U = (84.4 \times \delta\theta) + (84.4 \times 2\delta\theta) + (84.4 \times \delta\theta) = 337.6\delta\theta$$

so that

$$\lambda_p = 337.6/240 = 1.407.$$

The frame is only partially determinate when this mechanism forms, and so the member axial forces cannot be determined by statics alone. The axial force N_{358} determined by a computer first-order plastic analysis [13] is $N_{358} = 33.9$ kN and the axial force ratio is

$$(N/N_Y)_{358} = 33.9/(275 \times 32.3 \times 10^2/10^3) = 0.0382$$

which is less than 0.15, the approximate value at which N/N_Y begins to reduce M_p, and so λ_p is unchanged at $\lambda_p = 1.407$.

The collapse load factors calculated for the other possible mechanisms of:

- lower beam ($\lambda_p = 2 \times (155.9 + 85.0 + 246.1)/(80 \times 6) = 2.029$)
- lower storey sway ($\lambda_p = 2 \times 155.9/(30 \times 5) = 2.079$)
- upper storey sway ($\lambda_p = 2 \times (84.4 + 85.0)/(10 \times 5) = 6.776$)
- combined beam-sway

$$\left(\lambda_p = \frac{2 \times (84.4 + 84.4 + 246.1 + 85.0 + 155.9)}{\{(40 + 80) \times 6 + (20 \times 5 + 10 \times 10)\}} = 1.426\right)$$

are all greater than 1.407, and so $\lambda_p = 1.407$.

8.5.12 EXAMPLE 12 – ELASTIC MEMBER DESIGN

Problem.   Check the adequacy of the lower storey column 67 (of S275 steel) of the unbraced frame shown in Fig. 8.4 for the actions determined by elastic analysis. The column is fully braced against deflection out of the plane of the frame.

Design actions.
The design actions are $F^* = 136.7$ kN (Fig. 8.4) and $M_x^* = 142.3$ kNm (section 8.5.8).

Section capacity.

$$T = 12.5 \text{ mm}, p_y = 275 \text{ N/mm}^2 \qquad \text{T9}$$
$$\varepsilon = \sqrt{(275/275)} = 1.0$$
$$b/(t\varepsilon) = (203.9/2)/(12.5 \times 1.0) = 8.16 < 9 \qquad \text{T11}$$

and the flange is Class 1 plastic.

$$r_1 = \frac{136.7 \times 10^3}{(206.2 - 2 \times 12.5 - 2 \times 10.2) \times 8.0 \times 275} = 0.386 \qquad 3.5.5(a)$$

$$(1 + r_1)d/(t\varepsilon) = (1 + 0.386) \times (206.2 - 2 \times 12.5 - 2 \times 10.2)/(8.0 \times 1.0)$$
$$= 27.9 < 80 \text{ and the web is Class 1 plastic.} \qquad \text{T11}$$

Thus the section is Class 1 plastic and Annex J.2 can be used.

$$n = 136.7 \times 10^3/(66.4 \times 10^2 \times 275) = 0.0749 \qquad \text{J.2.1}$$
$$t(D - 2T)/A = 8.0 \times (206.2 - 2 \times 12.5)/(66.4 \times 10^2) \qquad \text{J.2.1}$$
$$= 0.218 > 0.0749 = n$$
$$S_{rx} = 567 \times 10^3 - (66.4 \times 10^2)^2 \times 0.0749^2/(4 \times 8.0) \qquad \text{J.2.1}$$
$$= 559 \times 10^3 \text{ mm}^3$$
$$M_{rx} = 275 \times 559 \times 10^3 \text{ Nmm} = 153.8 \text{ kNm} > 142.3 \text{ kNm} = M_x^* \quad \text{J.2.1}$$

and the section capacity is adequate.

Member resistance.
Because the member is continuously braced out-of-plane, beam lateral buckling and column minor axis buckling need not be considered.

$$L_E = 1.0 \times 5000 = 5000 \text{ mm} \qquad \text{5.6.4(b)}$$
$$\lambda_x = 5000/(8.90 \times 10) = 56.2 \qquad \text{4.7.2}$$

For x-axis buckling of a rolled H-section with $T < 40$ mm, use Strut Curve (b).
$$\text{T23}$$

$$p_{cx} = (227 \times 1.8 + 224 \times 0.2)/2 = 226.7 \text{ N/mm}^2 \qquad \text{T24(3)}$$
$$P_{cx} = 66.4 \times 10^2 \times 226.7 \text{ N} = 1505 \text{ kN} \qquad \text{4.7.4}$$
$$m = 0.60 \qquad \text{T26}$$

$$\frac{136.7 \times 10^3}{1505 \times 10^3} + \frac{0.60 \times 142.3 \times 10^6}{275 \times 510 \times 10^3} = 0.699 < 1.0 \qquad \text{4.8.3.3.1}$$

Also, using $L_E/L = 2.4$ (section 8.5.3)

$$\lambda_x = 2.4 \times 5000/(8.90 \times 10) = 134.8 \qquad \text{4.7.2}$$
$$p_{cx} = (95 \times 0.2 + 89 \times 4.8)/5 = 89.2 \text{ N/mm}^2 \qquad \text{T24(4)}$$
$$P_{cx} = 66.4 \times 10^2 \times 89.2 \text{ N} = 592.6 \text{ kN} > 136.7 \text{ kN} = F^* \qquad \text{4.7.4}$$

Thus the member resistance is adequate.

8.5.13 EXAMPLE 13 – PLASTIC DESIGN OF A BRACED FRAME

Problem.   Determine the plastic design capacity of the braced frame shown in Fig. 8.3.

Solution.   Using the solutions of sections 8.5.2 and 8.5.10

$$\lambda_c = 11.4 > 10 \qquad \text{5.5.2.2}$$
$$\lambda_p = 1.253 > 1.0 \qquad \text{5.5.2.2}$$

and the structure is satisfactory.

8.5.14 EXAMPLE 14 – PLASTIC DESIGN OF AN UNBRACED FRAME

Problem. Determine the plastic design capacity of the unbraced frame shown in Fig. 8.4.

Solution. Using the solution of sections 8.5.3 and 8.5.11,

$$\lambda_c = 6.15, \quad \lambda_p = 1.407,$$

$$\frac{0.95\lambda_c}{(\lambda_c - 1)} = \frac{0.95 \times 6.15}{(6.15 - 1)} = 1.134 < 1.407 = \lambda_p \qquad \text{5.7.3.2(b)}$$

and the structure is satisfactory.

8.6 Unworked examples

8.6.1 EXAMPLE 15 – TRUSS DESIGN

The members of the welded truss shown in Fig. 8.16a are all UC sections of S275 steel with their webs perpendicular to the truss. The member properties and the results of a first-order elastic analysis of the truss under the design loads are shown in Fig. 8.16b and c.

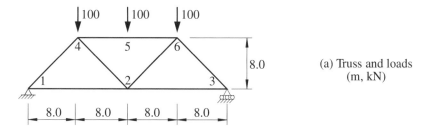

(a) Truss and loads
(m, kN)

Member	Section	I_y (cm^4)	A (cm^2)	p_y(N/mm^2)	
14, 36	305 x 305 UC 97	7272	123	275	(b) Member
45, 65	305 x 305 UC 158	12500	201	265	properties
12, 32	305 x 305 UC 97	7272	123	275	
42, 62	305 x 305 UC 97	7272	123	275	

Member	M_L^*(kNm)	M_R^*(kNm)	N^* (kN)	
14	16.8	-68.8	-217.4	
45	-150.3	249.7	-172.7	(c) Elastic
56	249.7	-150.4	-172.7	actions
12	-16.8	8.9	159.1	
42	81.5	-40.6	8.5	

Fig. 8.16 Example 15

Determine:

(i)   the effective length ratios k_e of the compression members and the frame elastic buckling load factor λ_c;

(ii) the amplified first-order design moments M^* for the compression members;

(iii) the adequacy of the members for the actions determined by elastic analysis;

(iv) the plastic collapse load factor λ_p;

(v)   the adequacy of the truss for plastic design.

8.6.2 EXAMPLE 17 – UNBRACED FRAME DESIGN

The members of the unbraced rigid-jointed two storey frame shown in Fig. 8.17a are all of S275 steel. The member properties and the results of a first-order elastic analysis of the frame under its design loads are shown in Fig. 8.17b and c. Determine:

(i)   the column effective length ratios k_e, the storey buckling load factors λ_{ms}, and the frame buckling load factor λ_c;

(ii) the amplified first-order design moments M^* for the columns;

(iii) the adequacy of the members for the actions determined by elastic analysis;

(iv) the plastic collapse load factor λ_p;

(v)   the adequacy of the frame for plastic design.

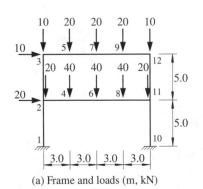

(a) Frame and loads (m, kN)

Member	M_L^*(kNm)	M_R^*(kNm)	N^* (kN)
1-2	-18.8	-13.7	-110.4
10-11	-65.4	79.5	-129.7
2-3	54.5	-48.8	-37.7
11-12	-77.4	75.9	-42.3
2-4	-68.2	89.7	1.7
4-6	89.7	127.5	1.7
6-8	127.5	45.3	1.7
8-11	45.3	-156.8	1.7
3-5	-48.8	34.5	-30.7
5-7	34.5	57.7	-30.7
7-9	57.7	20.9	-30.7
9-12	20.9	-75.9	-30.7

(b) Elastic actions

Member	Section	I_x (cm^4)	A (cm^2)	p_y(N/mm^2)
1-3, 10-12	152 x152 UC 37	2213	47.3	275
2-11	305 x 165 UB 40	8551	51.6	275
3-12	254 x 102 UB 25	3420	32.3	275

(c) Member properties

Fig. 8.17 Example 16

8.7 References

1. British Standards Institution (1969) *PD3343 Recommendations for Design* (Supplement No. 1 to BS 449), BSI, London.
2. Nethercot, D.A. (1986) The behaviour of steel frame structures allowing for semi-rigid joint action, *Steel Structures – Recent Research Advances and Their Applications to Design*, (ed. M.N. Pavlovic), Elsevier Applied Science Publishers, London, pp. 135–51.
3. Nethercot, D.A. and Chen, W-F. (1988) Effect of connections of columns, *Journal of Constructional Steel Research*, **10**, pp. 201–39.
4. Anderson, D.A., Reading, S.J. and Kavianpour, K. (1991) Wind moment design for unbraced frames, *SCI Publication P-082*, The Steel Construction Institute, Ascot.
5. Couchman, G.H. (1997) Design of semi-continuous braced frames, *SCI Publication P-183*, The Steel Construction Institute, Ascot.
6. Pippard, A.J.S. and Baker, J.F. (1968) *The Analysis of Engineering Structures*, 4th edition, Edward Arnold, London.
7. Norris, C.H., Wilburn, J.B. and Utku, S. (1976) *Elementary Structural Analysis*, 3rd edition, McGraw-Hill, New York.
8. Coates, R.C., Coutie, M.G. and Kong, F.K. (1988) *Structural Analysis*, 3rd edition, Van Nostrand Reinhold (UK), Wokingham, England.
9. Kleinlogel, A. (1931) *Mehrstielige Rahmen*, Ungar, New York.
10. Owens, G.W. and Knowles, P.R. (eds), (1992) *Steel Designers' Manual*, 5th edition, Blackwell Scientific Publications, Oxford.
11. Harrison, H.B. (1973) *Computer Methods in Structural Analysis*, Prentice-Hall, Englewood Cliffs, New Jersey.
12. Harrison, H.B. (1990) *Structural Analysis and Design, Parts 1 and 2*, 2nd edition, Pergamon Press, Oxford.
13. Hancock, G.J., Papangelis, J.P. and Clarke, M.J. (1995) *PRFSA User's Manual*, Centre for Advanced Structural Engineering, University of Sydney.
14. Computer Service Consultants (1999) *S-Frame Enterprise*, CSC (UK) Limited, Leeds.
15. Research Engineers Europe Limited (1999) *STAADpro-QSE Space Frame Analysis*, Bristol.
16. Hancock, G.J. (1984) Structural buckling and vibration analyses on microcomputers, *Civil Engineering Transactions*, Institution of Engineers, Australia, CE **24**, No. 4, pp. 327–332.
17. Bridge, R.Q. and Fraser, D.J. (1987) Improved G-factor method for evaluating effective lengths of columns, *Journal of Structural Engineering, ASCE*, **113**, No. 6, June, pp. 1341–56.
18. Wood, B.R., Beaulieu, D. and Adams, P.F. (1976) Column design by P-delta method, *Journal of the Structural Division, ASCE*, **102**, No. ST2, February, pp. 411–27.
19. Wood, B.R., Beaulieu, D. and Adams, P.F. (1976) Further aspects of design by P-delta method, *Journal of the Structural Division, ASCE*, **102**, No. ST3, March, pp. 487–500.
20. Column Research Committee of Japan (1971) *Handbook of Structural Stability*, Corona, Tokyo.

21. Davies, J.M. (1990) In-plane stability of portal frames, *The Structural Engineer*, **68**, No. 8, April, pp. 141–7.

22. Davies, J.M. (1991) The stability of multi-bay portal frames, *The Structural Engineer*, **69**, No. 12, pp. 223–9.

23. Bridge, R.Q. and Trahair, N.S. (1987) Limit state design rules for steel beam-columns, *Steel Construction*, Australian Institute of Steel Construction, **21**, No. 2, September, pp. 2–11.

24. Lai, S-M.A. and MacGregor, J.G. (1983) Geometric non-linearities in unbraced multistorey frames, *Journal of Structural Engineering*, *ASCE*, **109**, No. 11, pp. 2528–45.

25. Merchant, W. (1954) The failure load of rigid jointed frameworks as influenced by stability, *The Structural Engineer*, **32**, No. 7, July, pp. 185–90.

26. Levi, V., Driscoll, G.C. and Lu, L-W. (1965) Structural subassemblages prevented from sway, *Journal of the Structural Division*, *ASCE*, **91**, No. ST5, pp. 103–27.

27. Levi, V., Driscoll, G.C. and Lu, L-W. (1967) Analysis of restrained columns permitted to sway, *Journal of the Structural Division*, *ASCE*, **93**, No. ST1, pp. 87–108.

28. Hibbard, W.R. and Adams, P.F. (1973) Subassemblage technique for asymmetric structures, *Journal of the Structural Division*, *ASCE*, **99**, No. ST11, pp. 2259–68.

29. Driscoll, G.C., *et al.* (1965) *Plastic Design of Multi-Storey Frames*, Lehigh University, Pennsylvania.

30. American Iron and Steel Institute (1968) *Plastic Design of Braced Multi-Storey Steel Frames*, AISI, New York.

31. Driscoll, G.C., Armacost, J.O. and Hansell, W.C. (1970) Plastic design of multi-storey frames by computer, *Journal of the Structural Division*, *ASCE*, **96**, No. ST1, pp. 17–33.

32. Harrison, H.B. (1967) Plastic analysis of rigid frames of high strength steel accounting for deformation effects, *Civil Engineering Transactions*, Institution of Engineers, Australia, **CE9**, No. 1, April, pp. 127–36.

33. El-Zanaty, M.H. and Murray, D.W. (1983) Nonlinear finite element analysis of steel frames, *Journal of Structural Engineering*, *ASCE*, **109**, No. 2, February, pp. 353–68.

34. White, D.W. and Chen, W.F. (editors) (1993) *Plastic Hinge Based Methods for Advanced Analysis and Design of Steel Frames*, Structural Stability Research Council, Bethlehem, Pa.

35. Clarke, M.J., Bridge, R.Q., Hancock, G.J. and Trahair, N.S. (1993) Australian trends in the plastic analysis and design of steel building frames, *Plastic hinge based methods for advanced analysis and design of steel frames*, Structural Stability Research Council, Bethlehem, Pa, pp. 65–93.

36. Clarke, M.J. (1994) Plastic-zone analysis of frames, Chapter 6 of *Advanced Analysis of Steel Frames: Theory, Software, and Applications*, (eds. W.F. Chen and S. Toma), CRC Press, Inc., Boca Raton, Florida, pp. 259–319.

37. White, D.W., Liew, J.Y.R. and Chen, W.F. (1993) Toward advanced analysis in LRFD, *Plastic hinge based methods for advanced analysis and design of steel frames*, Structural Stability Research Council, Bethlehem, Pa, pp. 95–173.

38. Vacharajittiphan, P. and Trahair, N.S. (1975) Analysis of lateral buckling in plane frames, *Journal of the Structural Division*, *ASCE*, **101**, No. ST7, pp. 1497–516.

39. Vacharajittiphan, P. and Trahair, N.S. (1974) Direct stiffness analysis of lateral buckling, *Journal of Structural Mechanics*, **3**, No. 1, pp. 107–37.

40. Trahair, N.S. (1993) *Flexural–Torsional Buckling of Structures*, E. & F.N. Spon, London.
41. Papangelis, J.P., Trahair, N.S. and Hancock, G.J. (1995) Elastic flexural–torsional buckling of structures by computer, *Proceedings*, Sixth International Conference on Civil and Structural Engineering Computing, Cambridge, pp. 109–119.
42. Bradford, M.A., Cuk, P.E., Gizejowski, M.A. and Trahair, N.S. (1987) Inelastic buckling of beam-columns, *Journal of Structural Engineering, ASCE*, **113**, No. 11, pp. 2259–77.
43. Pi, Y.L. and Trahair, N.S. (1994) Nonlinear inelastic analysis of steel beam-columns – Theory, *Journal of Structural Engineering, ASCE*, **120**, No. 7, pp. 2041–61.
44. Pi, Y.L. and Trahair, N.S. (1994) Nonlinear inelastic analysis of steel beam-columns – Applications, *Journal of Structural Engineering, ASCE*, **120**, No. 7, pp. 2062–85.
45. Horne, M.R. (1956) The stanchion problem in frame structures designed according to ultimate carrying capacity, *Proceedings of the Institution of Civil Engineers, Part III*, **5**, pp. 105–46.
46. Horne, M.R. (1964) Safe loads on I-section columns in structures designed by plastic theory, *Proceedings of the Institution of Civil Engineers*, **29**, September, pp. 137–50.
47. Horne, M.R. (1964) The plastic design of columns, *Publication No. 23*, BCSA, London.
48. Institution of Structural Engineers and Institute of Welding (1971) *Joint Committee Second Report on Fully-Rigid Multi-Storey Steel Frames*, ISE, London.
49. Wood, R.H. (1973) Rigid-jointed multi-storey steel frame design: a state-of-the-art report, *Current Paper*, CP 25/73, Building Research Establishment, Watford, September.
50. Wood, R.H. (1974) *A New Approach to Column Design*. HMSO, London.

9 Connections

9.1 Introduction

Connections or joints are used to transfer the forces supported by a structural member to other parts of the structure or to the supports. They are also used to connect braces and other members which provide restraints to the structural member.

The components of a *connection* include connectors such as bolts, pins, rivets, or welds, and may include additional plates or cleats. The arrangement of the components is usually chosen to suit the type of action (force or moment) being transferred and the types of member (tension or compression member, beam, or beam-column) being connected. The arrangement should also be chosen to avoid excessive costs, since the manufacture and assembly of a connection is usually time consuming, and may require the use of specialized equipment. For example, it is often better to use more material in a connection if this will reduce the number of processes required for its manufacture.

A connection is designed by first analysing the method of force transfer from the member through the connection and its components to the other parts of the structure. Each component is then proportioned so that it has sufficient strength to resist the force that it is required to transmit. General guidance on connections is given in [1–6].

The components and arrangements of connections are discussed in sections 9.2 and 9.3. The methods used to analyse connections are summarized in section 9.4, including some common connections. The design of bolts, bolted plates, and welds in accordance with BS5950 is discussed in sections 9.5–9.7, respectively.

9.2 Connection components

9.2.1 BOLTS

Several different types of bolts may be used in structural connections, including ordinary structural bolts (i.e. commercial or precision bolts and black bolts), and high strength bolts. Turned close tolerance bolts are now rarely used. Bolts may transfer loads by shear and bearing, as shown in Fig. 9.1a, by friction between plates clamped together as shown in the preloaded friction-grip joint of Fig. 9.1b, or by tension as shown in Fig. 9.1c. The shear, bearing,

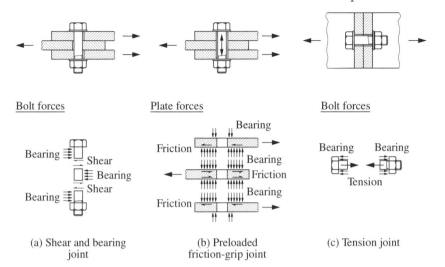

Bolt forces   Plate forces   Bolt forces

(a) Shear and bearing   (b) Preloaded   (c) Tension joint
joint   friction-grip joint

Fig. 9.1 Use of bolts in connections

and tension capacities of bolts and the slip capacities of preloaded friction-grip joints are discussed in section 9.5.

The use of bolts often facilitates the assembly of a structure, as only very simple tools are required. This is important in the completion of site connections, especially where the accessibility of a joint is limited, or where it is difficult to provide the specialized equipment required for other types of connectors. On the other hand, bolting usually involves a significant fabrication effort to produce the bolt holes and the associated plates or cleats. In addition, special but not excessively expensive procedures are required to ensure that the clamping actions required for preloaded friction-grip joints are achieved. Precautions may need to be taken to ensure that the bolts do not become undone, especially in situations where fluctuating loads may loosen them. Such precautions may involve the provision of special locking devices or the use of preloaded high strength bolts. Guidance on bolted (and on riveted) joints is given in [1–6].

9.2.2 PINS

Pin connections used to be provided in some triangulated frames where it was thought to be important to try to realize the common design assumption that these frames are pin-jointed. The cost of making a pin connection is high because of the machining required for the pin and its holes, and also because of difficulties in assembly. Pins are used in special architectural features where it is necessary to allow relative rotation to occur between the members being connected. In addition, a connection which requires a very large diameter fastener often uses a pin instead of a bolt.

9.2.3 RIVETS

In the past, hot-driven rivets were extensively used in structural connections. They were often used in the same way as ordinary structural bolts are used in shear and bearing and in tension connections. There is usually less slip in a riveted connection because of the tendency for the rivet holes to be filled by the rivets when being hot-driven. Shop riveting was cheaper than site riveting, and for this reason shop riveting was often combined with site bolting. However, riveting has been replaced by welding or bolting, except in some historical refurbishments.

9.2.4 WELDS

Structural connections between steel members are often made by arc-welding techniques, in which molten weld metal is fused with the parent metal of the members or joint plates being connected. Welding is used extensively in fabricating shops where specialized equipment is available and where control and inspection procedures can be exercised, ensuring the production of satisfactory welds. Welding is often cheaper than bolting because of the great reduction in the preparation required, while greater strength can be achieved, the members or plates no longer being weakened by bolt holes, and the strength of the weld metal being superior to that of the material connected. In addition, welds are more rigid than other types of load-transferring connectors. On the other hand, welding often produces distortion and high local residual stresses, and results in reduced ductility, while site welding may be difficult and costly.

Butt welds, such as that shown in Fig. 9.2a, may be used to splice tension members. A full penetration weld enables the full strength of the member to be developed, while the butting together of the members avoids any joint

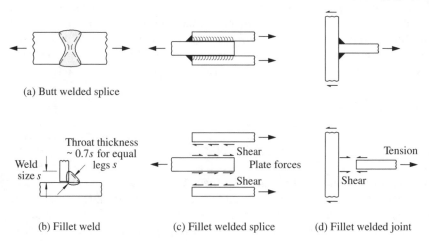

Fig. 9.2 Use of welds in connections

eccentricity. Butt welds often require some machining of the elements to be joined. Special welding procedures are usually needed for full strength welds between thick members to control the weld quality and ductility, while special inspection procedures may be required for critical welds to ensure their integrity. These butt welding limitations often lead to the selection of connections which use fillet welds.

Fillet welds (see Fig. 9.2b) may be used to connect lapped plates, as in the tension member splice (Fig. 9.2c), or to connect intersecting plates (Fig. 9.2d). The member force is transmitted by shear through the weld, either longitudinally or transversely. Fillet welds, although not as efficient as butt welds, require little if any preparation, which accounts for their extensive use.

9.2.5 PLATES AND CLEATS

Intermediate plates (or gussets), fin (or side) plates, and angle or tee cleats are frequently used in structural connections to transfer the forces from one member to another. Examples of flange cleats and plates are shown in the beam-to-column joints of Fig. 9.3, and an example of a gusset plate is shown in the truss joint of Fig. 9.4b. Stiffening plates, such as the seat stiffener (Fig. 9.3b) and the column web stiffeners (Fig. 9.3c), may also be used to help transfer the forces.

Plates are comparatively strong and stiff when they transfer the forces by in-plane actions, but are comparatively weak and flexible when they transfer the forces by out-of-plane bending. Thus the angle cleat and seat shown in Fig. 9.3a are flexible, and allow the relative rotation of the joint members, while the flange plates and web stiffeners (Fig. 9.3c) are stiff, and restrict the relative rotation.

The simplicity of welded joints and their comparative rigidity has often

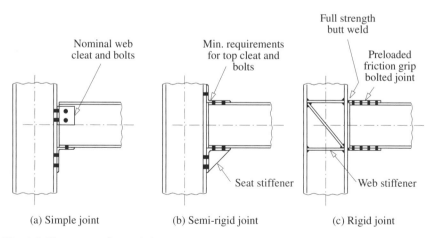

Fig. 9.3 Beam-to-column joints

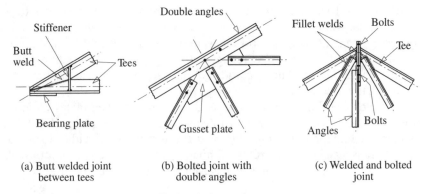

(a) Butt welded joint between tees	(b) Bolted joint with double angles	(c) Welded and bolted joint

Fig. 9.4 Joints between axially loaded members

resulted in the omission of stiffening plates when they are not required for strength purposes. Thus the rigid joint of Fig. 9.3c can be greatly simplified by butt welding the beam directly onto the column flange and by omitting the column web stiffeners. However, this omission will make the joint much more flexible since local distortions of the column flange and web will no longer be prevented.

9.3 Arrangement of connections

9.3.1 CONNECTIONS FOR FORCE TRANSMISSION

In many cases, a connection is only required to transmit a force, and there is no moment acting on the group of connectors. While the connection may be capable of also transmitting a moment, it will be referred to as a force connection.

Force connections are generally of two types. For the first, the force acts in the connection plane formed by the interface between the two plates connected, and the connectors between these plates act in shear, as in Fig. 9.1a. For the second type, the force acts out of the plane of the connection and the connectors act in tension, as in Fig. 9.1c.

Examples of force connections include splices in tension and compression members, truss joints, and shear splices and connections in beams. A simple shear and bearing bolted tension member splice is shown in Fig. 9.1a, and a friction-grip bolted splice in Fig. 9.1b. These are simpler than the tension bolt connection of Fig. 9.1c, and are typical of site connections.

Ordinary structural and high strength bolts are used in clearance holes (often 2 or 3 mm oversize to provide erection tolerances) as shown in Fig. 9.1a. The hole clearances lead to slip under service loading, and when this is undesirable, a preloaded friction-grip joint such as that shown in Fig. 9.1b may be used. In this connection, the transverse clamping action produced by preload-

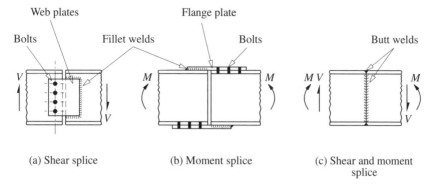

Fig. 9.5 Beam splices

ing the high strength bolts allows high frictional resistances to develop and transfer the longitudinal force. Preloaded friction-grip joints are often used to make site connections which need to be comparatively rigid.

The butt and fillet welded splices of Fig. 9.2a and c are typical of shop connections, and are of high rigidity. While they are often simpler to manufacture under shop conditions than the corresponding bolted joints of Fig. 9.1, special care may need to be taken during welding if these are critical joints.

The truss joint shown in Fig. 9.4b uses a gusset plate in order to provide sufficient room for the bolts. The use of a gusset plate is avoided in the joint of Fig. 9.4a, while end plates are used in the joint of Fig. 9.4c to facilitate the field-bolted connection of shop-welded assemblies.

The beam web shear splice of Fig. 9.5a shows a typical shop-welded and site-bolted arrangement. The common simple joint between a beam and column shown in Fig. 9.3a is often considered to transmit only shear from the beam to the column flange. Other examples of joints of this type are shown by the beam-to-beam connections of Fig. 9.6a and b. Common arrangements of beam splices are given in section 9.4.5 and in [1–6].

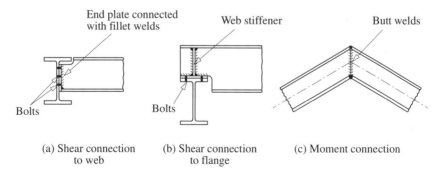

Fig. 9.6 Joints between beams

9.3.2 CONNECTIONS FOR MOMENT TRANSMISSION

While it is rare that a real connection transmits only a moment, it is not uncommon that the force transmitted by the connection is small enough to be neglected in design. Examples of connections which may be used when the force to be transmitted is negligible include the beam moment splice shown in Fig. 9.5b which combines site bolting with shop welding, and the welded moment connection of Fig. 9.6c. A moment connection is often capable of transmitting moderate transverse forces, as in the case of the beam-to-column joint shown in Fig. 9.3c.

9.3.3 FORCE AND MOMENT CONNECTIONS

A force and moment connection is required to transmit both force and moment, as in the seat for the beam-to-column connection shown in Fig. 9.7. Other examples include the semi-rigid beam-to-column connection of Fig. 9.3b, the full strength beam splice of Fig. 9.5c, and the beam joint of Fig. 9.6c. Common beam-to-column connections are given in [1–6], and are discussed in section 9.4.4.

In some instances, connections may actually transfer forces or moments which are not intended by the designer. For example, the cleats shown in Fig. 9.3a may transfer some horizontal forces to the column, even though the beam is designed as if simply supported, while a similar situation occurs in the shear splice shown in Fig. 9.5a.

9.4 Behaviour of connections

9.4.1 CONNECTIONS FOR FORCE TRANSMISSION

When shear and bearing bolts in clearance holes are used in an in-plane force connection, there is an initial slip when the shear is first applied to the connection

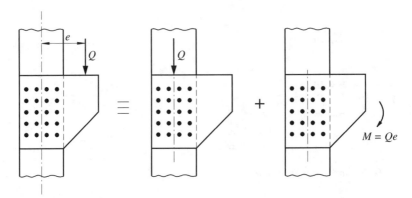

Fig. 9.7 Analysis of a force and moment connection

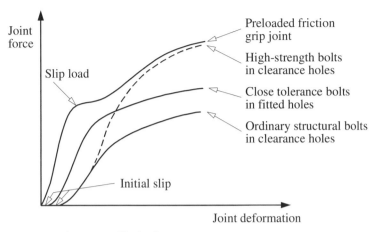

Fig. 9.8 Behaviour of bolted connections

as some of the clearances are taken up (Fig. 9.8). The connection then becomes increasingly stiff as more of the bolts come into play. At higher forces, the more highly loaded bolts start to yield, and the connection becomes less stiff. Later, a state may be reached in which each bolt is loaded to its maximum capacity.

It is not usual to analyse this complex behaviour, and instead it is commonly assumed that equal size bolts share equally in transferring the force as shown in Fig. 9.9b, even in the service load range. It is shown in section 9.8.1 that this is the case if there are no clearances and all the bolts fit perfectly, and if the members and connection plates act rigidly and the bolts elastically. If, however, the flexibilities of the members and plates are taken into account, then it is found that the forces transferred are highest in the end bolts of any line of bolts parallel to the connection force and lowest in the centre bolts, as shown in Fig. 9.9c. In long bolted connections, the end bolt forces may be so high as to lead to premature failure (before these forces can be redistributed by plastic action) and the subsequent 'unbuttoning' of the connection.

Shear and bearing connections using close tolerance bolts in fitted holes behave in a similar manner to connections with clearance holes, except that the bolt slips are greatly reduced. (It was noted earlier that fitted close tolerance bolts are now rarely used.) On the other hand, slip is not reduced in preloaded friction-grip bolted shear connections, but is postponed until the frictional resistance is overcome at the slip load as shown in Fig. 9.8. Again, it is commonly assumed that equal size bolts share equally in transferring the force.

This equal sharing of the force transfer is also often assumed for tension force connections in which the applied force acts out of the connection plane. It is shown in section 9.8.2 that this is the case for connections in which the plates act rigidly and the bolts act elastically.

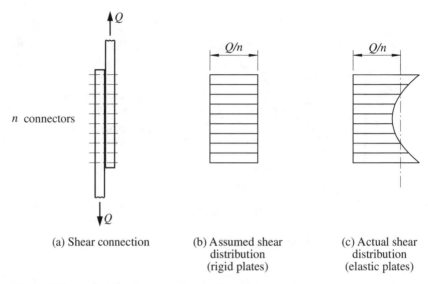

(a) Shear connection   (b) Assumed shear
distribution
(rigid plates)   (c) Actual shear
distribution
(elastic plates)

Fig. 9.9 Shear distribution in a force connection

Welded force connections do not slip, but behave as if almost rigid. Welds are often assumed to be uniformly stressed, whether loaded transversely or longitudinally. However, there may be significantly higher stresses at the ends of long welds parallel to the connection force, just as there are in the case of long bolted joints.

Due to the great differences in the stiffnesses of connections using different types of connectors, it is usual not to allow the connection force to be shared between slip and non-slip connectors, and instead, the non-slip connectors are required to transfer all the force. For example, when ordinary site bolting is used to hold two members in place at a joint which is subsequently site welded, the bolts are designed for the erection conditions only, while the welds are designed for the final total force. However, it is generally permissible to share the connection force between welds and preloaded friction-grip bolts which are designed against slip, and this is allowed in BS5950.

On the other hand, it is always satisfactory to use one type of connector to transfer the complete force at one part of a joint and a different type at another part. Examples of this are shown in Fig. 9.5a and b, where shop welds are used to join each connection plate to one member and field bolts to join it to the other.

When it is necessary to consider the total flexibility of a connection, this can be determined from the sum of the flexibilities of the components used at each link in the chain of force transfer through the connection. Thus the flexibilities of any plates or cleats used must be included with those of the connectors. Plates are comparatively stiff (and often assumed to be rigid) when loaded in their planes, but are comparatively flexible when bent out of their planes. In

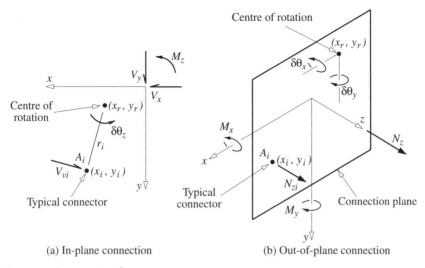

(a) In-plane connection   (b) Out-of-plane connection

Fig. 9.10 Connection forces

general, the overall behaviour of a connection can be assessed by determining the path by which force is transferred through the connection, and by synthesizing the responses of all the elements to their individual loads.

9.4.2 CONNECTIONS FOR MOMENT TRANSMISSION

Although a real connection is rarely required to transmit only a moment, the behaviour of a moment connection may be usefully discussed as an introduction to the consideration of connections which transmit both force and moment.

Connections where the moment acts in the plane of the connectors (as in Fig. 9.10a) so that the moment is transferred by connector shear are often analysed elastically [7] by assuming that all the connectors fit perfectly, and that each plate acts as if rigid, so that the relative rotation between them is $\delta\theta_z$. In this case, each connector transfers a shear force V_{vi} from one plate to the other. This shear force acts perpendicular to the radius r_i to the axis of rotation, and is proportional to the relative displacement $r_i\delta\theta_z$ of the two plates at the connector, whence

$$V_{vi} = k_v A_i r_i \delta\theta_z,$$ (9.1)

where A_i is the shear area of the connector, and the constant k_v depends on the shear stiffness of that type of connector. It is shown in section 9.8.1 by considering the equilibrium of the connector forces V_{vi} that the axis of rotation lies at the centroid of the connector group. The moment exerted by the connector force is $V_{vi}r_i$, and so the total moment M_z is given by

$$M_z = k_v \delta\theta_z \sum_i A_i r_i^2.$$ (9.2)

If this is substituted into equation 9.1, the connector force can be evaluated as

$$V_{vi} = \frac{M_z A_i r_i}{\sum_i A_i r_i^2}.$$ (9.3)

However, the real behaviour of in-plane moment connections is likely to be somewhat different, just as it is for the force connections discussed in section 9.4.1. This difference is due to the flexibility of the plates, the inelastic behaviour of the connections at higher moments, and the slip due to the clearances between any bolts and their holes.

The assumptions of elastic connectors and rigid plates may sometimes also be made for connections with moments acting normal to the plane of the connectors, as shown in Fig. 9.10b. It is shown in section 9.8.2 that the connector tension forces N_{zi} can be determined from

$$N_{zi} = \frac{M_x A_i y_i}{\sum A_i y_i^2} - \frac{M_y A_i x_i}{\sum A_i x_i^2}$$ (9.4)

in which x_i, y_i are the principal axis coordinates of the connector measured from the centroid of the connector group.

This result implies that the compression forces are transmitted only by the connectors, but in real bolted connections these are transmitted primarily through those portions of the connection plates which remain in contact. Thus the connector tension forces determined from equation 9.4 will be inaccurate when there is a significant difference between the centroid of the contact area and that of the assumed compression connectors.

Welded moment connections may be analysed by making the same assumptions as for bolted connections [7]. Thus equations 9.3 and 9.4 may be used, with the weld size substituted for the bolt area, and the summations replaced by integrals along the weld.

9.4.3 FORCE AND MOMENT CONNECTIONS

Connections which are required to transfer both force and moment may be analysed elastically by using the method of superposition, as shown in section 9.8. Thus, the individual bolt forces in the eccentrically loaded in-plane plate connections shown in Fig. 9.7 can be determined as the vector sum of the components caused by a concentric force Q and a moment Qe. Similarly, the individual connector forces in a connection loaded out of its plane as shown in Fig. 9.10b may be determined from the sum of the forces due to the out-of-plane force N_z and the principal axis moments M_x and M_y.

The connector forces in connections subjected to combined loadings may be analysed elastically by using superposition to combine the separate effects of in-plane and out-of-plane loading.

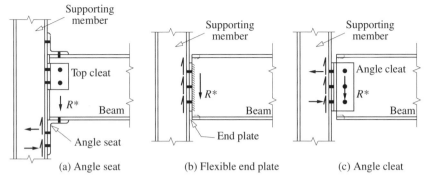

Fig. 9.11 Common flexible end connections

9.4.4 COMMON CONNECTIONS

9.4.4.1 General

There are many different connection arrangements used for force and moment transmission between members and supports, depending on the actions which are to be transferred. Fabrication and erection procedures may be simplified by standardizing a number of common connections for the more frequently occurring situations. Examples of some common connections are shown in Figs 9.11–9.15.

Flexible connections for joints in simple construction (for which any moment transfer can be neglected) are described in sections 9.4.4.2–9.4.4.5, a semi-rigid connection for joints in semi-continuous structures in section 9.4.4.6, and rigid connections for joints in fully-continuous structures (which are able to transfer moment as well as force) in sections 9.4.4.7 and 9.4.4.8. Welded and bolted splices are described in sections 9.4.4.9 and 9.4.4.10, and seats and base plates in sections 9.4.4.11 and 9.4.4.12, respectively.

Guidance on the arrangement and analysis of some of these connections are given in [1–6], while worked examples of a web side plate (fin plate) and a flexible end plate connection designed in accordance with BS5950 are given in section 9.9.

9.4.4.2 Angle seat connection

An angle seat connection (Fig. 9.11a) transfers a beam reaction force R^* to the supporting member through the angle seat. The top cleat is for lateral restraint only, and may be bolted either to the top of the web or to the top flange. The angle seat may be bolted or fillet welded to the supporting member. The connection has very little moment capacity, and is classified as a flexible end connection. It may be used for joints in simple construction.

The beam reaction is transferred by bearing, shear, and bending of the horizontal leg of the angle, by vertical shear through the connectors, and by horizontal forces in the connectors and between the vertical leg and the supporting member.

The beam is designed for zero end moment, and the supporting member for the eccentric beam reaction. The beam web may need to be stiffened to resist shear (section 4.7.4) and bearing (section 4.7.6). Although this connection is easily designed, its use is often discouraged because of erection difficulties associated with the close depth tolerances required at the top and bottom of the beam.

9.4.4.3 Flexible end plate connection

A flexible end plate connection (Figs 9.6c and 9.11b) also transfers a beam reaction R^* to the supporting member. The end plate is fillet welded to the beam web, and bolted to the supporting member. The flanges may be notched or coped, if required. This connection also has very little moment capacity, as there may be significant flexibility in the end plate, and is classified as a flexible end connection. It may be used for joints in simple construction.

The beam reaction is transferred by weld shear to the end plate, by shear and bearing to the bolts, and by shear and bearing to the supporting member.

The beam is designed for zero end moment, with the end plate augmenting the web shear and bearing capacity, while the supporting member is designed for the eccentric beam reaction.

9.4.4.4 Angle cleat connection

An angle cleat connection (Fig. 9.11c) also transfers a beam reaction R^* to the supporting member. One or two angle cleats may be used, and bolted to the beam web and to the supporting member. The flanges may be notched or coped, if required. This connection also has very little moment capacity, as there may be significant flexibility in the angle legs connected to the supporting member and in the bolted connection to the beam web. The connection is classified as a flexible end connection and may be used for joints in simple construction.

The beam reaction is transferred by shear and bearing from the web to the web bolts and to the angle cleats. These actions are transferred by the cleats to the supporting member bolts, and by these to the supporting member by shear, tension and compression.

The beam is designed for zero end moment, with the angles augmenting the web shear and bearing capacity, while the supporting member is designed for the eccentric beam reaction.

9.4.4.5 Web side plate (fin plate) connection

A web side plate (fin plate) connection (Fig. 9.12) transfers a beam reaction R^* to the supporting member, and can also transfer moment M^*. The fin plate is

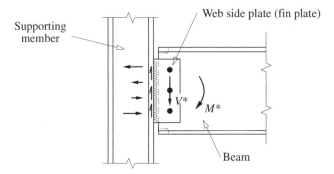

Fig. 9.12 Common semi-rigid end connection

fillet welded to the supporting member, and bolted to the beam web. The flanges may be notched or coped, if required. This connection has limited flexibility, and is classified as semi-rigid. It may be used for joints in simple construction, and for joints in semi-continuous construction provided that the degree of interaction between the members can be established.

The beam reaction R^* and moment M^* are transferred by shear through the bolts to the fin plate, by in-plane bending and shear to the welds, and by vertical and horizontal shear to the supporting member.

The beam and the supporting member are designed for the reaction R^* and moment M^*.

9.4.4.6 Welded moment connection

A fully welded moment connection (Fig. 9.13a) transfers moment M^*, axial force N^*, and shear V^* from one member to another. The welds may be fillet or butt welds, and erection cleats may be used to facilitate field welding. The

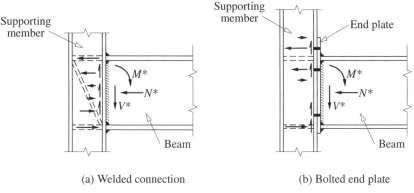

(a) Welded connection   (b) Bolted end plate

Fig. 9.13 Common rigid end connections

concentrated flange forces resulting from the moment M^* and axial force N^* may require the other member to be stiffened in order to transfer these forces. When suitably stiffened, such a connection acts as if rigid, and may be used in a continuous structure.

The shear V^* is transferred by shear through the welds from one member to the other. The flange forces are transferred by tension and compression (or shear) through the flange welds to the other member, while the web moment and the axial force are transferred by tension and compression (or shear) through the web welds to the other member.

9.4.4.7 Bolted moment end plate connection

A bolted moment end plate connection (Fig. 9.13b) also transfers moment M^*, axial force N^*, and shear V^* from one member to another. The end plate is fillet welded to the web and flanges of one member, and bolted to the other member. Concentrated flange forces may require the other member to be stiffened in order to transfer these forces. This connection is also classified as rigid, although it is less stiff than a welded moment connection, due to flexure of the end plate. BS5950 allows bolted moment end plate connections to be used in continuous construction.

The beam moment, axial force, and shear are transferred by tension and compression (or shear) through the flange welds and by shear through the web welds to the end plate, by bending and shear through the end plate to the bolts, and by bolt shear and tension and by horizontal plate reaction to the other member.

The bolt tensions are increased by prying actions resulting from bending of the end plate. Increases of up to 33% have been suggested. Prying actions must be considered in the design of bolts subjected to tension, and are discussed in section 9.5.2.

9.4.4.8 Welded splice

A fully welded splice (Fig. 9.14a) transfers moment M^*, axial force N^*, and shear V^* from one member to another concurrent member. The flange welds are butt welds, while the web welds are fillet welds, and erection plates may be used to facilitate site welding. A welded splice acts as a rigid connection between the members, and may be used in continuous construction.

The shear V^* is transferred by shear through the welds. The flange forces resulting from the moment M^* and axial force N^* are transferred by tension or compression through the flange welds, while the web moment and the axial force are transferred by tension or compression (or shear) through the web welds.

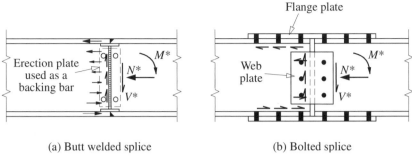

(a) Butt welded splice   (b) Bolted splice

Fig. 9.14 Common splices

9.4.4.9 Bolted splice

A fully bolted splice (Fig. 9.14b) also transfers moment M^*, axial force N^*, and shear V^* from one member to another concurrent member. Flange and web plates may be provided on one or both sides. This joint may be used in continuous construction.

The moment, axial force, and shear are transferred from one member by bearing and shear through the bolts to the plates, to the other bolts, and then to the other member. The flange plates only transfer the flange axial force components of the moment M^*, axial force N^*, while the web plates transfer all of the shear V^* together with the web components of M^* and N^*.

9.4.4.10 Beam seat

A beam seat (Fig. 9.15a) transfers a beam reaction R^* to its support. A seating plate may be fillet welded to the bottom flange to increase the support bearing

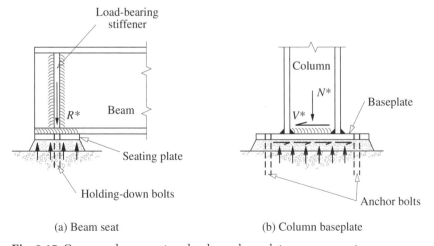

(a) Beam seat   (b) Column baseplate

Fig. 9.15 Common beam seat and column baseplate arrangements

area, while holding down bolts provide positive connections to the support. If a load-bearing stiffener is provided to increase the web bearing capacity, then this will effectively prevent lateral deflection of the top flange (see Fig. 6.19).

The beam reaction R^* is transferred by bearing through the seating plate to the support.

9.4.4.11 Base plate

A base plate (Fig. 9.15b) transfers a column axial force N^* and shear V^* to a support or a concrete foundation, and may also transfer a moment M^*. The base plate is fillet welded to the flanges and web of the column, while anchor or holding-down bolts provide connections to the support. Pinned bases used in simple construction often use four anchor bolts, and the flexibility of these and the baseplate limits the effective moment resistance.

An axial compression N^* is transferred from the column by end bearing or weld shear to the base plate, and then by plate bending, shear, and bearing to the support. An axial tension force N^* is transferred from the column by weld shear to the base plate, by plate bending and shear to the holding-down anchor bolts, and then by bolt tension to the support. The shear force V^* is transferred from the column by weld shear to the base plate, and then by shear and bearing through the holding down bolts to the support, or by friction. Specific guidance on the design of holding-down bolts is given in Clause 6.6 of BS5950, while guidance on the design of base plates is given in Clause 4.13.

9.5 Design of bolts

9.5.1 BEARING BOLTS IN SHEAR

The capacity P_s of a shearing bolt in shear (Fig. 9.1a) depends on the shear strength p_s of the bolt, the number of shear planes n, and the areas A_{sn} of the bolt in each shear plane (either the gross area, or the tensile stress area A_t through the threads, as appropriate). It can be expressed in the form

$$P_s = p_s \sum_n k_r A_{sn} \qquad (9.5)$$

in which k_r is a factor which allows for the overloading of end bolts that occurs in long connections.

BS5950 requires the factored shear force F_s^* to be limited by

$$F_s^* \le P_s, \qquad (9.6)$$

with the shear strength p_s of the bolt material being approximated [8] by the lesser of

$$p_s = 0.48\, U_b \qquad (9.7)$$

and

$$p_s = 0.69\, Y_b \qquad (9.8)$$

where U_b and Y_b are the ultimate tensile strength and the yield stress of the bolt material, respectively. For a bolt with $U_b/Y_b = 0.69/0.48 \approx 1.44$, these equations correspond to the use of a shear to tensile strength ratio of $\tau_{ub}/U_b = 0.60$ with a conservative factor of 0.8 (which ensures a higher safety level for bolts than for members). The ratio of $\tau_{ub}/U_b = 0.6$ is close to the theoretical yield value of $\tau_{yb}/Y_b = 1/\sqrt{3} \approx 0.577$ (see section 1.3.1) and the experimental value of 0.62 reported in [6]. Values of p_s are given in Table 30 of BS5950.

It is common and conservative to determine the area A_{sn} by substituting the tensile stress area A_t for the shank area of the bolt.

The joint length factor k_r is equal to 1.0 for lengths L_j less than 500 mm, and varies linearly for longer lengths according to

$$k_r = (5500 - L_j)/5000 \qquad (9.9)$$

9.5.2 BOLTS IN TENSION

The capacity of a bolt in tension (Fig. 9.1c) depends on the tensile strength f_{uf} of the bolt and the minimum cross-sectional area of the threaded length of the bolt. The design force F_t^* is limited by BS5950 to

$$F_t^* \leqslant P_t \qquad (9.10)$$

If any of the connecting plates is sufficiently flexible, then additional prying forces may be induced in the bolts. An example is given in Clause 6.3.4.3 of BS5950. When any prying actions can be calculated and are included in F_t^*, then

$$P_t = p_t A_t \qquad (9.11)$$

in which A_t is the tensile stress area of the bolt and the tensile strength of the bolt p_t is generally approximated by the lesser of 0.7 U_b and Y_b. For a bolt with $U_b/Y_b = 1.0/0.7 \approx 1.43$, these correspond to the use of a conservative factor of 0.7 to ensure a higher level of safety than for members. Values of p_t are given in Table 34 of BS5950.

Under some circumstances, the prying forces need not be calculated, provided P_t is reduced to

$$P_t = 0.8\, p_t A_t \qquad (9.12)$$

9.5.3 BEARING BOLTS IN SHEAR AND TENSION

Test results [6] for bearing bolts in shear and tension suggest a circular interaction relationship (Fig. 9.11) for the strength limit state, which can be expressed in the form of equations 9.6, 9.10, and

$$\frac{F_s^*}{P_s} + \frac{F_t^*}{P_t} \leq 1.4 \tag{9.13}$$

where P_s is the nominal shear capacity when there is no tension, and P_t the nominal tension capacity when there is no shear.

9.5.4 BOLTS IN BEARING

It is now commonly the case that bolt materials are of much higher strength than that of the steel plates or elements through which the bolts pass. As a result of this, bearing failure usually takes place in the plate material rather than in the bolt. The design of plates against bearing failure is discussed in section 9.6.2.

BS5950 requires the factored bearing force F_b^* on a bolt to be limited by

$$F_b^* \leq P_{bb} \tag{9.14}$$

where

$$P_{bb} = dt_p p_{bb} \tag{9.15}$$

in which t_p is the thickness of the connected element, d is the diameter of the bolt, and the bearing strength is generally given by

$$p_{bb} = 0.7(U_b + Y_b) \tag{9.16}$$

which allows for the enhancement of the bearing strength caused by the tri-axial stress state that exists in the bearing area of the bolt. Values of p_{bb} are given in Table 31 of BS5950.

9.5.5 PRELOADED FRICTION-GRIP BOLTS

9.5.5.1 Design as bearing bolts

Preloaded bolts may be used as bearing bolts, or as preloaded high strength (HSFG) bolts in friction-grip joints (Fig. 9.1b) which are designed against joint slip under service or factored loads. When preloaded bolts are used as bearing bolts, they should be designed as discussed in sections 9.5.1–9.5.4.

9.5.5.2 Design against slip under factored loads

For design against slip in a friction-grip joint under factored loads, the shear force on a preloaded bolt F_s^* determined from the factored loads must satisfy

$$F_s^* \leq P_{sL} \tag{9.17}$$

in which the slip resistance P_{sL} depends principally on the bolt preload and the slip factors between the surfaces in contact at each interface. For design against slip under factored loads, the slip resistance P_{sL} is given by BS5950 as

$$P_{sL} = 0.9K_s(\Sigma\mu_n)P_o \tag{9.18}$$

in which K_s is a coefficient which allows for the shape and size of the hole, values of the slip factors μ_n are given in Table 35 of BS5950, and the preload P_o is taken as the value of the minimum shank tension specified in BS4604 [9].

When the joint loading induces bolt tensions, these tend to reduce the friction clamping forces. BS5950 requires the factored shear design load F_s^* to satisfy

$$\frac{F_s^*}{P_{sL}} + \frac{F_t^*}{0.9P_o} \leqslant 1 \tag{9.19}$$

in which F_s^* is the total applied tension, including any prying forces.

9.5.5.3 Design against slip under service loads

Joint slip is often a serviceability limit state, in which case the joint must be designed against slip under the serviceability loads, and the bolts must be designed as bearing bolts under the factored loads as discussed in sections 9.5.1–9.5.4.

The procedure of BS5950 for design against slip in a friction-grip joint under service loads is presented as an approximate factored load equivalent, in that the shear force on a preloaded bolt F_s^* determined from the factored loads must satisfy equation 9.17, in which an increased value of P_{sL} is obtained from equation 9.18 by increasing the factor 0.9 to 1.1. This increase of $1.1/0.9 \approx 1.22$ provides an approximation for the ratio of the factored to the service loads.

When the joint loading induces bolt tensions, BS5950 requires the factored shear design load F_s^* to satisfy equation 9.19 with the factor 0.9 increased to 1.1.

9.6 Design of bolted plates

9.6.1 GENERAL

A connection plate used in a joint is required to transfer actions which may act in the plane of the plate, or out of it. These actions include axial tension and compression forces, shear forces, and bending moments, which themselves may include components induced by prying actions. The presence of bolt holes often weakens the plate, and failure may occur very locally by the bearing of a bolt on the surface of the bolt hole through the plate, or in an overall mode along a path whose position is determined by the positions of several holes and the actions transferred by the plate.

Generally, the proportions of the plates should be such as to ensure that there are no instability effects, in which case it will be conservative for the strength limit state to design against general yield, and satisfactory to design against local fracture.

The actual failure stress distributions in bolted connection plates are both uncertain and complicated. When design is to be based on general yielding, it is logical to take advantage of the ductility of the steel and to use a simple plastic analysis. A combined yield criterion such as

$$f_x^{*2} + f_y^{*2} - f_x^* f_y^* + 3\tau_{xy}^{*2} \leqslant f_y^2 \qquad (9.20)$$

(section 1.3.1) may then be used, in which f_x^* and f_y^* are the design normal stresses and τ_{xy}^* is the design shear stress.

When design is based on local fracture, an elastic analysis may be made of the stress distribution. Often approximate bending and shear stresses may be determined by elastic beam theory (Chapter 5). The failure criterion should then be tested at all potentially critical locations.

9.6.2 BEARING

Bearing failure of a plate may occur where a bolt bears against part of the surface of the bolt hole through the plate, as shown in Fig. 9.16a. After local yielding, the plate material flows plastically, increasing the circumference and thickness of the bearing area, and redistributing the contact force exerted by the bolt.

BS5950 requires the plate bearing force F_b^* due to the factored loads to be limited by

$$F_b^* \leqslant P_{bs} \qquad (9.21)$$

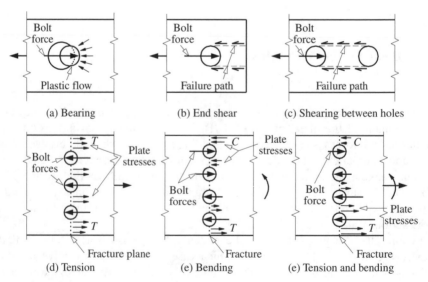

(a) Bearing   (b) End shear   (c) Shearing between holes

(d) Tension   (e) Bending   (e) Tension and bending

Fig. 9.16 Bolted plates in bearing, shear, tension, or bending

in which the bearing capacity of the plate is given by

$$P_{bs} = k_{bs}dt_pp_{bs} \leqslant 0.5k_{bs}et_pp_{bs} \tag{9.22}$$

in which k_{bs} is a coefficient whose value depends on the shape of the bolt hole, d is the bolt diameter, t_p is the plate thickness, e is the end distance (Fig. 9.16b) from the centre of the bolt hole to the end of the plate, and p_{bs} is the bearing strength of the plate. While strength test results [6] suggest that p_{bs}/Y_s should vary in the range of 4.5–4.9, large deformations occur at lower stresses. Because of this, BS5950 uses conservative values of p_{bs} which are generally approximated by

$$p_{bs} = 0.67(U_s + Y_s) \tag{9.23}$$

as an equivalent serviceability limitation. Values of p_{bs} are given in Table 32 of BS5950.

The bearing capacity P_{bs} is usually much higher than the shear strength of the bolts, and the limit state of bolt failure in shear often governs. However when a bolt is close to a plate edge, as in Fig. 9.16b, 'bearing' failure may be accompanied by plate tear-out by shearing. The BS5950 design method for this type of bearing failure is discussed in section 9.6.3.

9.6.3 SHEAR

Local in-plane shear failure may occur in a plate or element of a section at an end bolt hole, or where bolt holes in line are closely spaced, as shown in Fig. 9.16b and c. The shear yield capacity of the plate can be expressed as

$$P_r = tL_v\tau_y \tag{9.24}$$

in which L_v is the total length of the shear failure path and t is the thickness of the plate or element. If it is assumed that $\tau_y = p_y/\sqrt{3}$, then this becomes

$$P_r \approx 0.577\, p_ytL_v \tag{9.25}$$

BS5950 thus requires the factored shear force to satisfy

$$F_r^* \leqslant 0.6p_ytL_v \tag{9.26}$$

9.6.4 TENSION

Tension failure may be caused by tension actions as shown in Fig. 9.16d, or by in-plane bending actions as in Fig. 9.16e, or by combinations of these actions, as in Fig. 9.16f. Tension failures of tension members are treated in section 2.2, and it is logical to apply the BS5950 tension members method discussed in section 2.6.2 to the design of plates in tension. However, there are no specific limits given in BS5950.

A conservative method of designing connection plates against in-plane bending (Fig. 9.16c) would be to limit the design stress f_{bg}^* determined by an elastic beam analysis of the gross section by

$$f_{bg}^* \leq p_y \qquad (9.27)$$

and to limit the design stress f_{bn}^* determined by an elastic analysis of the net section by

$$f_{bn}^* \leq K_e p_y \qquad (9.28)$$

in which K_e is the effective net area coefficient discussed in section 2.6.2. This method is similar to the BS5950 method of designing tension members discussed in section 2.6.2.

Connection plates may be designed against combined tension and bending (Fig. 9.16f) by making elastic analyses of the gross and net sections to determine the maximum design local tension stresses, and using these in equations 9.27 and 9.28.

9.6.5 SHEAR AND TENSION

Plate sections may be subjected to simultaneous normal and shear stress, as in the case of the splice plates shown in Fig. 9.17a and b. These may be designed conservatively against general yield by using the shear and bending stresses determined by elastic analyses of the gross cross-section in the combined yield criterion of equation 9.20, and against fracture by using the stresses determined by elastic analyses of the net section in an ultimate version of equation 9.15, such as

$$f_x^{*2} + f_y^{*2} - f_x^* f_y^* + 3\tau_{xy}^{*2} = (K_e p_y)^2 \qquad (9.29)$$

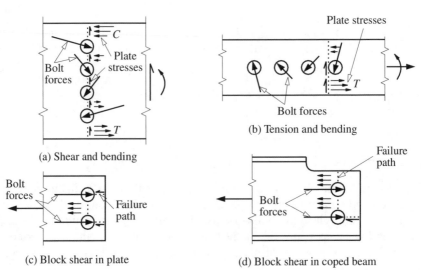

(a) Shear and bending

(b) Tension and bending

(c) Block shear in plate

(d) Block shear in coped beam

Fig. 9.17 Bolted plates in shear and tension

Block failure may occur in some connection plates as shown in Fig. 9.17c and d. In these failures, it may be assumed that the total resistance is provided partly by the tensile resistance across one section of the failure path, and partly by the shear resistance along another section of the failure path. This assumption implies considerable redistribution from the elastic stress distribution, which is likely to be very non-uniform. BS5950 conservatively uses the approximate shear strength $0.6\,p_y$ for both the shear and the tension strengths of the failure path sections. Thus equation 9.26 is used with the length of the shear failure path L_v replaced by the sum of L_v and K_e times the length of the tension failure path.

9.7 Design of welds

9.7.1 FULL PENETRATION BUTT WELDS

Full penetration butt welds are so made that their thicknesses and widths are not less than the corresponding lesser values for the elements joined. When the weld metal is of higher strength than that of the elements joined (and this is usually the case), the static capacity of the weld is greater than those of the elements joined. Hence, the design is controlled by the elements joined, and there are no design procedures required for the weld. BS5950 requires butt welds to have equal or superior properties to those of the elements joined.

Partial penetration butt welds have effective (throat) thicknesses which are less than those of the elements joined. The BS5950 value of the throat thickness is taken as equal to the minimum depth of penetration. The welds are designed using the relevant strength of the elements joined.

9.7.2 FILLET WELDS

Each of the fillet welds connecting the two plates shown in Fig. 9.18 is of length L, and its throat thickness a is inclined at $\alpha = \tan^{-1}(s_2/s_1)$. Each weld transfers a longitudinal shear V_L and transverse forces or shears V_{Tx} and V_{Ty} between the plates. The average normal and shear stresses f_w and τ_w on the weld throat may be expressed in terms of the forces

$$f_w La = V_{Tx} \sin \alpha + V_{Ty} \cos \alpha, \tag{9.30}$$

$$\tau_w La = \{(V_{Tx} \cos \alpha - V_{Ty} \sin \alpha)^2 + V_L^2\}^{1/2}. \tag{9.31}$$

In addition to these, there are local stresses in the weld arising from bending effects, shear lag, and stress concentrations, together with longitudinal normal stresses induced by compatibility between the weld and each plate.

It is customary to assume that the static strength of the weld is determined by the average throat stresses f_w and τ_w alone, and that the ultimate strength of the weld is reached when

$$\surd(f_w^2 + 3\tau_w^2) = f_{uw} \tag{9.32}$$

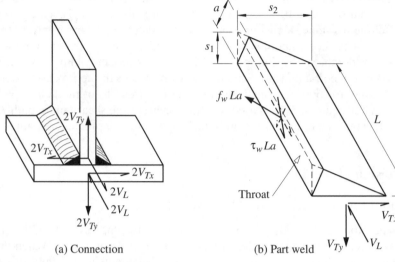

(a) Connection   (b) Part weld

Fig. 9.18 Fillet welded joint

in which f_{uw} is the ultimate tensile stress of the weld. This equation is similar to equation 1.1 used in section 1.3.1 to express the distortion energy theory for yield under combined stresses. Substituting equations 9.30 and 9.31 into equation 9.32 and rearranging leads to

$$3(V_{Tx}^2 + V_{Ty}^2 + V_L^2) - 2(V_{Tx} \sin \alpha + V_{Ty} \cos \alpha)^2 = (f_{uw}La)^2. \quad (9.33)$$

This is often simplified conservatively to

$$V_R/La = f_{uw}/\sqrt{3} \quad (9.34)$$

in which V_R is the vector resultant

$$V_R = \sqrt{\{V_{Tx}^2 + V_{Ty}^2 + V_L^2\}} \quad (9.35)$$

of the applied forces V_{Tx}, V_{Ty}, V_L.

BS5950 replaces the weld shear strength $f_{uw}/\sqrt{3}$ in equation 9.34 by the design strength p_w which is generally approximated by the lesser of $0.5\ U_e$ and $0.55\ U_s$, in which U_e and U_s are the minimum tensile strengths of the weld electrode and the parent metal, respectively. (Values of p_w are given in Table 37 of BS5950.) Thus in the simple design method of BS5950, the weld forces F_{Tx}^*, F_{Ty}^*, and F_L^* due to the factored loads are limited by

$$\sqrt{\{F_{Tx}^{*2} + F_{Ty}^{*2} + F_L^{*2}\}}/(La) \leq p_w \quad (9.36)$$

A less conservative directional method is provided by BS5950 which makes some allowance for the dependence of the weld strength on the direction of loading. This method is derived from a modification of equation 9.32 (with p_w

substituted for $f_{uw}/\sqrt{3}$), in which the transverse force stress components normal to the weld throat are approximated by decreasing the transverse forces to F_T^*/K, in which

$$K = 1.25\sqrt{\frac{1.5}{(1 + \cos^2 \theta)}} \qquad (9.37)$$

in which θ is the angle between the force F_T^* and the weld throat.

9.8 Appendix – Elastic analysis of connections

9.8.1 IN-PLANE CONNECTIONS

The in-plane connection shown in Fig. 9.10a is subjected to a moment M_z and to forces V_x, V_y acting through the centroid of the connector group (which may consist of bolts or welds) defined by (see section 5.9)

$$\left.\begin{array}{l} \Sigma A_i x_i = 0 \\ \Sigma A_i y_i = 0 \end{array}\right\} \qquad (9.38)$$

in which A_i is the area of the ith connector and x_i, y_i its coordinates.

It is assumed that the connection undergoes a rigid body relative rotation $\delta\theta_z$ between its plate components about a point whose coordinates are x_r, y_r. If the plate components are rigid and the connectors elastic, then it may be assumed that each connector transfers a force V_{vi}, which acts perpendicular to the line r_i to the centre of rotation (Fig. 9.10a) and which is proportional to the distance r_i, so that

$$V_{vi} = k_v A_i r_i \delta\theta_z \qquad (9.1)$$

in which k_v is a constant which depends on the elastic shear stiffness of the connector.

The centroidal force resultants V_x, V_y of the connector forces are

$$\begin{array}{l} V_x = -\Sigma V_{vi}(y_i - y_r)/r_i = k_v \delta\theta_z y_r \Sigma A_i \\ V_y = \Sigma V_{vi}(x_i - x_r)/r_i = -k_v \delta\theta_z x_r \Sigma A_i, \end{array} \qquad (9.39)$$

after using equations 9.38 and 9.1. The moment resultant M_z of the connector forces about the centroid is

$$M_z = \Sigma V_{vi}(y_i - y_r)y_i/r_i + \Sigma V_{vi}(x_i - x_r)x_i/r_i,$$

whence

$$M_z = k_v \delta\theta_z \Sigma A_i(y_i^2 + x_i^2), \qquad (9.40)$$

after using equations 9.38 and 9.1. This can be used to eliminate $k_v \delta\theta_z$ from equations 9.39 and these can then be rearranged to find the coordinates of the centre of rotation as

$$x_r = - \frac{V_y \Sigma A_i (x_i^2 + y_i^2)}{M_z \Sigma A_i},$$
(9.41)

$$y_r = \frac{V_x \Sigma A_i (x_i^2 + y_i^2)}{M_z \Sigma A_i}.$$
(9.42)

The term $k_v \delta \theta_z$ can also be eliminated from equation 9.1 by using equation 9.40 so that the connector force can be expressed as

$$V_{vi} = \frac{M_z A_i r_i}{\Sigma A_i (x_i^2 + y_i^2)}.$$
(9.43)

Thus the greatest connector shear stress V_{vi}/A_i occurs at the connector at the greatest distance r_i, from the centre of rotation x_r, y_r.

When there are n bolts of equal area, these equations simplify to

$$x_r = - \frac{V_y \Sigma (x_i^2 + y_i^2)/n}{M_z},$$
(9.41a)

$$y_r = \frac{V_x \Sigma (x_i^2 + y_i^2)/n}{M_z},$$
(9.42a)

$$V_{vi} = \frac{M_z r_i}{\Sigma (x_i^2 + y_i^2)}.$$
(9.43a)

When the connectors are welds, the summations in equations 9.41a–9.43a should be replaced by integrals, so that

$$x_r = - \frac{V_y (I_x + I_y)}{M_z A},$$
(9.41b)

$$y_r = \frac{V_x (I_x + I_y)}{M_z A},$$
(9.42b)

$$\tau_w = \frac{M_z r}{(I_x + I_y)}$$
(9.43b)

in which A is the area of the weld group given by the sum of the products of the weld lengths L and throat thicknesses a, I_x, I_y the second moments of area of the weld group about its centroid, and r the distance to the weld where the shear stress is τ_w. The properties of some specific weld groups are given in [4].

In the special case of a moment connection with V_x, V_y equal to zero, the centre of rotation is at the centroid of the connector group (equations 9.41 and 9.42), and the connector forces are given by

$$V_{vi} = \frac{M_z A_i r_i}{\Sigma A_i r_i^2}$$
(9.3)

with $r_i = \sqrt{(x_i^2 + y_i^2)}$. In the special case of a force connection with M_z, V_y

equal to zero, the centre of rotation is at $y_r = \infty$, and the connector forces are given by

$$V_{vi} = \frac{V_x A_i}{\Sigma A_i}. \tag{9.44}$$

In the special case of a force connection with M_z, V_x equal to zero, the centre of rotation is at $x_r = -\infty$, and the connector forces are given by

$$V_{vi} = \frac{V_y A_i}{\Sigma A_i}. \tag{9.45}$$

Equations 9.44 and 9.45 indicate that the connector stresses V_{vi}/A_i caused by forces V_x, V_y are constant throughout the connection.

It can be shown that the superposition by vector addition of the components of V_{vi} obtained from equations 9.3, 9.44, and 9.45 for the separate connection actions of V_x, V_y, and M_z leads to the general result given in equation 9.43.

9.8.2 OUT-OF-PLANE CONNECTIONS

The out-of-plane connection shown in Fig. 9.10b is subjected to a normal force N_z and moments M_x, M_y acting about the principal axes x, y of the connector group (which may consist of bolts or welds) defined by (section 5.9)

$$\left. \begin{array}{l} \Sigma A_i x_i = 0 \\ \Sigma A_i y_i = 0 \\ \Sigma A_i x_i y_i = 0 \end{array} \right\} \tag{9.46}$$

in which A_i is the area of the ith connector and x_i, y_i are its coordinates.

It is assumed that the plate components of the connection undergo rigid body relative rotations $\delta\theta_x$, $\delta\theta_y$ about axes which are parallel to the x, y principal axes and which pass through a point whose coordinates are x_r, y_r and that only the connectors transfer forces. If the plates are rigid and the connectors elastic, then it may be assumed that each connector transfers a force N_{zi} which acts perpendicular to the plane of the connection, and which has components which are proportional to the distances $(x_i - x_r)$, $(y_i - y_r)$ from the axes of rotation, so that

$$N_{zi} = k_t A_i (y_i - y_r)\delta\theta_x - k_t A_i (x_i - x_r)\delta\theta_y \tag{9.47}$$

in which k_t is a constant which depends on the axial stiffness of the connector.

The centroidal force resultant N_z of the connector forces is

$$N_z = \Sigma N_{zi} = k_t(-y_r\delta\theta_x + x_r\delta\theta_y)\Sigma A_i, \tag{9.48}$$

after using equations 9.46. The moment resultants M_x, M_y of the connector forces about the centroidal axes are

$$M_x = \Sigma N_{zi} y_i = k_t \delta\theta_x \Sigma A_i y_i^2, \tag{9.49}$$
$$M_y = -\Sigma N_{zi} x_i = k_t \delta\theta_y \Sigma A_i x_i^2, \tag{9.50}$$

after using equations 9.46.

Equations 9.48–9.50 can be used to eliminate k_t, $\delta\theta_x$, $\delta\theta_y$, x_r, y_r from equation 9.47, which then becomes

$$N_{zi} = \frac{N_z A_i}{\Sigma A_i} + \frac{M_x A_i y_i}{\Sigma A_i y_i^2} - \frac{M_y A_i x_i}{\Sigma A_i x_i^2}. \tag{9.51}$$

This result demonstrates that the connector force can be obtained by super-position of the separate components due to the centroidal force N_z and the principal axis moments M_x, M_y. When there are n bolts of equal area, this equation simplifies to

$$N_{zi} = \frac{N_z}{n} + \frac{M_x y_i}{\Sigma y_i^2} - \frac{M_y x_i}{\Sigma x_i^2}. \tag{9.51a}$$

When the connectors are welds, the summations in equation 9.47 should be replaced by integrals, so that

$$f_w = \frac{N_z}{A} + \frac{M_x y}{I_x} - \frac{M_y x}{I_y} \tag{9.51b}$$

in which x, y, are the coordinates of the weld where the stress is f_w.

The assumption that only the connectors transfer compression forces through the connection may be unrealistic when portions of the plates remain in contact. In this case the analysis above may still be used, provided that the values A_i, x_i, y_i used for the compression regions represent the actual contact areas between the plates.

9.9 Worked examples

9.9.1 EXAMPLE 1 – IN-PLANE ANALYSIS OF A BOLT GROUP

Problem.   The semi-rigid web side plate (fin plate) connection shown in Fig. 9.19a is to transmit factored design actions equivalent to a vertical down-wards force of Q^* kN acting at the centroid of the bolt group and a clockwise moment of $0.2Q^*$ kNm. Determine the maximum bolt shear force.

Solution.   For the bolt group

$$\Sigma(x_i^2 + y_i^2) = 4 \times (70^2 + 105^2) + 4 \times (70^2 + 35^2)$$
$$= 88\,200 \text{ mm}^2.$$

Using equation 9.41a

$$x_r = -\frac{(Q^* \times 10^3) \times 88\,200/8}{(-0.2Q^* \times 10^6)} \text{ mm} = 55.1 \text{ mm},$$

and so the most heavily loaded bolts are at the top and bottom of the right hand row. For these

$$r_i = \sqrt{\{(55.1 + 70)^2 + 105^2\}} = 163.3 \text{ mm},$$

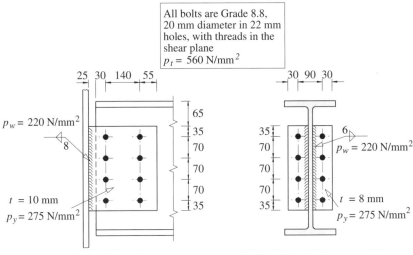

All bolts are Grade 8.8,
20 mm diameter in 22 mm
holes, with threads in the
shear plane
$p_t = 560$ N/mm^2

(a) Semi-rigid web fin plate connection   (b) Flexible end plate connection

Fig. 9.19 Examples 1–9

and the maximum bolt force may be obtained from equation 9.3 as

$$F_s^* = \frac{(-0.2Q^* \times 10^6) \times 163.3}{88\,200} \text{ N} = -0.370Q^* \text{ kN}.$$

9.9.2 EXAMPLE 2 – IN-PLANE DESIGN CAPACITY OF A BOLT GROUP

Problem.   Determine the design load capacity of the bolt group in the web fin plate connection shown in Fig. 9.19a.

Solution.   For 20 mm Grade 8.8 bolts, $A_s = A_t = 245$ mm^2
and $p_s = 375$ N/mm^2 T30
and so $P_s = 375 \times 245$ N $= 91.9$ kN   6.3.2.1
    Using this with the solution of example 1,

$$0.370Q^* \leqslant 91.9$$

so that $Q^* \leqslant 248.3$

9.9.3 EXAMPLE 3 – PLATE BEARING CAPACITY

Problem.   Determine the bearing load capacity of the web fin plate shown in Fig. 9.19a.

Bolt bearing capacity.

$$p_{bb} = 1000 \text{ N/mm}^2$$ T31
$$P_{bb} = 20 \times 10 \times 1000 \text{ N} = 200 \text{ kN}$$ 6.3.3.2

Plate bearing capacity.

$$k_{bs} = 1.0 \qquad \qquad \text{6.3.3.3}$$
$$p_{bs} = 460 \text{ N/mm}^2 \qquad \qquad \text{T32}$$
$$P_{bs} = 1.0 \times 20 \times 10 \times 460 \text{ N} = 92.0 \text{ kN} \leqslant 200 \text{ kN} = P_{bb} \qquad \text{6.3.3.3}$$

Using this with the solution of example 1, $Q_b^* \leqslant 92.0/0.370 = 248.6$ which is almost the same as $Q_s^* \leqslant 248.3$ for the bolt group in example 2.

9.9.4 EXAMPLE 4 – PLATE SHEAR AND TENSION CAPACITY

Problem. Determine the shear and tension capacity of the web fin plate shown in Fig. 9.19a.

Solution. If the yield criterion of equation 9.20 is used with the elastic stresses on the gross cross-section of the plate, then the elastic bending stress may be determined using an elastic section modulus of $bd^2/6$, so that

$$f_b^* = 0.2Q^* \times 10^6 \times 6/(10 \times 280^2) \text{ N/mm}^2 = 1.53Q^* \text{ N/mm}^2$$

and the average shear stress is

$$\tau_s^* = Q^* \times 10^3/(280 \times 10) \text{ N/mm}^2 = 0.357Q^* \text{ N/mm}^2.$$

Substituting into equation 9.20, $(1.53Q^*)^2 + 3 \times (0.357Q^*)^2 \leqslant 275^2$ so that $Q_{sty}^* \leqslant 166.6$.
   If the fracture criterion of equation 9.29 is used with the elastic stresses on the net cross-section of the plate, then

$$I_p = 280^3 \times 10/12 - 2 \times 22 \times 10 \times 105^2 - 2 \times 22 \times 10 \times 35^2 = 12.90 \times 10^6 \text{ mm}^4$$

so that

$$Z_p = 12.90/140 = 92.17 \times 10^3 \text{ mm}^3, \text{ and}$$
$$f_b^* = 0.2Q^* \times 10^6/(92.17 \times 10^3) \text{ N/mm}^2 = 2.17Q^* \text{ N/mm}^2$$

and the average shear stress is

$$\tau_s^* = Q^* \times 10^3/\{(280 - 4 \times 22) \times 10\} \text{ N/mm}^2 = 0.521Q^* \text{ N/mm}^2$$

Using $K_e = 1.2$ \qquad \qquad 3.4.3
and substituting into equation 9.29,

$$(2.17Q^*)^2 + 3 \times (0.521Q^*)^2 \leqslant (1.2 \times 275)^2$$

so that

$$Q_{stf}^* \leqslant 140.4 < 166.6 (\geqslant Q_{sty}^*) < 248.3 (\geqslant Q_s^*).$$

However, these calculations are conservative, since the maximum bending stresses occur at the top and bottom of the plate, where the shear stresses are zero. If the shear stresses are ignored, then $Q_y^* = 275/1.53 = 179.7$ and $Q_f^* = 1.2 \times 275/2.17 = 152.1$.

9.9.5 EXAMPLE 5 – FILLET WELD CAPACITY

Problem. Determine the load capacity of the two fillet welds shown in Fig. 9.19a.

Solution. At the welds, the design actions consist of a vertical shear of Q^* kN and a moment of

$$(-0.2Q^* - Q^* \times (25 + 30 + 70)/1000) \text{ kNm} = -0.325Q^* \text{ kNm}.$$

The average shear force per unit weld length can be determined from equation 9.51b as

$$F_L^*/L = (Q^* \times 10^3)/(2 \times 280) \text{ N/mm} = 1.79Q^* \text{ N/mm},$$

and the maximum bending force per unit weld length from equation 9.51b also as

$$\frac{F_T^*}{L} = \frac{(-0.325Q^* \times 10^6) \times 140}{2 \times 280^3/12} \text{ N/mm} = -12.44Q^* \text{ N/mm}.$$

For Class 35 electrodes, $p_w = 220$ N/mm^2 T37
  Substituting into equation 9.36,

$$\sqrt{\{(12.44Q^*)^2 + (1.79Q^*)^2\}/(8/\sqrt{2})} \leqslant 220$$ 6.8.7.2

so that

$$Q_w^* \leqslant 99.0 < 152.1 (\geqslant Q_f^*)$$

and so the connection capacity is governed by the shear and bending capacity of the welds.

9.9.6 EXAMPLE 6 – BOLT SLIP

Problem. If the bolts of the semi-rigid web fin plate connection shown in Fig. 9.19a are preloaded, then determine the value of Q at which the first bolt slip occurs, if the connection is designed to be non-slip in service.

Solution.

$$K_s = 1.0$$ 6.4.2
$$\mu = 0.4$$ T35
$$P_o = 144 \text{ kN}$$ BS4604[9]
$$P_{sL} = 1.1 \times 1.0 \times 0.4 \times 144 \text{ kN} = 63.4 \text{ kN}$$ 6.4.2

and using the solution of example 1,

$$Q_{sL} = 63.4/0.370 = 171.2$$

9.9.7 EXAMPLE 7 – OUT-OF-PLANE DESIGN CAPACITY OF A BOLT GROUP

Problem. The flexible end plate connection shown in Fig. 9.19b is to transmit factored design actions equivalent to a downwards force of Q^* kN acting at the centroid of the bolt group and an out-of-plane moment of $M_x = -0.05Q^*$ kNm. Determine the maximum bolt forces, and the design capacity of the bolt group.

Analysis. For the bolt group

$$\Sigma y_i^2 = 4 \times 105^2 + 4 \times 35^2 \text{ mm}^2 = 49\,000 \text{ mm}^2,$$

and using equation 9.51a, the maximum bolt tension is

$$F_t^* = \frac{(-0.05Q^* \times 10^6) \times (-35 - 70)}{49\,000} \text{ N} = 0.107Q^* \text{ kN}.$$

The average bolt shear is

$$F_s^* = (Q^* \times 10^3)/8 \text{ N} = 0.125Q^* \text{ kN}.$$

Capacity. Using the solution of example 2, $P_s = 91.9$ kN.
For 20 mm Grade 8.8 bolts, $A_t = 245$ mm^2 and

$p_t = 560$ N/mm^2	T34
$P_{nom} = 0.8 \times 560 \times 245$ N $= 109.8$ kN	6.3.4.2
$0.125Q^*/91.9 + 0.107Q^*/109.8 \leqslant 1.4$	6.3.4.4

so that

$$Q_{st}^* \leqslant 599.6$$

9.9.8 EXAMPLE 8 – PLATE BEARING CAPACITY

Problem. Determine the bearing load capacity of the end plate of example 7 shown in Fig. 9.19b.

Solution.

$k_{bs} = 1.0$	6.3.3.3
$p_{bs} = 460$ N/mm^2	T32
$k_{bs}d t_p p_{bs} = 1.0 \times 20 \times 8 \times 460$ N $= 73.6$ kN	6.3.3.3
$0.5k_{bs}e t_p p_{bs} = 0.5 \times 1.0 \times 35 \times 8 \times 460$ N $= 64.4$ kN < 73.6 kN	6.3.3.3

and so

$$P_{bs} = 64.4 \text{ kN}.$$

Using this with the analysis solution of example 7

$$0.125Q^* \leqslant 64.4$$

so that

$$Q_b^* \leqslant 515.2 < 599.6 (\geqslant Q_{st}^*)$$

9.9.9 EXAMPLE 9 – FILLET WELD CAPACITY

Problem.   Determine the load capacity of the two fillet welds shown in Fig. 9.19b.

Solution.   The average shear force per unit weld length can be determined from equation 9.51b as

$$F_L^*/L = (Q^* \times 10^3)/(2 \times 280) \text{ N/mm} = 1.79Q^* \text{ N/mm},$$

and the maximum bending force per unit weld length from equation 9.51b also as

$$\frac{F_T^*}{L} = \frac{(-0.05Q^* \times 10^6) \times 140}{2 \times 280^3/12} \text{ N/mm} = 1.91Q^* \text{ N/mm}.$$

Substituting into equation 9.36 with $p_w = 220 \text{ N/mm}^2$ for E35 electrodes   T37

$$\sqrt{\{(1.91Q^*)^2 + (1.79Q^*)^2\}/(6/\sqrt{2})} \leqslant 220 \qquad\qquad 6.8.7.2$$

so that

$$Q_w^* \leqslant 356.6 < 512.2 (\geqslant Q_b^*)$$

and so the connection capacity is governed by the shear and bending capacity of the welds.

9.10 Unworked examples

9.10.1 EXAMPLE 10 – BOLTED CONNECTION

Arrange and design a bolted connection to transfer a design force of 800 kN from the inclined member to the truss joint shown in Fig. 9.20a.

9.10.2 EXAMPLE 11 – TENSION MEMBER SPLICE

Arrange and design the tension member splice shown in Fig. 9.20b so as to maximize the tension capacity.

9.10.3 EXAMPLE 12 – TENSION MEMBER WELDS

Proportion the weld lengths shown in Fig. 9.20c to transfer the design tension force concentrically.

9.10.4 EXAMPLE 13 – CHANNEL SECTION BRACKET

Determine the design load capacity Q^* of the fillet welded bracket shown in Fig. 9.20d.

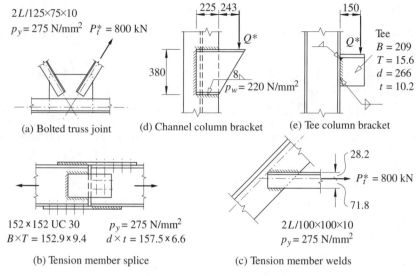

2L/125×75×10
p_y = 275 N/mm² P_t^* = 800 kN

(a) Bolted truss joint

(d) Channel column bracket

(e) Tee column bracket

Tee
B = 209
T = 15.6
d = 266
t = 10.2

152 × 152 UC 30 p_y = 275 N/mm²
B×T = 152.9 × 9.4   d × t = 157.5 × 6.6

(b) Tension member splice

2L/100×100×10
p_y = 275 N/mm²

(c) Tension member welds

Fig. 9.20 Examples 10–14

9.10.5 EXAMPLE 14 – TEE SECTION BRACKET

Design the fillet welds for the bracket shown in Fig. 9.20e.

9.11 References

1. The Steel Construction Institute and the British Constructional Steelwork Association (2001) *Joints in Steel Construction: Simple Connections*, SCI and BCSA, Ascot.
2. The Steel Construction Institute and the British Constructional Steelwork Association (1995) *Joints in Steel Construction: Moment Connections*, SCI and BCSA, Ascot.
3. Owens, G.W. and Cheal, B.D. (1989) *Structural Steelwork Connections*, Butterworths, London.
4. Owens, G.W. and Knowles, P.R. (eds) (1992) *Steel Designer's Manual*, 5th edition, Blackwell Scientific Publications, Oxford.
5. Morris, L.J. (1988) Connection design in the UK, *Steel Beam-to-Column Building Connections*, (ed. Chen, W.F.), Elsevier Applied Science, pp. 375–414.
6. Kulak, G.L., Fisher, J.W. and Struik, J.H.A. (1987) *Guide to Design Criteria for Bolted and Rivetted Joints*, 2nd edition, John Wiley and Sons, New York.
7. Harrison, H.B. (1980) *Structural Analysis and Design, Parts 1 and 2*, Pergamon Press, Oxford.
8. Nethercot, D.A. (2001) *Limit States Design of Structural Steelwork*, 3rd edition, E and FN Spon, London.
9. British Standards Institution (1970), *BS4604 Specification for the Use of High Strength Friction Grip Bolts in Structural Steelwork*, BSI, London.

10 Torsion members

10.1 Introduction

The resistance of a structural member to torsional loading may be considered to be the sum of two components. When the rate of change of the angle of twist rotation is constant along the member (see Fig. 10.1a), it is in a state of uniform torsion [1, 2], and the longitudinal warping deflections are also constant along the member. In this case, the torque acting at any cross-section is resisted by a single set of shear stresses distributed around the cross-section. The ratio of the torque acting to the twist rotation per unit length is defined as the torsional rigidity GJ of the member.

The second component of the resistance to torsional loading may act when the rate of change of the angle of twist rotation varies along the member (see Fig. 10.1b and c), so that it is in a state of non-uniform torsion [1]. In this case the warping deflections vary along the member, and an additional set of shear stresses may act in conjunction with those due to uniform torsion to resist the torque acting. The stiffness of the member associated with these additional shear stresses is proportional to the warping rigidity EI_w.

When the first component of the resistance to torsional loading completely dominates the second, the member is in a state of uniform torsion. This occurs when the torsion parameter $K = \sqrt{(\pi^2 EI_w/GJL^2)}$ is very small, as indicated in Fig. 10.2 which is adapted from [1]. Thin-walled closed section members, whose torsional rigidities are very large, behave in this way, as do members with narrow rectangular sections, and angle and tee sections, whose warping rigidities are negligible. If, on the other hand, the second component of the resistance to torsional loading completely dominates the first, the member is in a limiting state of non-uniform torsion referred to as warping torsion. This may occur when the torsion parameter K is very large, as indicated in Fig. 10.2, which is the case for some very thin-walled open sections (such as light gauge cold-formed sections) whose torsional rigidities are very small. Between these

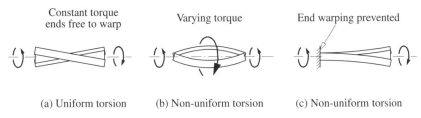

Constant torque ends free to warp	Varying torque	End warping prevented
(a) Uniform torsion	(b) Non-uniform torsion	(c) Non-uniform torsion

Fig. 10.1 Uniform and non-uniform torsion of an I-section member

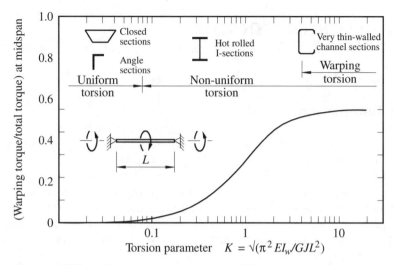

Fig. 10.2 Effect of cross-section on torsional behaviour

two extremes, the torsional loading is resisted by a combination of the uniform and warping torsion components, and the beam is in the general state of non-uniform torsion. This occurs for intermediate values of the parameter K, as shown in Fig. 10.2, which are appropriate for most hot-rolled I- or channel section members.

Whether a member is in a state of uniform or non-uniform torsion also depends on the loading arrangement and the warping restraints. If the torque resisted is constant along the member and warping is unrestrained (as in Fig. 10.1a), then the member will be in uniform torsion, even if the torsional rigidity is very small. If, however, the torque resisted varies along the length of the member (Fig. 10.1b), or if the warping displacements are restrained in any way (Fig. 10.1c), then the rate of change of the angle of twist rotation will vary, and the member will be in non-uniform torsion. In general, these variations must be accounted for, but in some cases they can be ignored, and the member analysed as if it were in uniform torsion. This is the case for members of very low warping rigidity and for members of high torsional rigidity, for which the rates of change of the angle of twist rotation vary only locally near points of concentrated torque and warping restraint. This simple method of analysis usually leads to satisfactory predictions of the angles of twist rotation, but may produce underestimates of the local stresses.

In this chapter, the behaviour and design of torsion members are discussed. Uniform torsion is treated in section 10.2, and non-uniform torsion in section 10.3, while the design of torsion members for strength and serviceability is treated in section 10.4.

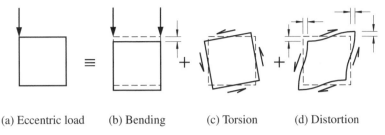

(a) Eccentric load   (b) Bending   (c) Torsion   (d) Distortion

Fig. 10.3 Eccentric loading of a thin-walled closed section

Structural members, however, are rarely used to resist torsion alone, and it is much more common for torsion to occur in conjunction with bending and other effects. For example, when the box section beam shown in Fig. 10.3a is subjected to an eccentric load, this causes bending (Fig. 10.3b), torsion (Fig. 10.3c), and distortion (Fig. 10.3d). There may also be interactions between bending and torsion. For example, the longitudinal stress due to bending may cause a change in the effective torsional rigidity. This type of interaction is of some importance in the flexural–torsional buckling of thin-walled open section members, as discussed in sections 3.6.5 and 6.7. Also, significant twist rotations of the member may cause increases in the bending stresses in thin-walled open section members. However, these interactions are usually negligible when the torsional rigidity is very high, as in thin-walled closed section members. The behaviour of members under combined bending and torsion is discussed in section 10.5.

The effects of distortion of the cross-section are only significant in very thin-walled open sections, and in thin-walled closed sections with high distortional loadings. Thus, for the box section beam shown in Fig. 10.3, the relatively flexible plates may bend out of their planes as indicated in Fig. 10.3d, causing the cross-section to distort. Due to this, the in-plane plate bending and shear stress distributions are changed, and significant out-of-plane plate bending stresses may be induced. The distortional behaviour of structural members is briefly discussed in section 10.6.

10.2 Uniform torsion

10.2.1 ELASTIC DEFORMATIONS AND STRESSES

10.2.1.1 General

In elastic uniform torsion, the twist rotation per unit length is constant along the length of the member. This occurs when the torque is constant and the ends of the member are free to warp, as shown in Fig. 10.1a. When a member twists in this way:

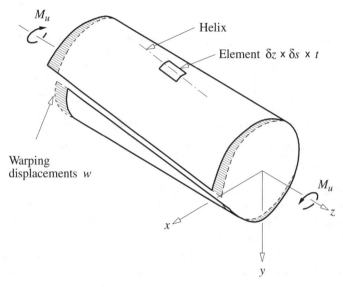

Fig. 10.4 Warping displacements *w* due to twisting

(a) lines which were originally parallel to the axis of twist become helices,
(b) cross-sections rotate ϕ as rigid bodies about the axis of twist,
(c) cross-sections warp out of their planes, the warping deflections *w* being
    constant along the length of the member (see Fig. 10.4).

The uniform torque M_u acting at any section induces shear stresses τ_{zx}, τ_{zy}
which act in the plane of the cross-section. The distribution of these shear
stresses can be visualized by using Prandtl's membrane analogy, as indicated in
Fig. 10.5 for a rectangular cross-section. Imagine a thin uniformly stretched
membrane which is fixed to the cross-section boundary and displaced by a
uniform transverse pressure. The contours of the displaced membrane coincide
with the shear stress trajectories, the slope of the membrane is proportional to
the shear stress, and the volume under the membrane is proportional to the
torque. The shear strains caused by these shear stresses are related to the twisting and warping deformations, as indicated in Fig. 10.6.

The stress distribution can be found by solving the equation [2]

$$\frac{\partial^2 \theta}{\partial x^2} + \frac{\partial^2 \theta}{\partial y^2} = -2G \frac{d\phi}{dz} \tag{10.1}$$

(where *G* is the shear modulus of elasticity) for the stress function θ (which
corresponds to the membrane displacement in Prandtl's analogy) defined by

$$\left.\begin{aligned} \tau_{zy} = \tau_{yz} = -\partial\theta/\partial x \\ \tau_{zx} = \tau_{xz} = \partial\theta/\partial y \end{aligned}\right\} \tag{10.2}$$

Sections through membrane

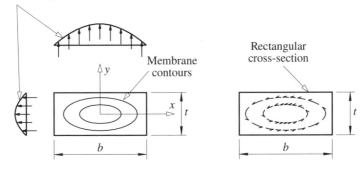

Membrane
contours

Rectangular
cross-section

(a) Transversely loaded membrane   (b) Distribution of shear stress

Fig. 10.5 Prandtl's membrane analogy for uniform torsion

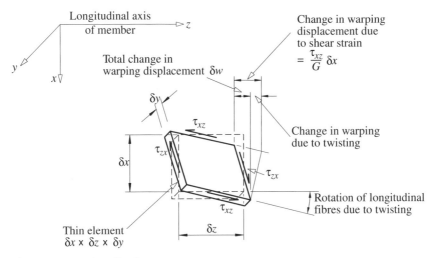

Longitudinal axis
of member

Change in warping
displacement due
to shear strain

$= \dfrac{\tau_{xz}}{G}\,\delta x$

Total change in
warping displacement δw

Change in warping
due to twisting

Rotation of longitudinal
fibres due to twisting

Thin element
$\delta x \times \delta z \times \delta y$

Fig. 10.6 Warping displacements

subject to the condition that

$$\theta = \text{constant}, \tag{10.3}$$

around the section boundary. The uniform torque M_u, which is the static equivalent of the shear stresses τ_{zx}, τ_{zy}, is given by

$$M_u = 2 \iint_{\text{section}} \theta \, dx \, dy. \tag{10.4}$$

10.2.1.2 Solid cross-sections

Closed form solutions of equations 10.1–10.3 are not generally available, except for some very simple cross-sections. For a solid circular section of radius R, the solution is

$$\theta = \frac{G}{2} \frac{d\phi}{dz} (R^2 - x^2 - y^2). \tag{10.5}$$

If this is substituted in equations 10.2, the circumferential shear stress τ_z at a radius $r = \sqrt{(x^2 + y^2)}$ is found to be

$$\tau_z = Gr \frac{d\phi}{dz}. \tag{10.6}$$

The torque effect of these shear stresses is

$$M_u = \int_0^R \tau_z (2\pi r) dr, \tag{10.7}$$

which can be expressed in the form

$$M_u = GJ \frac{d\phi}{dz}, \tag{10.8}$$

where the torsion section constant J is given by

$$J = \pi R^4 / 2, \tag{10.9}$$

which can also be expressed as

$$J = \frac{A^4}{4\pi^2 (I_x + I_y)}. \tag{10.10}$$

The same result can be obtained by substituting equation 10.5 directly into equation 10.4. Equation 10.6 indicates that the maximum shear stress occurs at the boundary and is equal to

$$\tau_m = \frac{M_u R}{J}. \tag{10.11}$$

For other solid cross-sections, equation 10.10 gives a reasonably accurate approximation for the torsion section constant, but the maximum shear stress depends on the precise shape of the section. When sections have re-entrant corners, the local shear stresses may be very large (as indicated in Fig. 10.7), although they can be reduced by increasing the radius of the fillet at the re-entrant corner.

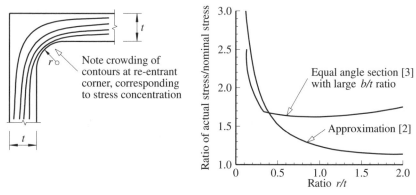

(a) Membrane contours at re-entrant corner

Note crowding of contours at re-entrant corner, corresponding to stress concentration

(b) Increased shear stresses

Equal angle section [3] with large b/t ratio

Approximation [2]

Fig. 10.7 Stress concentrations at re-entrant corners

10.2.1.3 Rectangular cross-sections

The stress function θ for a very narrow rectangular section of width b and thickness t is approximated by

$$\theta \approx G \frac{d\phi}{dz}\left(\frac{t^2}{4} - y^2\right). \tag{10.12}$$

The shear stresses can be obtained by substituting equation 10.12 into equations 10.2, whence

$$\tau_{zx} = -2y\, G\, \frac{d\phi}{dz}, \tag{10.13}$$

which varies linearly with y as shown in Fig. 10.8c. The torque effect of the τ_{zx} shear stresses can be obtained by integrating their moments about the z axis, whence

$$-\int_{-t/2}^{t/2} \tau_{zx} by\, dy = G\, \frac{bt^3}{6}\, \frac{d\phi}{dz}, \tag{10.14}$$

which is one half of the total torque effect. The other half arises from the shear stresses τ_{zy}, which have their greatest effects near the ends of the thin

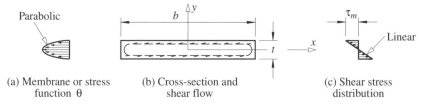

Parabolic

(a) Membrane or stress function θ

b

y

t

x

(b) Cross-section and shear flow

τ_m

Linear

(c) Shear stress distribution

Fig. 10.8 Uniform torsion of a thin rectangular section

rectangle. Although they act there over only a small length of the order of t (compared with b for the τ_{zx} stresses), they have a large lever arm of the order of $b/2$ (instead of $t/2$), and they make an equal contribution. The total torque can therefore be expressed in the form of equation 10.8 with

$$J \approx bt^3/3. \tag{10.15}$$

The same result can be obtained by substituting equation 10.12 directly into equation 10.4. Equation 10.13 indicates that the maximum shear stress occurs at the centre of the long boundary, and is given by

$$\tau_m \approx \frac{M_u t}{J}. \tag{10.16}$$

For stockier rectangular sections, the stress function θ varies significantly along the width b of the section, as indicated in Fig. 10.5, and these approximations may not be sufficiently accurate. In this case, the torsion section constant J and the maximum shear stress τ_m can be obtained from Fig. 10.9.

10.2.1.4 Thin-walled open cross-sections

The stress distributions in thin-walled open cross-sections are very similar to that in a narrow rectangular section, as shown in Fig. 10.10a. Thus, the shear stresses are parallel to the walls of the section, and vary linearly across the thickness t, this pattern remaining constant around the section except at the

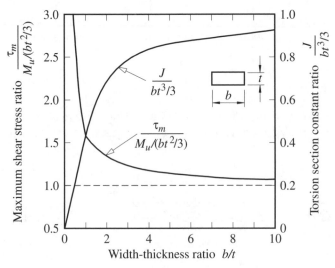

Fig. 10.9 Torsion properties of rectangular sections

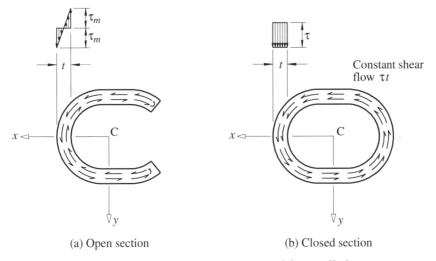

| (a) Open section | (b) Closed section |

Fig. 10.10 Shear stresses due to uniform torsion of thin-walled sections

ends. Because of this similarity, the torsion section constant J can be closely approximated by

$$J \approx \Sigma bt^3/3, \tag{10.17}$$

where b is the developed length of the midline and t the thickness of each thin-walled element of the cross-section, while the summation is carried out for all such elements.

The maximum shear stress away from the concentrations at re-entrant corners can also be approximated as

$$\tau_{m} \approx \frac{M_u t_m}{J}, \tag{10.18}$$

where t_m is the maximum thickness. When more accurate values for the torsion section constant J are required, the formulae developed in [3] can be used. The values given in [4–6] for British hot-rolled steel sections have been calculated in this way.

10.2.1.5 Thin-walled closed cross-sections

The uniform torsional behaviour of thin-walled closed section members is quite different from that of open section members, and there are dramatic increases in the torsional stiffness and strength. The shear stress no longer varies linearly across the thickness of the wall, but is constant as shown in Fig. 10.10b, and there is a constant shear flow around the closed section.

This shear flow is required to prevent any discontinuities in the longitudinal warping displacements of the closed section. To show this, consider the slit

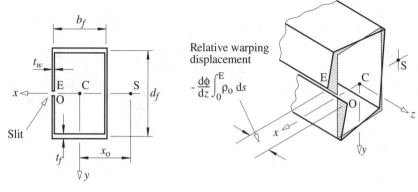

(a) Cross-section of a slit tube   (b) Warping displacements

Fig. 10.11 Warping of a slit tube

rectangular tube shown in Fig. 10.11a. The mid-thickness surface of this open section is unstrained, and so the warping displacements of this surface are entirely due to the twisting of the member. The distribution of the warping displacements caused by twisting (see section 10.3.1.2) is shown in Fig. 10.11b. The relative warping displacement at the slit is equal to

$$w_E - w_0 = \frac{-d\phi}{dz} \int_0^E \rho_0 \, ds \tag{10.19}$$

in which ρ_0 is the distance (see Fig. 10.12a) from the tangent at the mid-thickness line to the axis of twist (which passes through the shear centre of the section, as discussed in sections 10.3.1.2 and 5.4.3).

However, when the tube is not slit, this relative warping displacement cannot occur. In this case, the relative warping displacement due to twisting shown in Fig. 10.13a is exactly balanced by the relative warping displacement

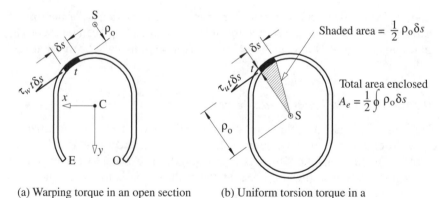

(a) Warping torque in an open section   (b) Uniform torsion torque in a
                                                                        closed section

Fig. 10.12 Torques due to shear flows in thin-walled sections

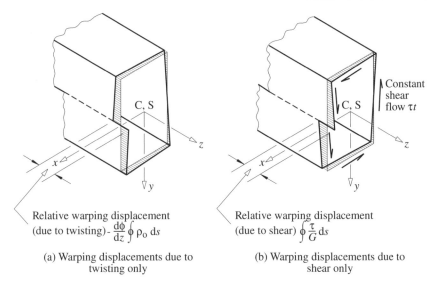

Relative warping displacement
(due to twisting)- $\dfrac{d\phi}{dz}\displaystyle\oint \rho_0\, ds$

Relative warping displacement
(due to shear) $\displaystyle\oint \dfrac{\tau}{G}\, ds$

(a) Warping displacements due to
twisting only

(b) Warping displacements due to
shear only

Fig. 10.13 Warping of a closed section

caused by shear straining (see Fig. 10.6) of the mid-thickness surface shown in Fig. 10.13b, so that the total relative warping displacement is zero. It is shown in section 10.7.1 that the required shear straining is produced by a constant shear flow τt around the closed section which is given by

$$\tau t = \frac{M_{\mathrm{u}}}{2A_{\mathrm{e}}}, \tag{10.20}$$

where A_{e} is the area enclosed by the section, and that the torsion section constant is given by

$$J = \frac{4A_{\mathrm{e}}^2}{\oint(1/t)ds}. \tag{10.21}$$

The membrane analogy can also be used for thin-walled closed sections, by imagining that the inner section boundary is fixed to a weightless horizontal plate supported at a distance equivalent to τt above the outer section boundary by the transverse pressures and the forces in the membrane stretched between the boundaries, as shown in Fig. 10.14b. Thus the slope $(\tau t)/t$ of the membrane is substantially constant across the wall thickness, and is equivalent to the shear stress τ, while twice the volume $A_{\mathrm{e}}\tau t$ under the membrane is equivalent to the uniform torque M_{u} given by equation 10.20.

The membrane analogy is used in Fig. 10.14 to illustrate the dramatic increases in the stiffnesses and strengths of thin-walled closed sections over those of open sections. It can be seen that the torque (which is proportional to the volume under the membrane) is much greater for the closed section when the maximum shear stress is the same. The same conclusion can be reached by

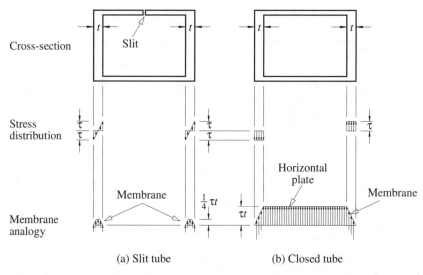

Cross-section

Stress
distribution

Membrane
analogy

(a) Slit tube   (b) Closed tube

Fig. 10.14 Comparison of uniform torsion of open and closed sections

considering the effective lever arms of the shear stresses, which are of the same
order as the overall dimensions of the closed section, compared with those of
the same order as the wall thickness of the open section.

The behaviour of multi-cell closed section members in uniform torsion can be
determined by using the warping displacement continuity condition for each cell,
as indicated in section 10.7.1. A general matrix method of carrying out this analysis
by computer has been given in [7], and a computer program has been developed
[8]. Alternatively, the membrane analogy can be extended by considering a set of
horizontal plates at different heights, one for each cell of the cross-section.

10.2.2 ELASTIC ANALYSIS

10.2.2.1 Statically determinate members

Some members in uniform torsion are statically determinate, such as the cantilever shown in Fig. 10.15. In these cases, the distribution of the total torque
$M_z = M_u$ can be determined from statics, and the maximum stresses can be
evaluated from equation 10.11, 10.16, 10.18 or 10.20, or from Fig. 10.7 or 10.9.
The twisted shape of the member can be determined by integrating equation
10.8, using the sign conventions of Fig. 10.16a. Thus, the angle of twist rotation
in a region of constant torque M_z will vary linearly in accordance with

$$\phi = \phi_0 + \frac{M_z z}{GJ}, \tag{10.22}$$

as indicated in Fig. 10.15c.

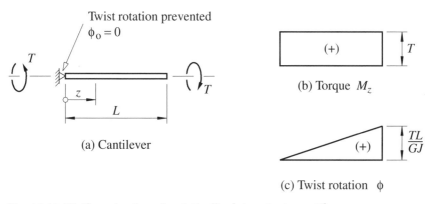

(a) Cantilever

(b) Torque M_z

(c) Twist rotation ϕ

Fig. 10.15 Uniform torsion of a statically determinate cantilever

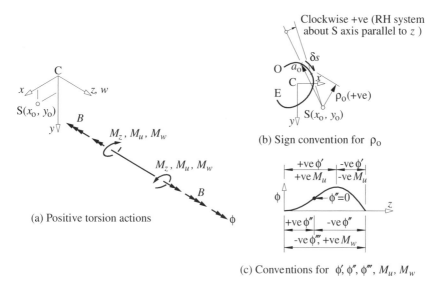

(a) Positive torsion actions

(b) Sign convention for ρ_o

(c) Conventions for ϕ', ϕ'', ϕ''', M_u, M_w

Fig. 10.16 Torsion sign conventions

10.2.2.2 Statically indeterminate members

Many torsion members are statically indeterminate, such as the beam shown in Fig. 10.17, and the distribution of torque cannot be determined from statics alone. The redundant quantities can be determined by substituting the corresponding compatibility conditions into the solution of equation 10.8. Once these have been found, the maximum torque can be evaluated by statics, and the maximum stress can be determined. The maximum angle of twist can also be derived from the solution of equation 10.8.

As an example of this procedure, the indeterminate beam shown in

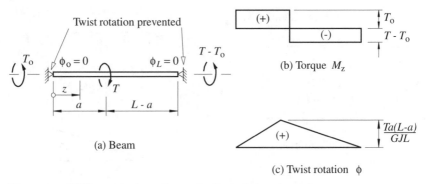

Fig. 10.17 Uniform torsion of a statically indeterminate beam

Fig. 10.17 is analysed in section 10.7.2. The theoretical twisted shape of this beam is shown in Fig. 10.17c, and it can be seen that there is a jump discontinuity in the twist per unit length $d\phi/dz$ at the loaded point. This discontinuity is a consequence of the use of the uniform torsion theory to analyse the behaviour of the member. In practice such a discontinuity does not occur, because an additional set of local warping stresses is induced. These additional stresses, which may have high values at the loaded point, have been investigated in [9, 10]. High local stresses can usually be tolerated in static loading situations when the material is ductile, whether the stresses are induced by concentrated torques or by other concentrated loads. In members for which the use of the uniform torsion theory is appropriate, such as those with high torsional rigidity and low warping rigidity, these additional stresses decrease rapidly as the distance from the loaded point increases. Because of this, the angles of twist predicted by the uniform torsion analysis are of sufficient accuracy.

10.2.3 PLASTIC COLLAPSE ANALYSIS

10.3.2.1 Fully plastic stress distribution

The elastic shear stress distributions described in the previous subsections remain valid while the shear yield stress τ_y is not exceeded. The uniform torque at nominal first yield M_{uY} may be obtained from equation 10.11, 10.16 or 10.18 by using $\tau_m = \tau_y$. Yielding commences at M_{uY} at the most highly stressed regions, and then spreads until the section is fully plastic.

The fully plastic shear stress distribution can be visualized by using the 'sand heap' modification of the Prandtl membrane analogy. Because the fully plastic shear stress distribution is constant, the slope of the Prandtl membrane is also constant, and its contours are equally spaced in the same way as are those of a heap formed by pouring sand on to a base area of the same shape as the member cross-section until the heap is fully formed.

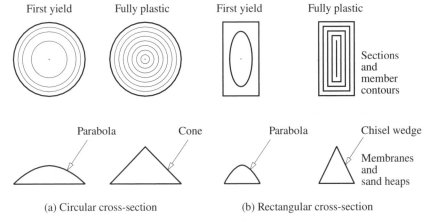

(a) Circular cross-section   (b) Rectangular cross-section

Fig. 10.18 First yield and fully plastic uniform torsion shear stress distributions

This is demonstrated in Fig. 10.18a for a circular cross-section for which the sand heap is conical. Its fully plastic uniform torque may be obtained from

$$M_{up} = \int_0^R \tau_y (2\pi r) r \; dr, \tag{10.23}$$

whence

$$M_{up} = (2\pi R^3/3)\tau_y. \tag{10.24}$$

The corresponding first yield uniform torque may be obtained from equations 10.9 and 10.11 as

$$M_{uY} = (\pi R^3/2)\tau_y, \tag{10.25}$$

and so the uniform torsion plastic shape factor is $M_{up}/M_{uY} = 4/3$.

The sand heap for a fully plastic rectangular section $b \times t$ is chisel shaped as shown in Fig. 10.18b. In this case the fully plastic uniform torque is given by

$$M_{up} = \frac{bt^2}{2}\left(1 - \frac{t}{3b}\right)\tau_y, \tag{10.26}$$

and for a narrow rectangular section, this becomes $M_{up} \approx (bt^2/2)\tau_y$. The corresponding first yield uniform torque is

$$M_{uY} \approx (bt^2/3)\tau_y, \tag{10.27}$$

and so the uniform torsion plastic shape factor is $M_{up}/M_{uY} = 3/2$.

For a fully plastic I-section, the fully plastic uniform torque is

$$M_{up} = (b_f t_f^2 - t_f^3/3 + b_w t_w^2/2 + t_w^3/6)\tau_y. \tag{10.28}$$

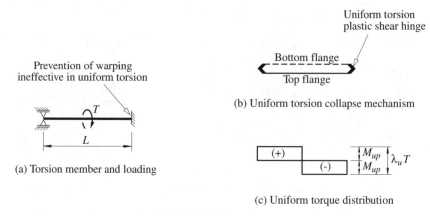

(a) Torsion member and loading

(b) Uniform torsion collapse mechanism

(c) Uniform torque distribution

Fig. 10.19 Uniform torsion plastic collapse

10.2.3.2 Plastic analysis

A member in uniform torsion will collapse plastically when there is a sufficient number of fully plastic cross-sections to transform the member into a mechanism, as shown for example in Fig. 10.19. Each fully plastic cross-section can be thought of a shear 'hinge' which allows increasing torsional rotations under the uniform plastic torque M_{up}. The plastic shear hinges usually form at the supports where there are reaction torques, as indicated in Fig. 10.19. In general, the plastic shear hinges develop progressively until the collapse mechanism is formed. Examples of uniform torsion plastic collapse mechanisms are shown in Figs 10.20 and 10.21.

At plastic collapse, the mechanism is statically determinate, and can be analysed by equilibrium to determine the plastic collapse torques. For the member shown in Fig. 10.19, the mechanism forms when there are plastic shear hinges at each support. The total applied torque $\lambda_u T$ at plastic collapse must be in equilibrium with these plastic uniform torques, and so

$$\lambda_u T = 2M_{up}, \tag{10.29}$$

so that the uniform torsion plastic collapse load factor is given by

$$\lambda_u = 2M_{up}/T. \tag{10.30}$$

Values of the uniform torsion plastic collapse load factor λ_u for a number of example torsion members can be obtained from Figs 10.20 and 10.21.

Member and Loading	Plastic Collapse				Rotations	
	Uniform Torsion		Warping Torsion		Uniform	Warping
	Mechanism	$\lambda_u T/M_{up}$	Mechanism	$\lambda_w TL/M_{fp}\,d_f$	$\phi_{um}GJ/TL$	$\phi_{wm}EI_w/TL^3$
		1.0		0	1.0	∞
		1.0		1.0	1.0	0.333
		1.0		1.0	1.0	0.667
		1.0		1.0	1.0	0.583
		2.0		4.0	0.25	0.0208
		2.0		6.0	0.25	0.00931
		2.0		8.0	0.25	0.00521
		2.0		6.0	0.25	0.0150
		2.0		6.0	0.25	0.0142
		2.0		8.0	0.25	0.00751
		2.0		8.0	0.25	0.00721

×   Free to warp   ○   Warping torsion frictionless hinge

    Warping prevented   •   Warping torsion plastic hinge

——   Top flange   ❮ ❯   Uniform torsion shear hinges

·······   Bottom flange

Fig. 10.20 Plastic collapse and maximum rotation – concentrated torque

10.3 Non-uniform torsion

10.3.1 ELASTIC DEFORMATIONS AND STRESSES

10.3.1.1 General

In elastic non-uniform torsion, both the rate of change of the angle of twist rotation $d\phi/dz$ and the longitudinal warping deflections w vary along the length of the member. The varying warping deflections induce longitudinal strains and

Member and Loading	Plastic Collapse				Rotations	
	Uniform Torsion		Warping Torsion		Uniform	Warping
	Mechanism	$\lambda_u\,T/M_{up}$	Mechanism	$\lambda_w\,TL/M_{fp}d_f$	$\phi_{um}\,GJ/TL$	$\phi_{wm}\,EI_w/TL^3$
		1.0		0	0.5	∞
		1.0		2.0	0.5	0.139
		1.0		2.0	0.5	0.292
		1.0		2.0	0.5	0.250
		2.0		8.0	0.125	0.0130
		2.0		11.66	0.125	0.00541
		2.0		16.0	0.125	0.00260
		2.0		11.66	0.125	0.00915
		2.0		11.66	0.125	0.00860
		2.0		16.0	0.125	0.00417
		2.0		16.0	0.125	0.00397

$\times$	Free to warp	$\circ$	Warping torsion frictionless hinge
	Warping prevented	$\bullet$	Warping torsion plastic hinge
——	Top flange	$\langle\,\rangle$	Uniform torsion shear hinges
⋯⋯	Bottom flange		

Fig. 10.21 Plastic collapse and maximum rotation – uniformly distributed torque

stresses f_w. When these warping normal stresses also vary along the member, there are associated warping shear stresses τ_w distributed around the cross-section, and these may act in conjunction with the shear stresses due to uniform torsion to resist the torque acting at the section [1, 11]. In the following subsections, the warping deformations, stresses, and torques in thin-walled open section members of constant cross-section are treated. The local warping stresses induced by the non-uniform torsion of closed sections are beyond the scope of this book, but are treated in [9, 10], while some examples of the non-uniform torsion of tapered open section members are discussed in [12, 13].

10.3.1.2 Warping deflections and stresses

In thin-walled open section members, the longitudinal warping deflections w due to twist are much greater than those due to shear straining (see Fig. 10.6), and the latter are usually neglected. The warping deflections due to twist arise because the lines originally parallel to the axis of twist become helical locally, as indicated in Figs 10.4, 10.22 and 10.23. In non-uniform torsion, the axis of twist is the locus of the shear centres (see section 5.4.3), since otherwise the shear centre axis would be deformed, and the member would be bent as well as twisted. It is shown in section 10.8.1 that the warping deflections due to twist (see Fig. 10.23) can be expressed as

$$w = (\alpha_n - \alpha) \frac{d\phi}{dz}, \tag{10.31}$$

where

$$\alpha = \int_0^s \rho_0 \, ds, \tag{10.32}$$

and

$$\alpha_n = \frac{1}{A} \int_0^E \alpha t \, ds, \tag{10.33}$$

where ρ_0 is the perpendicular distance from the shear centre S to the tangent to the mid-line of the section wall (see Figs 10.16b and 10.23).

In non-uniform torsion, the warping displacements w vary along the length of the member, as shown for example in Fig. 10.1b and c. As a result of this,

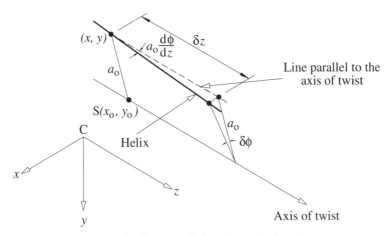

Fig. 10.22 Rotation of a line parallel to the axis of twist

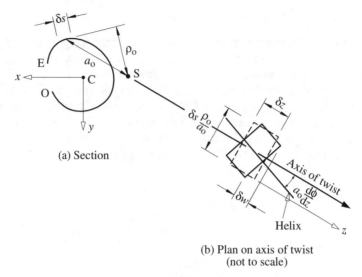

(a) Section

(b) Plan on axis of twist
(not to scale)

Fig. 10.23 Warping of an element $\delta z \times \delta s \times t$ due to twisting

longitudinal strains are induced, and the corresponding longitudinal warping normal stresses are

$$f_w = E(\alpha_n - \alpha)\frac{d^2\phi}{dz^2}. \tag{10.34}$$

These stresses define the bimoment stress resultant

$$B = -\int_0^E f_w(\alpha_n - \alpha)t\, ds. \tag{10.35}$$

Substituting equation 10.34 leads to

$$B = -EI_w \frac{d^2\phi}{dz^2} \tag{10.36}$$

in which

$$I_w = \int_0^E (\alpha_n - \alpha)^2 t\, ds \tag{10.37}$$

is the warping section constant. Substituting equation 10.36 into equation 10.34 allows the warping normal stresses to be expressed as

$$f_w = -\frac{B(\alpha_n - \alpha)}{I_w}. \tag{10.38}$$

Values of the warping section constant I_w and the warping normal stresses f_w

for structural sections are given in [5, 6], while a general computer program for their evaluation has been developed [8].

When the warping normal stresses f_w vary along the length of the member, warping shear stresses τ_w are induced, in the same way that variations in the bending normal stresses in a beam induce bending shear stresses (see section 5.4). The warping shear stresses can be expressed in terms of the shear flow

$$\tau_w t = -E \frac{d^3\phi}{dz^3} \int_0^s (\alpha_n - \alpha) t \, ds. \tag{10.39}$$

Information for calculating the warping shear stresses in structural sections is also given in [5, 6], while a general computer program for their evaluation has been developed [8].

The warping shear stresses exert a warping torque

$$M_w = \int_0^E \rho_0 \tau_w t \, ds, \tag{10.40}$$

about the shear centre, as shown in Fig. 10.12a. It can be shown [14, 15] that this reduces to

$$M_w = -EI_w \frac{d^3\phi}{dz^3}, \tag{10.41}$$

after substituting for τ_w and integrating by parts. The warping torque M_w given by equation 10.41 is related to the bimoment B given by equation 10.36 through

$$M_w = \frac{dB}{dz}. \tag{10.42}$$

Sign conventions for M_w and B are illustrated in Fig. 10.16.

A somewhat simpler explanation of the warping torque M_w and bimoment B can be given for the I-section illustrated in Fig. 10.24. This is discussed in section 10.8.2 where it is shown that for an equal flanged I-section, the bimoment B is related to the flange moment M_f through

$$B = d_f M_f, \tag{10.43}$$

and the warping section constant I_w is given by

$$I_w = \frac{I_y d_f^2}{4} \tag{10.44}$$

in which d_f is the distance between the flange centroids.

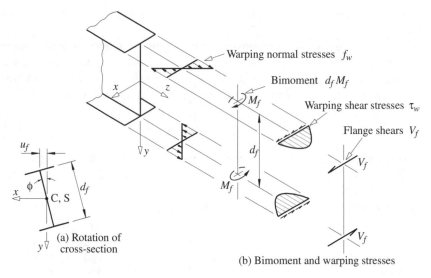

Fig. 10.24 Bimoment and stresses in an I-section member

10.3.2 ELASTIC ANALYSIS

10.3.2.1 Uniform torsion

Some thin-walled open-section members, such as thin rectangular sections and angle and tee sections, have very small warping section constants. In such cases it is usually sufficiently accurate to ignore the warping torque M_w, and to analyse the member as if it were in uniform torsion, as discussed in section 10.2.2.

10.3.2.2 Warping torsion

Very thin-walled open-section members have very small torsion section constants J, while some of these, such as I- and channel sections, have significant warping section constants I_w. In such cases, it is usually sufficiently accurate to analyse the member as if the applied torque were resisted entirely by the warping torque, so that $M_z = M_w$. Thus, the twisted shape of the member can be obtained by solving

$$-EI_w \frac{d^3\phi}{dz^3} = M_z, \tag{10.45}$$

as

$$-EI_w\phi = \int_0^z \int_0^z \int_0^z M_z \, dz \, dz \, dz + \frac{A_1 z^2}{2} + A_2 z + A_3, \tag{10.46}$$

where A_1, A_2, A_3 are constants of integration whose values depend on the boundary conditions.

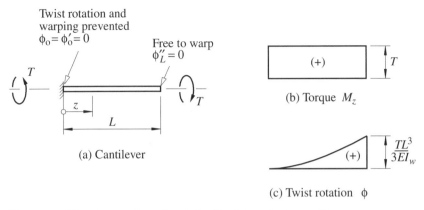

(a) Cantilever

(b) Torque M_z

(c) Twist rotation ϕ

Fig. 10.25 Warping torsion of a statically determinate cantilever

For statically determinate members, there are three boundary conditions from which the three constants of integration can be determined. As an example of three commonly assumed boundary conditions, the statically determinate cantilever shown in Figs 10.1c and 10.25 is analysed in section 10.8.3. The theoretical twisted shape of the cantilever is shown in Fig. 10.25c, and it can be seen that the maximum angle of twist rotation occurs at the loaded end and is equal to $TL^3/3EI_w$.

For statically indeterminate members such as the beam shown in Fig. 10.26, there is an additional boundary condition for each redundant quantity involved. These redundants can be determined by substituting the additional boundary conditions into equation 10.46. As an example of this procedure, the beam shown in Fig. 10.26 is analysed in section 10.8.4.

The warping torsion analysis of equal flanged I-sections can also be carried out by analysing the flange bending model shown in Fig. 10.27. Thus the

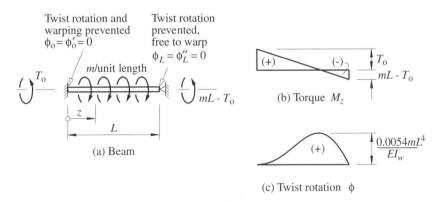

(a) Beam

(b) Torque M_z

(c) Twist rotation ϕ

Fig. 10.26 Warping torsion of a statically indeterminate beam

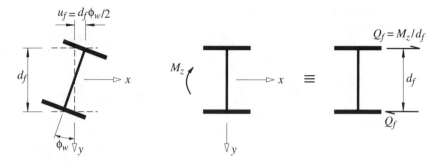

(a) Flange deflection u_f (b) Equivalent flange shears Q_f

Fig. 10.27 Flange bending model of warping torsion

warping torque $M_w = M_z$ acting on the I-section is replaced by two transverse shears

$$V_f = M_z/d_f, \tag{10.47}$$

one on each flange. The bending deflections u_f of each flange can then be determined by elastic analysis or by using available solutions [4], and the twist rotations calculated from

$$\phi_w = 2u_f/d_f. \tag{10.48}$$

10.3.2.3 Non-uniform torsion

When neither the torsional rigidity nor the warping rigidity can be neglected, as in hot-rolled steel I- and channel sections, the applied torque is resisted by a combination of the uniform and warping torques, so that

$$M_z = M_u + M_w. \tag{10.49}$$

In this case, the differential equation of non-uniform torsion is

$$M_z = GJ \frac{d\phi}{dz} - EI_w \frac{d^3\phi}{dz^3}, \tag{10.50}$$

the general solution of which can be written as

$$\phi = p(z) + A_1 e^{z/a} + A_2 e^{-z/a} + A_3, \tag{10.51}$$

where

$$a^2 = \frac{EI_w}{GJ} = \frac{K^2 L^2}{\pi^2}. \tag{10.52}$$

The function $p(z)$ in this solution is a particular integral whose form depends on the variation of the torque M_z along the beam. This can be determined by

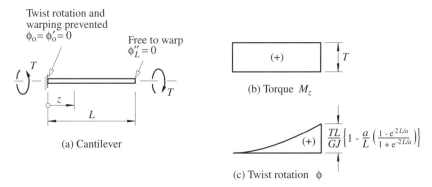

Fig. 10.28 Non-uniform torsion of a cantilever

the standard techniques used to solve ordinary linear differential equations (see [16] for example). The values of the constants of integration A_1, A_2, A_3 depend on the form of the particular integral and on the boundary conditions.

Complete solutions for a number of torque distributions and boundary conditions have been determined, and graphical solutions for the twist ϕ and its derivatives $d\phi/dz$, $d^2\phi/dz^2$, $d^3\phi/dz^3$ are available [5, 6]. As an example, the non-uniform torsion of the cantilever shown in Fig. 10.28 is analysed in section 10.8.5.

Sometimes the accuracy of the method of solution described above is not required, in which case a much simpler method may be used. The maximum angle of twist rotation ϕ_m may be approximated by

$$\phi_m = \frac{\phi_{um}\phi_{wm}}{\phi_{um} + \phi_{wm}} \tag{10.53}$$

in which ϕ_{um} is the maximum uniform torsion angle of twist rotation obtained by solving equation 10.8 with $M_u = M_z$, and ϕ_{wm} is the maximum warping torsion angle of twist rotation obtained by solving equation 10.45. Values of ϕ_{um} and ϕ_{wm} can be obtained from Figs 10.20 and 10.21 for a number of different torsion members.

This approximate method tends to overestimate the true maximum angle of twist rotation, partly because the maximum values ϕ_{um} and ϕ_{wm} often occur at different locations along the member. In the case of the cantilever of Fig. 10.28, the approximate solution

$$\phi_m = \frac{TL/GJ}{1 + 3EI_w/GJL^2} \tag{10.54}$$

obtained using Figs 10.15c and 10.25c in equation 10.53 has a maximum error of about 12%. [17].

10.3.3 PLASTIC COLLAPSE ANALYSIS

10.3.3.1 Fully plastic bimoment

The warping normal stresses f_w developed in elastic warping torsion are usually much larger than the warping shear stresses τ_w. Thus the limit of elastic behaviour is often reached when the bimoment is close to its nominal first yield value for which the maximum value of f_w is equal to f_y.

For the equal flanged I-section shown in Fig. 10.24, the first yield bimoment B_Y corresponds to the flange moment M_f reaching the yield value

$$M_{fY} = f_y b_f^2 t_f / 6 \tag{10.55}$$

so that

$$B_Y = f_y d_f b_f^2 t_f / 6 \tag{10.56}$$

in which b_f is the flange width, t_f the flange thickness, and d_f the distance between flange centroids. As the bimoment increases above B_Y, yielding spreads from the flange tips into the cross-section, until the flanges become fully yielded at

$$M_{fp} = f_y b_f^2 t_f / 4, \tag{10.57}$$

which corresponds to the fully plastic bimoment

$$B_p = f_y d_f b_f^2 t_f / 4. \tag{10.58}$$

The plastic shape factor for warping torsion of the I-section is therefore given by $B_p/B_Y = 3/2$.

10.3.3.2 Plastic analysis of warping torsion

An equal flanged I-section member in warping torsion will collapse plastically when there is a sufficient number of warping hinges (frictionless or plastic) to transform the member into a mechanism, as shown for example in Fig. 10.29. In general, the warping hinges develop progressively until the collapse mechanism forms.

A frictionless warping hinge occurs at a point where warping is unrestrained ($\phi'' = 0$) so that there is a frictionless hinge in each flange, while at a plastic warping hinge, the bimoment is equal to the fully plastic value B_p (equation 10.58) and the moment in each flange is equal to the fully plastic value M_{fp} (equation 10.57), so that there is a flexural plastic hinge in each flange. Warping torsion plastic collapse therefore corresponds to the simultaneous plastic collapse of the flanges in opposite directions, with a series of flexural hinges (frictionless or plastic) in each flange. Thus warping torsion plastic collapse can be analysed by using the methods discussed in section 5.5.5 to analyse the flexural plastic collapse of the flanges.

Warping torsion plastic hinges often form at supports or at points of concen-

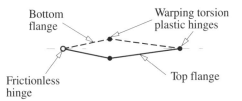

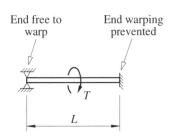

(a) Torsion member and loading

(b) Warping torsion collapse mechanism

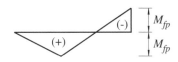

(c) Top flange moment distribution

Fig. 10.29 Warping torsion plastic collapse

trated torque, as indicated in Fig. 10.29. Examples of warping torsion plastic collapse mechanisms are shown in Figs 10.20 and 10.21.

At plastic collapse, each flange becomes a mechanism which is statically determinate, so that it can be analysed by equilibrium to determine the plastic collapse torques. For the member shown in Fig. 10.29, each flange forms a mechanism with plastic hinges at the concentrated torque and the right hand support and a frictionless hinge at the left hand support where warping is unrestrained. If the concentrated torque $\lambda_w T$ acts at mid-length, then the flanges collapse when

$$\lambda_w (T/d_f) L/4 = M_{fp} + M_{fp}/2 \tag{10.59}$$

so that the warping torsion plastic collapse load factor is given by

$$\lambda_w = 6M_{fp}d_f/TL. \tag{10.60}$$

Values of the warping torsion plastic collapse load factor λ_w for a number of example torsion members are given in Figs 10.20 and 10.21.

10.3.3.3 Plastic analysis of non-uniform torsion

It has not been possible to develop a simple but rigorous model for the analysis of the plastic collapse of members in non-uniform torsion where both uniform and warping torsion are important. This is because:

(a) different types of stress (shear stress τ_u and normal stress f_w) are associated with uniform and warping torsion collapse,
(b) the different stresses τ_u and f_w are distributed differently across the section,

(c) the uniform torque M_u and the warping torque M_w are distributed differently along the member.

A number of approximate theories have been proposed, the simplest of which is the Merchant approximation [18], according to which the plastic collapse load factor λ_z is the sum of the uniform and warping torsion collapse load factors, so that

$$\lambda_z = \lambda_u + \lambda_w. \qquad (10.61)$$

This approximation assumes that there is no interaction between uniform and warping torsion at plastic collapse, so that the separate plastic collapse capacities are additive.

If strain-hardening is ignored and only small twist rotations are considered, then the errors arising from this approximation are on the unsafe side, but are small because: the warping torsion shear stresses τ_w are small; the yield interactions between normal and shear stresses (see section 1.3.1) are small; and cross-sections which are fully plastic due to warping torsion often occur at different locations along the member than those which are fully plastic due to uniform torsion. These small errors are more than compensated for by the conservatism of ignoring strain-hardening and second-order longitudinal stresses that develop at large rotations. Test results [19, 20] and numerical studies [21] have shown that these cause significant strengthening at large rotations as indicated in Fig. 10.30, and that the approximation of equation 10.61 is conservative.

An example of the plastic collapse analysis of non-uniform torsion is given in section 10.9.6.

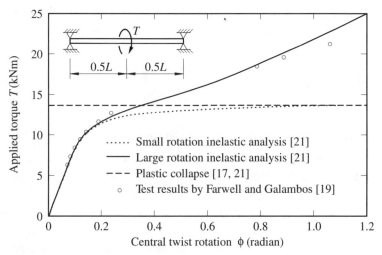

Fig. 10.30 Comparison of analysis and test results

10.4 Torsion design

10.4.1 GENERAL

Although very few design codes give guidance for designing against torsion, it is possible to formulate procedures which are consistent with those used in BS5950 for the design of beams against bending and shear. Such a set of procedures is given in the following subsections.

First the member cross-section is classified as plastic, compact, semi-compact, or slender, in much the same way as is a beam cross-section (see sections 4.7.2 and 5.6.1.2), and then the effective section capacities for uniform and warping torsion are determined. Following this, an appropriate method of torsion analysis (plastic or elastic) is selected, and then an appropriate method (plastic, first hinge, first yield, or local buckling) is used for strength design. Finally, a method of serviceability design is discussed.

10.4.2 SECTION CLASSIFICATION

Cross-sections of torsion members need to be classified according to the extent by which local buckling effects may reduce their cross-section capacities. Cross-sections which are capable of reaching and maintaining plasticity while a torsion plastic hinge collapse mechanism develops may be called plastic, as are the corresponding cross-sections of beams. Compact sections are capable of developing a first hinge, but inelastic local buckling may prevent the development of a plastic collapse mechanism. Semi-compact sections are capable of reaching the nominal first yield before local buckling occurs, while slender section members buckle locally before the nominal first yield is reached.

For closed cross-sections, the only significant stresses that develop during torsion are uniform torsion shear stresses τ_u that are constant across the wall thickness t and along the length b of any rectangular element of the cross-section. This stress distribution is very similar to that caused by a bending shear force in the web of an equal-flanged I-section. Hence the section classification may be based on the same width–thickness limits. Thus the elements of a plastic or compact hollow section would satisfy

$$\lambda_e \leqslant 62 \tag{10.62}$$

in which λ_e is the element slenderness given by

$$\lambda_e = (b/t)\sqrt{(p_y/275)}. \tag{10.63}$$

Sections which do not satisfy this condition would be classified as slender.

For open cross-sections, the warping shears stresses τ_w are usually very small, and can be neglected without serious error. The uniform shear stresses τ_u change sign and vary linearly across the wall thickness t, and so can be considered to have no effect on local buckling. The flange elastic and plastic warping normal stress distributions are similar to those due to bending in the plane of the flange, and so the section classification may be based on the same width–thickness limits. Thus the flange outstands of a plastic open section would satisfy

$$\lambda_e \leq \lambda_{ep} \qquad (10.64)$$

in which λ_{ep} is the appropriate plastic limit of Table 11 of BS5950 (for example, $\lambda_{ep} = 9$ for the flange outstands of hot-rolled sections). Compact open sections would satisfy

$$\lambda_{ep} < \lambda_e \leq \lambda_{ec} \qquad (10.65)$$

in which λ_{ec} is the appropriate compact limit (for example, $\lambda_{ec} = 10$ for the flange outstands of hot-rolled sections). Semi-compact open sections would satisfy

$$\lambda_{ec} < \lambda_e \leq \lambda_{es} \qquad (10.66)$$

in which λ_{es} is the appropriate semi-compact limit (for example, $\lambda_{es} = 15$ for the flange outstands of hot-rolled sections). Slender open sections would satisfy

$$\lambda_{es} < \lambda_e. \qquad (10.67)$$

10.4.3 UNIFORM TORSION SECTION CAPACITY

The uniform torsion section capacity of a plastic or compact hollow section M_{up} can be determined by assuming that the complete cross-section reaches the approximate shear yield stress

$$\tau_y \approx 0.6p_y, \qquad (10.68)$$

and finding the torque resultant M_{up} of these stresses.

The effective uniform torsion section capacity of a slender hollow section M_{ue} can be approximated by using equation 10.20 with $\tau = \tau_y$ to find the first yield uniform torque

$$M_{uY} = 1.2f_y A_e t_{min} \qquad (10.69)$$

in which t_{min} is the least thickness of the cross-section, and then reducing it to

$$M_{ue} = M_{uY}(\lambda_{es}/\lambda_e). \qquad (10.70)$$

The effective uniform torsion section capacity of a plastic or compact open section M_{up} can be determined by using the 'sand heap' analogy discussed in section 10.2.3.1.

The effective uniform torsion section capacity of a semi-compact or slender open section can be approximated by using

$$M_{ue} = M_{uY}. \qquad (10.71)$$

in which M_{uY} is the first yield uniform torque discussed in section 10.2.3.1.

10.4.4 BIMOMENT SECTION CAPACITY

The bimoment section capacity of a plastic or compact equal-flanged I-section B_p is given by equation 10.58. The bimoment section capacity of a semi-compact equal-flanged I-section can be approximated by using

$$B_e = B_Y \tag{10.72}$$

in which B_Y is the first yield bimoment given by equation 10.56. The bimoment section capacity of a slender equal-flanged I-section can be approximated by using

$$B_e = B_Y(\lambda_{es}/\lambda_e). \tag{10.73}$$

The bimoment section capacity of other plastic, compact, or semi-compact open sections can be conservatively approximated using equation 10.72, unless there is specific information available [22].

The bimoment section capacity of other slender open sections can be approximated using equation 10.73.

10.4.5 PLASTIC DESIGN

The use of plastic design should be limited to plastic members which have sufficient ductility to reach the plastic collapse mechanism. Plastic design should be carried out by using plastic analysis to determine the plastic collapse load factor λ_z (see section 10.3.3.3), and then checking that

$$1 \leq \lambda_z \tag{10.74}$$

10.4.6 FIRST HINGE DESIGN

The use of first hinge design should be limited to plastic or compact members in which the first hinge of a collapse mechanism can form. The maximum design uniform torques M_u^* and bimoments B^* can be determined by using elastic analysis, and the member is satisfactory if

$$M_u^* \leq M_{up}, \tag{10.75}$$

$$B^* \leq B_p. \tag{10.76}$$

The first hinge method is more conservative than plastic design, and more difficult to use because elastic member analysis is much more difficult than plastic analysis.

10.4.7 FIRST YIELD DESIGN

The use of first yield design is appropriate for semi-compact members. The maximum design uniform torques M_u^* and bimoments B^* can be determined by elastic analysis. The member is satisfactory if

$$\left(\frac{M_u^*}{M_{uY}}\right)^2 + \left(\frac{B^*}{B_Y}\right)^2 \leqslant 1, \tag{10.77}$$

which is often conservative when M_u^* and B^* occur at different locations along the member. The first yield method is just as difficult to use as the first hinge method.

10.4.8 LOCAL BUCKLING DESIGN

Slender members should be designed against local buckling. The maximum design uniform torque M_u^* and B^* can be determined by elastic analysis, and the member is satisfactory if

$$\left(\frac{M_u^*}{M_{ue}}\right)^2 + \left(\frac{B^*}{B_e}\right)^2 \leqslant 1 \tag{10.78}$$

is satisfied, in which the effective torsion capacity M_{ue} is now given by equation 10.71 and the effective bimoment capacity B_e by equation 10.73.

10.4.9 DESIGN FOR SERVICEABILITY

Serviceability design usually requires the estimation of the deformations under part or all of the serviceability loads. Because these loads are usually significantly less than the factored combined loads used for strength design, the member remains largely elastic, and the serviceability deformations can be evaluated by linear elastic analysis.

There are no specific serviceability criteria for satisfactory deformations of a torsion member given in BS5950. However, general recommendations such as 'the deformations of a structure and its component members should be appropriate to the location, loading and function of the structure and the component members', may be applied to torsion members, so that the onus is placed on the designer to determine what are appropriate magnitudes of the twist rotations.

Because of the lack of precise serviceability criteria, highly accurate predictions of the serviceability rotations are not required. Thus the maximum twist rotations of members in non-uniform torsion may be approximated by using equation 10.53.

Usually, the twist rotations of thin-walled closed section members are very small, and can be ignored (although in some cases, the distortional deformations may be quite large, as discussed in section 10.6). On the other hand, thin-walled open section members are comparatively flexible, and special measures may be required to limit their rotations.

10.5 Torsion and bending

10.5.1 BEHAVIOUR

Pure torsion occurs very rarely in steel structures. Most commonly, torsion occurs in combination with bending actions. The torsion actions may be classified as primary or secondary, depending on whether the torsion action is required to transfer load (primary torsion), or whether it arises as a secondary action. Secondary torques may arise as a result of differential twist rotations compatible with the joint rotations of primary frames, as shown in Fig. 10.31, and are predicted by three-dimensional analysis programs. They are not unlike the secondary bending moments which occur in rigid-jointed trusses, but which are usually ignored (a procedure justified by many years of satisfactory experience based on the long-standing practice of analysing a rigid-jointed truss as if pin-jointed). Secondary torques are usually small when there are alternative load paths of high stiffness, and may often be ignored.

Primary torsion actions may be classified as being restrained, free, or destabilizing, as shown in Fig. 10.32. For restrained torsion, the member ABC applying the torsion action also applies a restraining action to the member DCE resisting the torsion, as shown in Fig. 10.32b. In this case, the structure is redundant, and compatibility between the members must be satisfied in the analysis if the magnitudes of the torques and other actions are to be determined correctly. Free torsion occurs when the member ABC applying the torsion action does not restrain the twisting of the torsion member DCE, as shown in Fig. 10.32c but does prevent its lateral deflection. Destabilizing

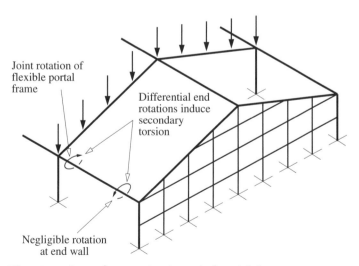

Fig. 10.31 Secondary torsion in an industrial frame

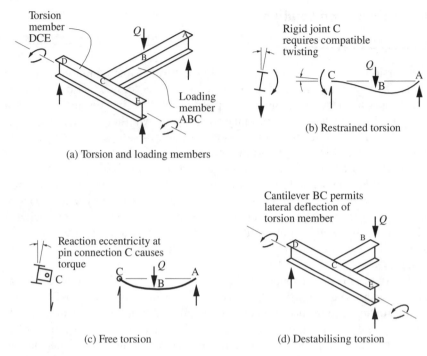

(a) Torsion and loading members

(b) Restrained torsion

(c) Free torsion

(d) Destabilising torsion

Fig. 10.32 Torsion actions

torsion may occur when the member BC applying the torsion action does not restrain either the twisting or the lateral deflection of the torsion member DCE, as shown in Fig. 10.32d. In this case, lateral buckling actions (Chapter 6) caused by the in-plane loading of the torsion member amplify the torsion and out-of-plane bending behaviour.

The inelastic non-linear bending and torsion of fully braced, centrally braced, and unbraced I-beams with central concentrated loads (see Fig. 10.33) has been analysed [23], and interaction equations developed for predicting their strengths. It was found that while circular interaction equations are appropriate for short length braced beams, these provide unsafe predictions for beams subject to destabilizing torsion, where lateral buckling effects become important. On the other hand, destabilizing interactions between lateral buckling and torsion tend to be masked by the favourable effects of the secondary axial stresses that develop at large rotations, and linear interaction equations based on plastic analyses provide satisfactory strength predictions, as shown in Fig. 10.34. Proposals based on these findings are made below for the analysis and design of members subject to combined torsion and bending.

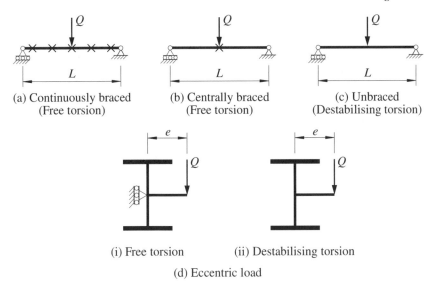

Fig. 10.33 Beams in torsion and bending

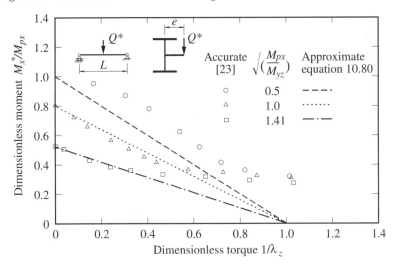

Fig. 10.34 Interaction between lateral buckling and torsion for unbraced beams

10.5.2 PLASTIC DESIGN OF FULLY BRACED MEMBERS

The member must be a plastic I-section. Plastic collapse analyses should be used to determine the plastic collapse load factors λ_i for in-plane bending (section 5.5) and λ_z for torsion (section 10.3.3.3).

The member should satisfy the circular interaction equation

$$1/\lambda_i^2 + 1/\lambda_z^2 \leq 1.0 \tag{10.79}$$

10.5.3 DESIGN OF MEMBERS WITHOUT FULL BRACING

A member without full bracing has its bending resistance reduced by lateral buckling effects (Chapter 6). Its bending should be analysed elastically, and the resulting moment distribution used to determine its bending resistance M_b/m_{LT} according to BS5950.

An elastic torsion analysis should be used to find the values of the design uniform torque M_u^* and bimoment B^*. An appropriate torsion design method should be selected from section 10.4.5 and used to find the load factor λ_z by which the design torque and bimoment should be multiplied so that the appropriate torsion design equation (equations 10.75–10.78) is just satisfied.

The member should then satisfy the linear interaction equation

$$\frac{m_{LT}M_x^*}{M_b} + \frac{1}{\lambda_z} \leqslant 1. \tag{10.80}$$

There is no need to amplify the moment because of the finding that linear interaction equations are satisfactory, even for unbraced beams, as shown in Fig. 10.34.

10.6 Distortion

Twisting and distortion of a flexural member may be caused by the local distribution around the cross-section of the forces acting, as shown in Fig. 10.3. If the member responds significantly to either of these actions, then the bending stress distribution may depart considerably from that calculated in the usual way, while additional distortional stresses may be induced by out-of-plane bending of the wall (see Fig. 10.3d). To avoid possible failure, the designer must either increase the strength of the member, which may be uneconomic, or must limit both twisting and distortion.

The resistance to distortion of a thin-walled member depends on the arrangement of the cross-section. Members with triangular closed cells have high resistances because of the truss-like action of the triangulated cross-section which causes the distortional loads to be transferred by in-plane bending and shear of the cell walls. On the other hand, the walls of the members with rectangular or trapezoidal cells bend out of their planes when under distortional loading and when the resistances to distortion are low [24], while members of open cross-section are even more flexible. Concentrated distortional loads can be distributed locally by providing stiff diaphragms which reduce or prevent local distortion. However, isolated diaphragms are not fully effective when the distortional loads are distributed or can move along the member. In such cases, it is necessary to analyse the distortion of the section and the stresses induced by the distortion.

The resistance to torsion of a thin-walled open section is comparatively small, and the dominant mode of deformation is twisting of the member. Because this twisting is usually accounted for, the importance of distortion lies in its modifying influence on the torsional behaviour. Significant distortions occur only in very thin-walled members, for which the distortional rigidity, which varies as the cube of the wall thickness, reduces at a much faster rate than the warping rigidity, which varies directly with the thickness [25]. These distortions may induce significant wall bending stresses, and may increase the angles of twist and change the warping stress distributions [26, 27].

Thin-walled closed section members are very stiff torsionally, and the twisting deformations due to uniform torsion are not usually of great importance except in curved members. On the other hand, rectangular and trapezoidal members are not very resistant to distortional loads, and the distortional deformations may be large, while the stresses induced by the distortional loads may dominate the design [24].

The distortional behaviour of single cell rectangular or trapezoidal members can be classified as uniform or non-uniform. Uniform distortion occurs when the applied distortional loading is uniformly distributed along a member of constant cross-section which has no diaphragms. In this case the distortional loading is resisted solely by the out-of-plane bending rigidity of the plate elements of the section. In non-uniform distortion, the distortions vary along the member, and the plate elements bend in their planes, producing in-plane shear stresses which help to resist the distortional loading.

Perhaps the simplest method of analysing non-uniform distortion is by using an analogy to the problem [28] of a beam on an elastic foundation (BEF analogy). In this analogy, which is shown diagrammatically in Fig. 10.35, the distortional loading corresponds to the applied beam loads in the analogy, and the out-of-plane resistance of the plate elements corresponds to the stiffness k of the elastic foundation, while the in-plane resistance of these elements corresponds to the flexural rigidity EI of the analogous beam. The BEF analogy can account for concentrated and distributed distortional loads, for rigid diaphragms which either permit or prevent warping, and for flexible diaphragms and stiffening rings, as indicated in Fig. 10.35. Many solutions for problems of beams on elastic foundations have been obtained [29], and these can be used to find the out-of-plane bending stresses in the plate elements (which are related to the foundation reactions), and the warping and normal shear stresses due to non-uniform distortion (which are related to the bending moment and shear force in the beam on the elastic foundation). The distributions of the warping stresses in a rectangular box section are shown in Fig. 10.36.

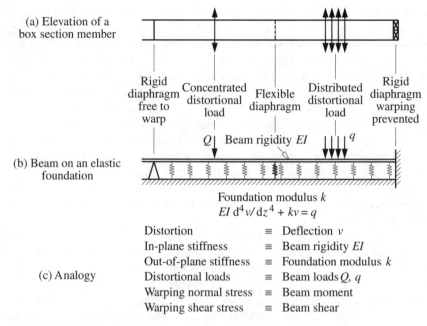

(a) Elevation of a box section member

Rigid diaphragm free to warp — Concentrated distortional load — Flexible diaphragm — Distributed distortional load — Rigid diaphragm warping prevented

Q | Beam rigidity EI q

(b) Beam on an elastic foundation

Foundation modulus k

$$EI\, \mathrm{d}^4 v / \mathrm{d}z^4 + kv = q$$

(c) Analogy

Distortion	$\equiv$	Deflection v
In-plane stiffness	$\equiv$	Beam rigidity EI
Out-of-plane stiffness	$\equiv$	Foundation modulus k
Distortional loads	$\equiv$	Beam loads Q, q
Warping normal stress	$\equiv$	Beam moment
Warping shear stress	$\equiv$	Beam shear

Fig. 10.35 BEF analogy for box section distortion

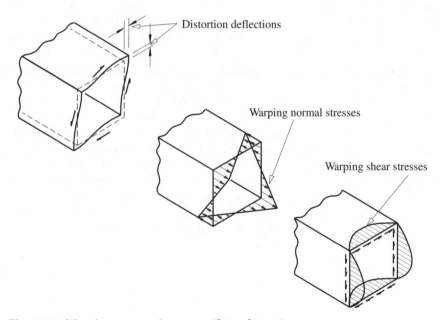

Distortion deflections

Warping normal stresses

Warping shear stresses

Fig. 10.36 Warping stresses in non-uniform distortion

10.7 Appendix – uniform torsion

10.7.1 THIN-WALLED CLOSED SECTIONS

The shear flow τt around a thin-walled closed section member in uniform torsion can be determined by using the condition that the total change in the warping displacement w around the complete closed section must be zero for continuity. The change in the warping displacement due to the twist rotations ϕ is equal to $-(d\phi/dz)\oint \rho_0\,ds$ (see section 10.2.1.5 and Fig. 10.13a), while the change in the warping displacement due to the shear straining of the mid-surface is equal to $\oint (\tau/G)ds$ (see Figs 10.6 and 10.13b). Thus

$$\oint \frac{\tau}{G}\,ds = \frac{d\phi}{dz}\oint \rho_0\,ds.$$

The shear flow τt is constant around the closed section (otherwise the corresponding longitudinal shear stresses would not be in equilibrium), while $\oint \rho_0\,ds$ is equal to twice the area A_e enclosed by the section, as shown in Fig. 10.12b. Thus,

$$\tau t \oint \frac{1}{t}\,ds = 2GA_e \frac{d\phi}{dz}. \tag{10.81}$$

The moment resultant of the shear flow around the section is equal to the torque acting, and so

$$M_u = \oint \tau t \rho_0\,ds, \tag{10.82}$$

whence

$$\tau t = \frac{M_u}{2A_e}. \tag{10.20}$$

Substituting this into equation 10.81 leads to

$$M_u = G\frac{4A_e^2}{\oint(1/t)ds}\frac{d\phi}{dz},$$

and so the torsion section constant is

$$J = \frac{4A_e^2}{\oint(1/t)ds}. \tag{10.21}$$

The torsion section constants and the shear flows in multicell closed section members can be determined by extending this method. If the member consists of n junctions linked by m walls, then there are $m - n + 1$ independent cells. The longitudinal equilibrium conditions require there to be m constant shear flows τt, one for each wall, and $n - 1$ junction equations of the type

$$\sum_{\text{junction}} (\tau t) = 0. \tag{10.83}$$

There are $m - n + 1$ independent cell warping continuity conditions of the type

$$\sum_{\text{cell}} \int_{\text{wall}} (\tau/G)ds = 2A_e \frac{d\phi}{dz}, \tag{10.84}$$

where A_e is the area enclosed by the cell. Equations 10.83 and 10.84 can be solved simultaneously for the m shear flows τt. The total uniform torque M_u can be determined as the torque resultant of these shear flows, and the torsion section constant J can then be found from

$$J = \frac{M_u}{Gd\phi/dz}.$$

10.7.2 ANALYSIS OF STATICALLY INDETERMINATE MEMBERS

The uniform torsion of the statically indeterminate beam shown in Fig. 10.17a may be analysed by taking the left hand reaction torque T_0 as the redundant quantity. The distribution of torque M_z is therefore as shown in Fig. 10.17b, and equation 10.8 becomes

$$GJ \frac{d\phi}{dz} = T_0 - T\langle z - a \rangle^0,$$

where the value of the second term is taken as zero when the value inside the Macaulay brackets $\langle \rangle$ is negative.

The solution of this equation which satisfies the condition that the angle of twist at the left hand support is zero is

$$GJ\phi = T_0 z - T\langle z - a \rangle. \tag{10.85}$$

The other condition which must be satisfied is that the angle of twist at the right hand support must be zero. Using this in equation 10.85,

$$0 = T_0 L - T(L - a),$$

and so

$$T_0 = \frac{T(L - a)}{L}.$$

Thus the maximum torque is T_0 when a is less than $L/2$, while the maximum angle of twist rotation occurs at the loaded point, and is equal to

$$\phi_a = \frac{Ta(L - a)}{GJL}.$$

10.8 Appendix – non-uniform torsion

10.8.1 WARPING DEFLECTIONS

When a member twists, lines originally parallel to the axis of twist become helical, and any element $\delta z \times \delta s \times t$ of the section wall rotates $a_0 \, d\phi/dz$ about the line a_0 from the shear centre S, as shown in Figs 10.4 and 10.22. The increase δw in the warping displacement of the element is shown in Fig. 10.23, and is equal to

$$\delta w = -\rho_0 \frac{d\phi}{dz} \, \delta s,$$

where ρ_0 is the perpendicular distance from the shear centre S to the tangent to the mid-line of the wall. The sign convention for ρ_0 is demonstrated in Fig. 10.16b, which shows a cross-section viewed by looking in the positive z direction. The value of ρ_0 is positive when the line a_0 from the shear centre rotates in a clockwise sense when the distance s increases, which is the case for Fig. 10.16b. Thus the warping displacement is given by

$$w = (\alpha_n - \alpha) \frac{d\phi}{dz}, \tag{10.31}$$

where

$$\alpha = \int_0^s \rho_0 \, ds, \tag{10.32}$$

and the quantity α_n is chosen to be

$$\alpha_n = \frac{1}{A} \int_0^E \alpha t \, ds, \tag{10.33}$$

so that the average warping displacement of the section is zero.

10.8.2 WARPING TORQUE AND BIMOMENT IN AN I-SECTION

When an equal flanged I-section member twists ϕ as shown in Fig. 10.24a, the flanges deflect laterally

$$u_f = \frac{d_f}{2} \, \phi.$$

Thus the flanges may have curvatures

$$\frac{d^2 u_f}{dz^2} = \frac{d_f}{2} \frac{d^2 \phi}{dz^2}$$

in opposite directions, as for example in the cantilever shown in Fig. 10.1c. These curvatures can be thought of as being produced by the equal and opposite flange bending moments

$$M_f = EI_f \frac{d^2 u_f}{dz^2} = \frac{EI_y}{2} \frac{d_f}{2} \frac{d^2\phi}{dz^2}$$

shown in Fig. 10.24b. This pair of equal and opposite flange moments M_f, spaced d_f apart, form the bimoment

$$B = d_f M_f. \tag{10.43}$$

The bending stresses induced by the bimoment are the warping normal stresses

$$f_w = \mp \frac{M_f x}{I_f} = \pm E \frac{d_f}{2} x \frac{d^2\phi}{dz^2}$$

shown in Fig. 10.24b. These can also be obtained from equation 10.34.

When the flange moments M_f vary along the member, they induce equal and opposite flange shears

$$V_f = - \frac{dM_f}{dz} = - \frac{EI_y}{2} \frac{d_f}{2} \frac{d^3\phi}{dz^3},$$

as shown in Fig. 10.24b. The warping shear stresses

$$\tau_w = \pm \frac{V_f}{b_f t_f} \left(1.5 - \frac{6x^2}{b_f^2} \right)$$

(in which b_f is the flange width) which are statically equivalent to these shear forces are also shown in Fig. 10.24b. They can also be obtained from equation 10.39.

The equal and opposite flange shears V_f are statically equivalent to the warping torque

$$M_w = V_f d_f = - \frac{EI_y d_f^2}{4} \frac{d^3\phi}{dz^3}.$$

If this is compared with equation 10.41, it can be deduced that the warping section constant for the I-section is

$$I_w = \frac{I_y d_f^2}{4} \tag{10.44}$$

This result can also be obtained from equation 10.37.

10.8.3 WARPING TORSION ANALYSIS OF STATICALLY DETERMINATE MEMBERS

The twisted shape of a statically determinate member in warping torsion is given by equation 10.46. For the example of the cantilever shown in Fig. 10.25a, the torque M_z is constant and equal to the applied torque T, and so the twisted shape is given by

$$-EI_w\phi = \frac{Tz^3}{6} + \frac{A_1 z^2}{2} + A_2 z + A_3. \tag{10.86}$$

The constants of integration A_1, A_2, and A_3 depend on the boundary conditions. At the support, it is assumed that twisting is prevented, so that $\phi_0 = 0$ and that warping is also prevented, so that (see equation 10.31)

$$\left(\frac{d\phi}{dz}\right)_0 = 0.$$

At the loaded end, the member is free to warp (i.e. the warping normal stresses f_w are zero), so that (see equation 10.34)

$$\left(\frac{d^2\phi}{dz^2}\right)_L = 0.$$

By substituting these conditions into equation 10.86, the constants of integration can be determined as

$$A_1 = -TL, \quad A_2 = 0, \quad A_3 = 0.$$

If these are substituted into equation 10.86, the complete solution for the twisted shape is obtained as

$$-EI_w\phi = \frac{Tz^3}{6} - \frac{TLz^2}{2}.$$

The maximum angle of twist occurs at the loaded end and is equal to

$$\phi_L = \frac{TL^3}{3EI_w}.$$

The warping shear stress distribution (see equation 10.39) is constant along the member, but the maximum warping normal stress (see equation 10.34) occurs at the support where $d^2\phi/dz^2$ reaches its highest value.

10.8.4 WARPING TORSION ANALYSIS OF STATICALLY INDETERMINATE MEMBERS

The warping torsion of the statically indeterminate beam shown in Fig. 10.26a may be analysed by taking the left hand reaction torque T_0 as the redundant. The distribution of torque M_z is then given by (see Fig. 10.26b)

$$M_z = T_0 - mz,$$

and equation 10.46 becomes

$$-EI_w\phi = \frac{T_0z^3}{6} - \frac{mz^4}{24} + \frac{A_1z^2}{2} + A_2z + A_3.$$

After using the boundary conditions $\phi_0 = (d\phi/dz)_0 = 0$, the constants A_2 and A_3 are determined as

$$A_2 = A_3 = 0,$$

while the condition that $\phi_L = 0$ requires that

$$A_1 = -\frac{T_0 L}{3} + \frac{mL^2}{12}.$$

The additional boundary condition is $(d^2\phi/dz^2)_L = 0$, and so

$$T_0 = 5mL/8.$$

Thus the complete solution is

$$-EI_w\phi = m\left(\frac{-z^4}{24} + \frac{5Lz^3}{48} - \frac{L^2z^2}{16}\right).$$

The maximum angle of twist occurs near $z = 0.58L$, and is approximately equal to $0.0054mL^4/EI_w$. The warping shear stresses are greatest at the left hand support, where the torque M_z has its greatest value of $5mL/8$. The warping normal stresses are greatest at $z = 5L/8$, where $-EI_w\, d^2\phi/dz^2$ has its maximum value of $9mL^2/128$.

10.8.5 NON-UNIFORM TORSION ANALYSIS

The twisted shape of a member in non-uniform torsion is given by equation 10.51. For the example of the cantilever shown in Fig. 10.28a, the torque M_z is constant and equal to the applied torque T. In this case, a particular integral is

$$p(z) = \frac{Tz}{GJ},$$

while the boundary conditions require that

$$\left.\begin{array}{l} 0 = A_1 + A_2 + A_3 \\[2mm] 0 = \dfrac{T}{GJ} + \dfrac{A_1}{a} - \dfrac{A_2}{a} \\[2mm] 0 = \dfrac{A_1}{a^2}\, e^{L/a} + \dfrac{A_2}{a^2}\, e^{-L/a} \end{array}\right\},$$

so that

$$\phi = \frac{Tz}{GJ} - \frac{Ta}{GJ(1 + e^{-2L/a})}\left[e^{(z-2L)/a} - e^{-z/a} + 1 - e^{-2L/a}\right].$$

At the fixed end, the uniform torque is zero because $d\phi/dz = 0$. Because of this the torque M_z is resisted solely by the warping torque, and the maximum warping shear stresses occur at this point. The value of $d^2\phi/dz^2$ is greatest at the fixed end, and therefore the value of the warping normal stress is also greatest there. The value of $d\phi/dz$ is greatest at the loaded end of the canti-

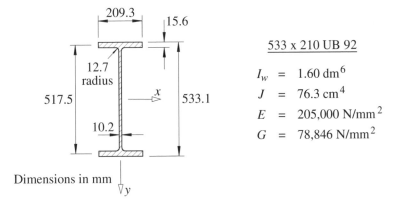

Fig. 10.37 Examples 1–7

lever, and therefore the value of the shear stress due to uniform torsion is also greatest there. The warping torque at the loaded end is

$$M_{wL} = T \frac{2e^{-L/a}}{1 + e^{-2L/a}},$$

which decreases from T to zero as L/a increases from 0 to ∞. Thus the warping shear stresses at the loaded end are not zero, but are less than those at the fixed end.

10.9 Worked examples

10.9.1 EXAMPLE 1 – APPROXIMATIONS FOR THE SERVICEABILITY TWIST ROTATION

Problem. The cantilever shown in Fig. 10.28 is 5 m long and has the properties shown in Fig. 10.37. If the serviceability end torque T is 3 kNm, determine approximate values of the serviceability end twist rotation either by:

(a) assuming that the cantilever is in uniform torsion ($EI_w \equiv 0$), or
(b) assuming that the cantilever is in warping torsion ($GJ \equiv 0$), or
(c) using the approximation of equation 10.53.

Solution.

(a) If $EI_w = 0$, then the end twist rotation (see section 10.2.2.1) is

$$\phi_{um} = \frac{TL}{GJ} = \frac{3 \times 10^6 \times 5000}{78\,846 \times 76.3 \times 10^4}$$

$$= 0.249 \text{ rad} = 14.3°.$$

(b) If $GJ = 0$, then the end twist rotation is (see section 10.3.2.2)

$$\phi_{wm} = TL^3/3EI_w$$

$$= \frac{3 \times 10^6 \times 5000^3}{3 \times 205\,000 \times 1.60 \times 10^{12}}$$

$$= 0.381 \text{ rad} = 21.8°.$$

(c) Using the results of (a) and (b) above in the approximation of equation 10.53

$$\phi_m = \frac{0.249 \times 0.381}{0.249 + 0.381}$$

$$= 0.151 \text{ rad} = 8.6°.$$

10.9.2 EXAMPLE 2 – SERVICEABILITY TWIST ROTATION

Problem. Use the solution obtained in section 10.8.5 to determine an accurate value of the serviceability end twist rotation of the cantilever of example 1.

Solution. Using equation 10.52,

$$a = \sqrt{\left(\frac{EI_w}{GJ}\right)} = \sqrt{\left(\frac{205\,000 \times 1.6 \times 10^{12}}{78\,846 \times 76.3 \times 10^4}\right)} = 2235 \text{ mm.}$$

Therefore

$$e^{-2L/a} = e^{-2 \times 5000/2235} = 0.0114.$$

Using section 10.8.5,

$$\phi_L = \frac{TL}{GJ}\left[1 - \frac{a}{L}\frac{(1 - e^{-2L/a})}{(1 + e^{-2L/a})}\right]$$

$$= \frac{3 \times 10^6 \times 5000}{78\,846 \times 76.3 \times 10^4}\left[1 - \frac{2235}{5000}\frac{(1 - 0.0114)}{(1 + 0.0114)}\right]$$

$$= 0.140 \text{ rad} = 8.0°,$$

which is about 9% less than the approximate value of 8.6° obtained in section 10.9.1c.

10.9.3 EXAMPLE 3 – ELASTIC UNIFORM TORSION SHEAR STRESS

Problem. For the cantilever of example 2, determine the maximum elastic uniform torsion shear stress caused by a strength design end torque of 5 kNm.

Solution. An expression for the nominal maximum elastic uniform torsion shear stress can be obtained by combining equations 10.8 and 10.18, whence

$$\tau_{\mathrm{m}} \approx G\left(\frac{d\phi}{dz}\right)_{\mathrm{m}} t_{\mathrm{m}}.$$

An expression for $d\phi/dz$ can be obtained from the solution in section 10.8.5 for ϕ, whence

$$\frac{d\phi}{dz} = \frac{T}{GJ}\left\{1 - \frac{[e^{(z-2L)/a} + e^{-z/a}]}{(1 + e^{-2L/a})}\right\}.$$

The maximum value of this occurs at the free end $z = L$ and is given by

$$\left(\frac{d\phi}{dz}\right)_{\mathrm{m}} = \frac{T}{GJ}\left[1 - \frac{2e^{-L/a}}{(1 + e^{-2L/a})}\right].$$

Adapting the solution of example 2,

$$\left(\frac{d\phi}{dz}\right)_{\mathrm{m}} = \frac{T}{GJ}\left[1 - \frac{2 \times \sqrt{(0.0114)}}{1.0114}\right] = 0.789\,\frac{T}{GJ}\,.$$

Thus

$$\tau_{\mathrm{m}} \approx 0.789\,\frac{Tt_{\mathrm{m}}}{J},$$

$$\approx \frac{0.789 \times 5 \times 10^6 \times 15.6}{76.3 \times 10^4}\,\mathrm{N/mm^2}$$

$$\approx 80.7\,\mathrm{N/mm^2}.$$

This nominal maximum shear stress occurs at the centre of the long edge of the flange. A similar value may be calculated for the opposite edge, but the actual stresses near this point are increased because of the re-entrant corner at the junction of the web and flange. An approximate estimate of the stress concentration factor can be made by using Fig. 10.7b with

$$\frac{r}{t} = \frac{12.7}{15.6} = 0.814.$$

Thus the actual local maximum stress is approximately $\tau_{\mathrm{m}} = 1.7 \times 80.7 = 137\,\mathrm{N/mm^2}$.

10.9.4 EXAMPLE 4 – ELASTIC WARPING SHEAR AND NORMAL STRESSES

Problem.   For the cantilever of example 3, determine the maximum elastic warping shear and normal stresses.

Solution.   The maximum warping shear stress τ_{w} occurs at the fixed end ($z = 0$) of the cantilever (see section 10.8.5) where the uniform torque is zero. Thus the total torque there is resisted only by the warping torque, and so

$$V_f = T/d_{\mathrm{f}},$$

where V_f is the flange shear (see section 10.8.2). The shear stresses in the flange vary parabolically (see Fig. 10.24b), and so

$$\tau_w = \frac{1.5 V_f}{b_f t_f}.$$

Thus

$$\tau_w = \frac{1.5 \times (5 \times 10^6)/517.5}{209.3 \times 15.6} \ \text{N/mm}^2$$

$$= 4.4 \ \text{N/mm}^2$$

The maximum warping normal stress f_w occurs where $d^2\phi/dz^2$ is a maximum (see equation 10.34). An expression for $d^2\phi/dz^2$ can be obtained from the solution in section 10.8.5 for ϕ, whence

$$\frac{d^2\phi}{dz^2} = \frac{T}{aGJ}\left[\frac{(-e^{(z-2L)/a} + e^{-z/a})}{(1 + e^{-2L/a})}\right].$$

The maximum value occurs at the fixed end $z = 0$ and is given by

$$\left(\frac{d^2\phi}{dz^2}\right)_m = \frac{T}{aGJ}\ \frac{(1 - e^{-2L/a})}{(1 + e^{-2L/a})}.$$

The corresponding maximum warping normal stress (see section 10.8.2) is

$$f_w = \frac{Ed_t b_f}{2 \times 2}\left(\frac{d^2\phi}{dz^2}\right)_m$$

$$= \frac{205\,000 \times 517.5 \times 209.3 \times 5 \times 10^6}{2 \times 2 \times 2235 \times 78\,846 \times 76.3 \times 10^4} \times \frac{(1 - 0.0114)}{(1 + 0.0114)}\ \text{N/mm}^2$$

$$= 201.8 \ \text{N/mm}^2.$$

10.9.5 EXAMPLE 5 – FIRST YIELD DESIGN TORQUE CAPACITY

Problem.   If the cantilever of example 2 has a yield stress of 275 N/mm², then use section 10.4.7 to determine the torque capacity which is consistent with BS5950 and a first yield strength criterion.

Solution.   In this case, the maximum uniform torsion shear and warping torsion shear and normal stresses occur either at different points along the cantilever, or at different points around the cross-section. Because of this, each stress may be considered separately.

For first yield in uniform torsion, equation 10.77 reduces to

$$\tau_u^*/\tau_y \leqslant 1$$

in which $\tau_y \approx 0.6 \times 275 = 165$ N/mm². Thus

$$\tau_u^* \leqslant 165 \ \text{N/mm}^2.$$

Using section 10.9.3, a torque of $T^* = 5$ kNm causes a shear stress of $\tau_u^* = 80.7$ N/mm^2, if the stress concentration at the flange–web junction is ignored. Thus the uniform torsion torque capacity is

$$T_u = 5 \times 165/80.7 = 10.22 \text{ kNm.}$$

The maximum warping shear stress τ_w^* caused by a torque of $T^* = 5$ kNm (see section 10.9.4) is 4.4 N/mm^2, which is much less than the corresponding maximum uniform torsion shear stress $\tau_u^* = 80.7$ N/mm^2. Because of this, the torque capacity is not limited by the warping shear stress.

For first yield in warping torsion, equation 10.77 reduces to

$$f_w^*/f_y \leqslant 1.$$

Thus

$$f_w^* \leqslant 275 \text{ N/mm}^2.$$

Using section 10.9.4, a torque of $T^* = 5$ kNm causes a warping normal stress of $f_w^* = 201.8$ N/mm^2. Thus the warping torsion torque capacity is

$$T_w = 5 \times 275/201.8 = 6.81 \text{ kNm.}$$

This is less than the value of 10.22 kNm determined for uniform torsion, and so the torque capacity is 6.81 kNm.

10.9.6 EXAMPLE 6 – PLASTIC COLLAPSE ANALYSIS

Problem. Determine the plastic collapse torque of the cantilever of example 5.

Uniform torsion plastic collapse. Using equation 10.28,

$$M_{up} = \left(209.3 \times 15.6^2 - \frac{15.6^3}{3} + \frac{501.9 \times 10.2^2}{2} + \frac{10.2^3}{6} \right) \times (0.6 \times 275) \text{ Nmm}$$

$$= 12.53 \text{ kNm.}$$

Using Fig. 10.20, $\lambda_u T = M_{up} = 12.53$ kNm.

Warping torsion plastic collapse. Using equation 10.57,

$$M_{fp} = 275 \times 209.3^2 \times 15.6/4 \text{ Nmm} = 46.98 \text{ kNm.}$$

Using Fig. 10.20, $\lambda_w T = M_{fp} d_f/L = 46.98 \times 517.5/5000 = 4.86$ kNm.

Non-uniform torsion plastic collapse. Using equation 10.61,

$$\lambda_z T = 12.53 + 4.86 = 17.39 \text{ kNm.}$$

10.9.7 EXAMPLE 7 – PLASTIC DESIGN

Problem.   Determine the plastic design torque resistance of the cantilever of example 5.

Section classification.

$$T = 15.6 \text{ mm}, p_y = 275 \text{ N/mm}^2 \qquad \text{T9}$$
$$\varepsilon = \sqrt{(275/275)} = 1.0$$
$$b/(T\varepsilon) = (209.3/2)/(15.6 \times 1.0)$$
$$= 6.71 < 9 \qquad \text{T11}$$

and so the section is plastic, and plastic design can be used.

Plastic design torque resistance.   Using the result of section 10.9.6,

$$T = 17.39 \text{ kNm},$$

which is significantly higher than the first yield torque capacity of 6.81 kNm determined in section 10.9.5.

10.10 Unworked examples

10.10.1 EXAMPLE 8 – UNIFORM TORSION OF A BOX SECTION

Determine the torsion section constant J for the box section shown in Fig. 10.38a, and find the maximum elastic shear stress caused by a uniform torque of 10 kNm.

10.10.2 EXAMPLE 9 – WARPING TORSION SECTION CONSTANT OF A CHANNEL SECTION

Determine the warping torsion section constant I_w for the channel section shown in Fig. 10.38b.

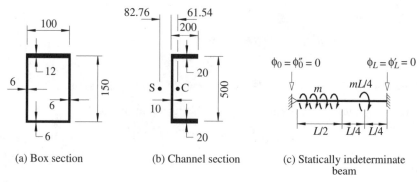

(a) Box section	(b) Channel section	(c) Statically indeterminate beam

Fig. 10.38 Examples 8–13

10.10.3 EXAMPLE 10 – APPROXIMATION FOR THE MAXIMUM TWIST ROTATION

The steel beam shown in Fig. 10.38c is 5 m long, has the cross-section shown in Fig. 10.38b, and the torque per unit length is $m = 1000$ Nm/m. Determine approximate values of the maximum angle of twist rotation by assuming:

(a) that the beam is in uniform torsion ($EI_w \equiv 0$), or
(b) that the beam is in warping torsion ($GJ \equiv 0$), or
(c) the approximation of equation 10.53.

10.10.4 EXAMPLE 11 – ACCURATE TWIST ROTATION

If the steel beam shown in Fig. 10.26 is 5 m long and has the cross-section shown in Fig. 10.38b, determine an accurate value for the maximum angle of twist rotation.

10.10.5 EXAMPLE 12 – ELASTIC STRESSES

Use the solutions of examples 9 and 11 to determine:

(a) the maximum nominal uniform torsion shear stress,
(b) the maximum warping shear stress,
(c) the maximum warping normal stress.

10.10.6 EXAMPLE 13 – DESIGN

Determine the maximum factored design torque per unit length that should be permitted on the beam of examples 11 and 12 if the beam has a yield stress of $p_y = 275$ N/mm^2.

10.11 References

1. Kollbrunner, C.F. and Basler, K. (1969) *Torsion in Structures*, 2nd edition, Springer-Verlag, Berlin.
2. Timoshenko, S.P. and Goodier, J.N. (1970) *Theory of Elasticity*, 3rd edition, McGraw-Hill, New York.
3. El Darwish, I.A. and Johnston, B.G. (1965) Torsion of structural shapes, *Journal of the Structural Division, ASCE*, **91**, No. ST1, pp. 203–27.
4. Owens, G.W. and Knowles, P.R. (eds), (1992) *Steel Designers' Manual*, 5th edition, Blackwell Scientific Publications, Oxford.
5. Terrington, J.S. (1968) Combined bending and torsion of beams and girders, *Publication No. 31* (First Part), British Construction Steelwork Association, London.
6. Nethercot, D.A., Salter, P.R. and Malik, A.S. (1989) Design of members subject to

combined bending and torsion, *SCI Publication 057*, The Steel Construction Institute, Ascot.

7. Hancock, G.J. and Harrison, H.B. (1972) A general method of analysis of stresses in thin-walled sections with open and closed parts, *Civil Engineering Transactions*, Institution of Engineers, Australia, CE**14**, No. 2, pp. 181–8.

8. Papangelis, J.P. and Hancock, G.J. (1995) *THIN-WALL – Cross Section Analysis and Finite Strip Buckling Analysis of Thin-Walled Structures – Users Manual*, Centre for Advanced Structural Engineering, University of Sydney.

9. Von Karman, T. and Chien, W-Z. (1946) Torsion with variable twist, *Journal of the Aeronautical Sciences*, **13**, No. 10, pp. 503–10.

10. Smith, F.A., Thomas, F.M. and Smith, J.O. (1970) Torsion analysis of heavy box beams in structures, *Journal of the Structural Division, ASCE*, **96**, No. ST3, pp. 613–35.

11. Galambos, T.V. (1968) *Structural Members and Frames*, Prentice-Hall, Englewood Cliffs, New Jersey.

12. Kitipornchai, S. and Trahair, N.S. (1972) Elastic stability of tapered I-beams, *Journal of the Structural Division, ASCE*, **98**, No. ST3, pp. 713–28.

13. Kitipornchai, S. and Trahair, N.S. (1975) Elastic behaviour of tapered mono-symmetric I-beams, *Journal of the Structural Division, ASCE*, **101**, No. ST8, pp. 1661–78.

14. Vlasov, V.Z. (1961) *Thin Walled Elastic Beams*, 2nd edition, Israel Program for Scientific Translations, Jerusalem.

15. Zbirohowski-Koscia, K. (1967) *Thin Walled Beams*, Crosby Lockwood and Son, Ltd, London.

16. Piaggio, H.T.H. (1962) *Differential Equations*, 3rd edition, G. Bell, London.

17. Trahair, N.S. and Pi, Y-L. (1996) Simplified torsion design of compact I-beams, *Steel Construction, AISC*, **30**, No. 1, pp. 2–19.

18. Dinno, K.S. and Merchant, W. (1965) A procedure for calculating the plastic collapse of I-sections under bending and torsion, *The Structural Engineer*, **43**, No. 7, pp. 219–221.

19. Farwell, C.R. and Galambos, T.V. (1969) Non-uniform torsion of steel beams in inelastic range, *Journal of the Structural Division, ASCE*, **95**, No. ST12, pp. 2813–29.

20. Aalberg, A. (1995) An experimental study of beam-columns subjected to combined torsion, bending, and axial actions, *Dr.ing. thesis*, Department of Civil Engineering, Norwegian Institute of Technology, Trondheim.

21. Pi, Y.L. and Trahair, N.S. (1995) Inelastic torsion of steel I-beams, *Journal of Structural Engineering, ASCE*, **121**, No. 4, pp. 609–20.

22. Trahair, N.S. (1999) Plastic torsion analysis of mono- and point-symmetric beams, *Journal of Structural Engineering, ASCE*, **125**, No. 2, pp. 175–82.

23. Pi, Y.L. and Trahair, N.S. (1994) Inelastic bending and torsion of steel I-beams, *Journal of Structural Engineering, ASCE*, **120**, No. 12, pp. 3397–417.

24. Subcommittee on Box Girders of the ASCE-AASHO Task Committee on Flexural Members (1971) Progress report on steel box bridges, *Journal of the Structural Division, ASCE*, **97**, No. ST4, pp. 1175–86.

25. Vacharajittiphan, P. and Trahair, N.S. (1974) Warping and distortion at I-section joints, *Journal of the Structural Division, ASCE*, **100**, No. ST3, pp. 547–64.

26. Goodier, J.N. and Barton, M.V. (1944) The effects of web deformation on the torsion of I-beams, *Journal of Applied Mechanics, ASME*, **2**, pp. A-35–A-40.

27. Kubo, G.G., Johnston, B.G. and Eney, W.J. (1956) Non-uniform torsion of plate girders, *Transactions*, *ASCE*, **121**, pp. 759–85.
28. Wright, R.N., Abdel-Samad, S.R. and Robinson, A.R. (1968) BEF analogy for analysis of box girders, *Journal of the Structural Division*, *ASCE*, **94**, No. ST7, pp. 1719–43.
29. Hetenyi, M. (1946) *Beams on Elastic Foundations*, University of Michigan Press.

Index